Plants, People, and Environment

Plants, People,

Peter B. Kaufman
Department of Botany/University of Michigan

J. Donald LaCroix
Department of Biology/University of Detroit

and Environment

MACMILLAN PUBLISHING CO., INC.
New York

COLLIER MACMILLAN PUBLISHERS
London

Macmillan Publishing Co., Inc.
866 Third Avenue, New York, New York 10022

Collier Macmillan Canada, Ltd.

Library of Congress Cataloging in Publication Data

Plants, people, and environment.

   Bibliography: p.
   Includes index.
   1. Botany.   2. Human ecology.   3. Ecology.
4. Conservation of natural resources.   5. En-
vironmental protection.   I. Kaufman, Peter B.
II. LaCroix J. Donald, (date)
QK47.P63   301.31   77–28594
ISBN 0–02–362120–6

Printing: 1 2 3 4 5 6 7 8   Year: 9 0 1 2 3 4 5

Preface

In our present-day civilization, we have witnessed an ever-accelerating rush toward overpopulation, exhaustion of our natural resources, and a gobbling up of our wilderness and natural areas by sprawling suburbs and agricultural and industrial enterprises. The awful consequences of this are documented daily. We are enmeshed in a synthetic environment, and people are being caught up in it where they work and where they live. Little wonder people are embarking on a new "green revolution"—one that embraces green spaces, plants in their homes, paintings and posters of nature, and gardens that may be no bigger than a windowbox! But we must seek other solutions—solutions on a much larger scale. These are necessary if we are to survive long enough to provide our children's children with a fit place in which to live.

Solutions to the crisis facing us today can be appreciated and understood only by obtaining a basic background in botany and ecology. Therefore, we have devoted an entire section of the book to plants and their way of life, and another section to probing the nature of our ecological environment. Our colleagues and students have suggested that we get right down to basic solutions of our environmental problems. Many books and articles have documented what the specific problems are, but few present practical, workable solutions.

We hope, after reading this book, that you will join us in our efforts to fight to conserve our natural heritage, to learn from nature, and to participate in working out solutions to our environmental crises. From such efforts, we can develop a more sane environment for all of us to live in, where people can be free from rubbish, fumes, and concrete jungles. We desperately need a new life-style in which we conserve our energy, use our plants more effectively, recycle our industrial products, and put the brakes on the population explosion. We hope that our book will provide a framework for such a life-style through its emphasis on plants, ecology, and solutions to all the many types of environmental problems confronting us today.

We would like to express our profound appreciation and thanks to the following people who helped in the preparation of this book: typing—Martha M. Jones, Karen White, and Suzanne Weller; photos and art work—Larry Mellichamp, David Bay, and Linda Kaufman; research work for preparation of the text—Kathy Rybarz, Steve Amatangelo,

v

Tom Gross, Kathy Cavanagh, Janet Eary, Joan Stroud, Irene Lee, Michael Ginsburg, and Robert Mistiatis; and preparation of the glossary—Keith Heller, Diane Finneren, and Ed Conley.

P. K.
D. LaC.

Contributors

Van D. Baldwin, a North Carolina native, is pursuing a Ph.D. in botany at the University of Michigan, Ann Arbor. He is currently involved in a study of the morphology of the inferior ovary in certain flowering plants.

Charles R. Barnes received his degree in wildlife ecology from the University of Michigan in 1973. After graduation, he served as Interpretive Naturalist for the Sarett Nature Center in Benton Harbor, Michigan, and is currently Director-naturalist for the Love Creek Nature Center and the Berrien County Parks and Recreation Commission.

Nancy Buckingham is a researcher and technical writer for The Nature Conservancy at its national office in Arlington, Virginia. She received her degree in natural resource conservation from the University of Connecticut in 1974.

Eugene V. Coan has an undergraduate degree from the University of California at Santa Barbara and a doctorate from Stanford University, with a speciality in marine zoology. He worked for one year for Zero Population Growth and has worked for the Sierra Club for over three years, specializing in energy and international conservation programs.

P. Dayanandan is a graduate student in the Department of Botany, University of Michigan, working with Dr. Peter Kaufman. Dayanandan's interests range from developmental physiology of plants to electron microscopy. He has written and lectured on hallucinogenic plants and Indian art history.

George F. Estabrook is Associate Professor of Botany and Research Scientist in the University Herbarium at the University of Michigan, Ann Arbor. His major research activities are in phylogenetic systematics (the classification of organisms based on relative recency of their common ancestry) and evolutionary ecology.

Richard I. Ford is Director of the Museum of Anthropology and head of the Ethnobotanical Laboratory at the University of Michigan. He combined training in anthropology and botany at Oberlin College, where he earned an A.B. degree; and at the University of Michigan, which awarded him a Ph.D. in anthropology in 1968. He has conducted ethnobotanical and ecological research in North American Indian communities, Ann Arbor, and Mexico. His interest in past environments and in the evolution of human-dominated ecosystems has stimulated paleo-ethnobotanical studies of plant remains from archaeological sites in the United States, Mexico, Poland, and Tunisia.

Dale M. Grimes is a Professor of Electrical Engineering, whose interests are in the application of modern electrical technology to things useful to man. He is the author of 2 books and about 25 technical papers.

Spencer W. Havlick is Assistant Dean and Professor of Environmental Design at the University of Colorado, Boulder. In the early 1970s, Dr. Havlick taught at San José State University in the areas of environmental advocacy and ecological assessment of urbanization. He serves as consultant to several federal agencies and the Athens Center of Ekistics. Spencer's many published works include *The Urban Organism,* published by the Macmillan Publishing Co., Inc., in 1974.

Frederick V. Hebard is a graduate student in plant physiology at the University of Michigan. He is currently working on plant tissue culture to develop new strains of American chestnut that are blight-resistant.

John W. Humke has been associated with The Nature Conservancy for the past 13 years. His experience with the organization includes the establishment of the Conservancy's Midwestern Regional Office, the development of the professional staff at the state chapter level, supervision of the organization's national land conservation program as Director of Operations, and the development of natural area preservation programs in government. He earned his masters degree at Michigan State University with studies in resource development and urban planning.

Hugh H. Iltis is Professor of Botany and Director of the Herbarium of the University of Wisconsin in Madison. His work has been an important influence in spreading awareness of the social responsibilities of evolutionists and taxonomists as guardians of the world's rapidly dwindling wilderness and plant and animal diversities.

Dan Janzen studies the interactions of animals and plants in tropical habitats. His primary concern is understanding how the traits of tropical animals and plants are the evolutionary outcome of interactions.

Hazel S. Kaufman obtained her degree from East Stroudsburg State Teachers College in Pennsylvania. She has a master's degree in physiology from the University of Michigan and did cancer research at the Kresge Cancer Research Institute in the University of Michigan Medical School. In addition, she served as Meteorological Officer in the U.S. Navy for 10 years.

Peter B. Kaufman did his undergraduate work in horticulture at Cornell University and completed his Ph.D. in botany at the University of California, Davis. He has worked as a Professional Botanist at the University of California, Davis; University of Michigan; Lund University in Sweden; University of Colorado in Boulder; and the University of Saskatchewan. His research interests include hormonal control of plant growth; silicification in plants; scanning electron microscopy of rice, *Cannabis,* and hops; stomatal functioning in plants; herbicide damage to rice; and geotropic responses in plants following lodging. Currently, he does collaborative research with the University of Michigan, Flint; University of Calgary; du Pont Chemical Company; University of California, Santa Cruz; University of Colorado; University of Washington; University of Leningrad; University of Oulu, Finland; Michigan State University; University of Detroit; Eastern Michigan University; and Cranbrook Institute of Science, Bloomfield Hills, Michigan.

J. Donald LaCroix studied botany at Purdue University. His wide experience in research includes work as a Kellogg Fellow at the University of Michigan, a position as Research Associate with Argonne National Laboratories, and work in mycology with Parke Davis. He has authored over 20 articles for professional journals based on his interests in electron probe analysis of silicification patterns in plants and scanning electron microscopy. He is a member of several scientific organizations and presently is a Professor of Biology at the University of Detroit and a Research Associate in Botany at the University of Michigan.

Richard MacMath is a partner of Sunstructures, Inc., a partnership of architectural designers and engineers in Ann Arbor, Michigan, who actively promote and utilize alternate, nonpolluting sources of energy.

Timothy Mahoney is an iconoclastic New Englander, who has spent a fair amount of time in the White Mountains. Presently, he is studying ecology and doing conservation work in the West.

Charles Nelson received his degree in Natural Resource Management from the University of Wisconsin, Madison. He is currently Director-Naturalist at Sarett Nature Center, an environmental education facility and wildlife sanctuary, and extension lecturer for Michigan State University.

Susan Schick is a biologist-illustrator for the Kentucky Nature Conservancy in Frankfort. She attended Earlham College and has served as a naturalist for Sarett Nature Center.

Richard W. Snyder was a member of the student Environmental Law Society while attending the University of Michigan Law School. In addition to specializing in environmental law, he still finds time to give special lectures to undergraduate students of botany.

Ellen Elliot Weatherbee teaches at the Extension Department of the University of Michigan. In addition to leading field trips and lecturing, she is in charge of the University's course in Edible Wild Plants.

James A. Weber is a plant physiologist who attended the University of Michigan. He is presently working with David M. Gates on biophysical ecology.

James R. Wells is a native of Ohio who has studied botany at the University of Tennessee, the University of Michigan, and Ohio State University. He is Adjunct Associate Professor at Oakland University and Botanist for Cranbrook Institute of Science.

Contents

SECTION I

*Plants: Their Way
of Life*

7

SECTION II

*Probing the Nature
of Our Environment*

77

SECTION III

*Constructive Action
and Solutions to
Our Environmental
Problems*

271

Plants, People, and Environment

The Environmental Landscape—Untouched, Touched, and Retouched
Peter B. Kaufman

1

Questions for Consideration

1. How does one distinguish between a wilderness area and other types of natural areas? Why are wilderness areas important to set aside and preserve?
2. What are some of the ways in which we can live with nature instead of squandering and destroying our natural resources? What lessons can we learn from American Indians and other peoples?
3. What kinds of landscapes in our environment are being destroyed most rapidly today? What solutions must we invoke to prevent this needless destruction?

In this introduction to *Plants, People, and Environment,* we first need to see what has been happening to our natural environment. To do this, we shall explore some of our unspoiled environments or wilderness areas. Then we shall look at the multitude of ways in which people have been tampering with their environment. Finally, we look at ways in which people have, and are now, attempting to live with their natural environment. We hope that you will perceive that we have reached a crisis on this planet in the overutilization of natural resources, including energy; in overpopulating large areas of the world; and in polluting our environment to such an extent that in many industrialized societies, people suffer from chronic diseases and greatly increased death rates. All of these problems can be alleviated. It is the aim of this book to point out ways in which this can be done.

Unspoiled Environments or Wilderness Areas

We shall define a *wilderness area,* "in the ideal sense, as a large, primitive landscape in an otherwise essentially people-shaped environment. It is thus an extensive area of rugged, undeveloped land and water that, among other things, offers the opportunity for contemporary people to experience the challenges of a hostile, rugged, natural environment and opportunities for solitude or an escape from masses of people" (Paul Rasmussen, Michigan Department of Natural Resources).

On planet earth, the number of unspoiled, wilderness areas is diminishing rapidly. The pressure to set aside and dedicate those that remain has increased remarkably, especially with greater environmental awareness and education, and concerted action by environmental groups.

FIGURE 1-1.
View of the Lake of the Clouds and escarpment in the Porcupine Mountains State Wilderness Area of Michigan. (Courtesy of the Michigan Department of Natural Resources.)

One such area that has been set aside in Michigan is the Porcupine Mountains State Wilderness Area (Figures 1-1 to 1-3). Fortunately, this area was set aside in the 1930s, before it was logged off or intruded on to any extent by "development" interests. Other states have set aside equally beautiful wilderness areas.

It is essential to point out that attempts to set aside all types of natural areas as wilderness areas is neither feasible nor in agreement with the true wilderness concept, as defined above. Many natural areas set aside today include nonwilderness uses, such as nature interpretive centers, snowmobiling, marinas, dune buggies, vacation homes, villages, and scenic roads. Some environmentalists would like to "lock up" all such areas as "wilderness areas." As just pointed out, these types of human activities and structures completely violate the wilderness areas that are left in populous areas of the eastern United States, as well as those that are more vast and commonplace (but equally vulnerable) in the western United States.

FIGURE 1-2.
Aerial view of Mirror Lake in the Porcupine Mountains State Wilderness Area. (Courtesy of the Michigan Department of Natural Resources.)

Why is it important to save wilderness areas? Several salient reasons have been promulgated by various national and state conservation organizations. They include the following:

1. To do ecological research.
2. To preserve endangered species of plants and animals.
3. To maintain the "gene pool"—to have available species of plants that can be used for crop improvement or as a source of wild plants that can be used to breed new, more desirable plants for people's food, fiber, and enjoyment.
4. To provide an "escape" for people from the urban ghetto.
5. To obtain exercise, as in hiking (or backpacking), cross-country skiing, and horseback riding.
6. To pursue hobbies, such as mushroom hunting, photography, and observing birds and other animals.
7. To study nature education.

All these goals are vital to human physical and mental health. Wilderness areas give us a sense of "ecological therapy," just as gardening is a kind of "horticultural therapy." We predict that without wilderness areas, people will become completely estranged from nature, suffer many more physical and mental maladies, and will drown in their own synthetic quagmire.

How People Harm the Natural Environment

Ever since people have existed on this planet, they have, in all kinds of ways, with increasing frequency, been tampering with the environment. Let us look at some of the most dramatic cases.

1. *Man-made fires,* leading to the loss of timberland, as has occurred so dramatically in the California Sierra chaparral vegetation and in the once-magnificent forests of Douglas fir and other trees west of the Cascade Mountains in Oregon and Washington.

2. *Slash-and-burn agriculture in the tropical areas* throughout the world, leading to serious soil erosion and the loss of valuable cropland and natural wildlife habitats.

3. *Timber clearing and cutting* (especially clear-cutting), resulting in nearly complete loss of forests and wilderness areas. A good case in point is seen with relic tree stumps of white and red pine that once dominated

FIGURE 1-3.
Aerial view of the mouth of the Carp River in the Porcupine Mountains State Wilderness Area. (Courtesy of the Michigan Department of Natural Resources.)

central and northern lower and upper Michigan before the 1800s—all virtually logged off by early 1900.

4. *Mining minerals,* especially in the eastern United States, leading to ugly scars from strip-mining operations on hills and mountain slopes. In the western United States, many mines of the 1800s have left denuded areas that have *not* become revegetated to any great extent.

5. *Dumping of sewage and industrial wastes* into lakes, rivers, and oceans, leading to eutrophication of lakes (see Chapter 8), denuding of lakes and rivers of vegetation and animal life (e.g., Lake Erie, parts of the Rhine River), and hazards to human health from polluted drinking water (witness the dumping of takonite, containing asbestos fibers, into Lake Superior from an iron ore operation at Silver Bay, Minnesota).

6. *Intrusion of new modes of transportation and transportation systems into natural areas,* particularly new highways, snowmobile trails, motorcycle trails, dune-mobile areas.

7. *Devastation of the landscape by herbicides,* as has so clearly been documented for mangrove swamps and rubber plantations in South Vietnam through aerial application of herbicides such as 2,4,5-T and cacodylic acid.

8. *Scars upon the landscape as a result of conflict* between nations and tribes, as witnessed by huge bomb craters and relics of weapons left all over the landscape, both in aquatic and land habitats.

9. *Invasion of agricultural and natural areas by sprawling shopping centers and suburbs* at an ever-accelerating pace, especially because of the lack of land-use planning and adequate zoning restrictions.

10. *Introduction of pests into new areas,* including disease-causing bacteria and fungi, insect pests, and weeds (over 80 percent of the weeds in California have been introduced from other areas, particularly the Mediterranean).

11. *Overgrazing, coupled with drought,* especially dramatic in northern Africa, where there has been a tragic loss of vital food supplies for both humans and livestock.

12. *The pressure of people on the environment as a result of overpopulation,* especially in Japan, India, Mexico, and the eastern United States.

These are but a few of the ways by which people have been destroying our natural environment. Most of these are of major importance. Not all of these are irreversible; indeed, today we are witnessing a long-overdue attempt to prevent these types of destructive activities.

People's Attempt to Live with Nature: Restoration at Work

It is refreshing to see all the positive steps now under way in our country to correct many of our environmental disasters and wrongdoings. We shall go into these steps in more detail in later chapters, but for now, let us list a few of these:

1. Conservation of heat, food, fuel, water, timber, and other sources of energy.
2. Use of alternative energy sources, such as wind, geothermal, and solar energy.

3. Recycling of wastes: paper, glass, metal, tires, garbage, compost, chips, and sawdust.
4. Establishment of wilderness areas: greenbelts and parks in urban areas.
5. Enforced (by law) revegetation of strip-mined areas.
6. Establishment of land-use laws.
7. Population planning and control.
8. Use of wild, edible foods.
9. Establishment of nature centers and environmental education programs in our schools.
10. Curtailment of industrial pollution by law and imposition of stiff fines for violation of antipollution laws.
11. Use of alternative modes of transportation: mass transit, bicycles (and bicycle paths), cross-country skis; banning of ORVs (off-road vehicles) from wilderness areas.
12. Use of multiple-crop rotation systems to prevent erosion, add nutrients to the soil, and increase yields (as opposed to continuous monoculture systems).
13. Use of plants to improve environmental quality on streets, around city buildings and homes, and on school properties.

You should now get a glimpse of what this book will be about. We shall continually reiterate the central theme, that people *must live with* and *respect* their natural environment, much as American Indians have done in the New World, as cultures in India and China have done for centuries, and as environmentally concerned people in many countries are trying to do today. We have no other alternative, as we may find that we are on a collision course with nature if we continue in our destructive ways, magnified by an ever-burgeoning world population with ever-increasing numbers of mouths to feed.

Plants:
Their Way
of Life

The theme of our book emphasizes that people must live in harmony with their environment. Plants, for centuries, have been living in harmony with nature through various types of associations and interactions. Humans, in contrast, have for centuries tried to manipulate their environment, and in so doing, have often brought on irreparable damage to the environment and to themselves. Many "panaceas" developed by people turn out, in retrospect, to be boondoggles. To avoid these tragedies, people must try to attain as complete an understanding of their environment as possible. This requires a great amount of open-mindedness and experience. There is, therefore, a crying need for all of us to learn something more about plants and their way of life. In this first section of Plants, People, and Environment, *we introduce you to the various kinds of plants on our planet, then see how they function at various stages in their life history, and finally, learn something about the ways plants have benefited people economically. Once you see how plants work, how people use them, and what kinds exist in our environment, you should gain a much better appreciation of how essential plants are in our everyday lives and what a central role they play in our attempts to live with nature.*

On the Types of Plants
Van D. Baldwin

2

Questions for Consideration

1. Which groups of plants serve as decomposers in ecosystems? What are the essential roles of decomposers in the ecosystem?
2. Which groups of plants serve as pathogens in ecosystems? What are their essential roles?
3. Which groups of plants serve as primary producers in the ecosystem? What are their primary roles in the ecosystem?
4. How are the following groups of plants distinguished from each other: Monera, Fungi, and Plantae?
5. What do we mean by common names and scientific names for plants? What advantages do scientific names have over common names? What disadvantages do they have?
6. What are some of the major economic uses of the plants or groups of plants: blue-green algae, mushrooms (club fungi), bacteria, sphagnum moss, brown algae, conifers, grasses, legumes, cacti and succulents, and composites (members of the thistle family: e.g., dandelions, asters, zinnias, chicory, marigolds, and sunflowers)?

The process and principles of classification are inherently obvious to the trained biologist. To the nonbotanist, the business of naming biological entities with vague-sounding epithets may seem quite mysterious. This is not so. Man, as a species, is by necessity and by nature a nomenclaturist: one who names things. Without the use of names, the identification of even the most common items would be impossible. One must realize that the use of names is primarily a *means of communication.* These names evoke the identity of an object by their mere mention.

Classification goes beyond the business of naming. It tries not only to identify, but also to account for the different types of objects or organisms in the universe. The logic behind biological classification is based on a comparison of organisms. By comparing the morphology (both of living and fossil specimens), mode of nutrition, physiology, ecology, and other features of plants and animals, one may hopefully discern which organisms are related and which ones are not related. To this comparative method, modern classification adds evolutionary interpretation in an attempt to understand how different groups of organisms changed, gave rise to new groups over time, and how they are now evolving.

With the preceding explanation in mind, let us stress again that the listing of the groups of plants below is not an attempt to depict all of the major groups. This listing proposes to focus your attention on the different types of the nearly 350,000 species that at one time or another have been described as plants. For a breakdown of the number of species in different groups of plants, see Table 2-1.

Although it is certainly presumptuous to assume that the only value of a group of plants or animals is to benefit mankind, a note concerning the *economic* or *ecological* importance of the group is included, in keeping with the topics presented in this book.

The most familiar breakdown of living things is to differentiate between plants and animals. This assumes that there are these two great groups, and all organisms are included either in one group or the other. In fact, such a division should not separate plants from animals: rather, it

TABLE 2-1 / Types and Numbers of Plants: Their Ecological Characteristics and Economic Roles

Kingdom	Group	Number of Genera	Number of Species
Monera	Blue-green algae (Cyanophyta)	150	2,500
	Bacteria (Schizomycota)	122	1,335
	Viruses (−)	−	−
Fungi	Slime molds (Myxomycota)	63	450
	Algal-like fungi (Phycomycota) ⎫		
	Sac fungi (Ascomycota) ⎬	3,385	40,000
	Club fungi (Basidiomycota) ⎭		
	Lichens (−)	−	15,000
Plantae			
Algae			
	Red algae (Rhodophyta)	400	2,500
	Brown algae (Phaeophyta)	190	1,000
	Green algae (Chlorophyta)	400	6,900
	Stoneworts (Charophyta)	6	250
Other groups (certain Protista)		460	7,035
Bryophytes			
	Liverworts (Hepaticae)	175	8,500
	Hornworts (Anthocerotae)	5	320
	Mosses (Musci)	660	14,000
Tracheophytes (Vascular plants)			
Lower vascular plants			
	Lycopsids (Lycopsida)	5	944
	Sphenopsids (Sphenopsida)	1	25
Ferns			
	Ferns (Pteropsida)	260	9,280
Primitive seed plants			
	Cycads (Cycadophyta)	9	100
	Conifers (Coniferophyta)	50	550
	Ginkgo (Ginkgophyta)	1	1
	Gnetes (Gnetophyta)	3	71
Flowering plants			
	Dicots (Dicotyledoneae)	9,500	200,000
	Monocots (Monocotyledoneae)	3,000	50,000

should serve to distinguish between major groups of organisms which go about the business of life in *different ways.* Recent conceptual advances now recognize at least five of these major groups (or kingdoms). The animal kingdom (Animalia) has been retained for all multicellular forms of that group. The unicellular and colonial members have been placed in a separate kingdom, the *Protista.* Certain unicellular forms, once classified as plants (i.e., certain "algae"), are placed in the Protista as well. For further details on these groups, a general biology or zoology text will be helpful.

The plant kingdom, in the older sense, has been divided into three kingdoms, the Monera, the Fungi, and the Plantae.

Nutritional Habit	Ecological Role	Economic Role
Autotrophic	Primary producer	—
Heterotrophic	Decompcser or pathogen	Disease, dairy industry
Heterotrophic	Pathogen	Disease
Heterotrophic	Decomposer or pathogen	—
Heterotrophic	Decomposer or pathogen	—
Heterotrophic	Decomposer or pathogen ⎫	Edible species; crop
Heterotrophic	Decomposer or pathogen ⎭	loss due to pathogenicity
Functionally autotrophic	Soil builder, colonizer	Food, dyestuffs
Autotrophic	Primary producers	—
Autotrophic	Primary producers	Kelp eaten
Autotrophic	Primary producers	—
Autotrophic	Primary producers	—
Autotrophic	Primary producers	—
Autotrophic	Primary producers	—
Autotrophic	Primary producers	—
Autotrophic	Primary producers	Sphagnum industry
Autotrophic	Primary producers	—
Autotrophic	Primary producers	—
Autotrophic	Primary producers	Cultivated, food
Autotrophic	Primary producers	Ornamentals
Autotrophic	Primary producers	Many uses
Autotrophic	Primary producers	Cultivated
Autotrophic	Primary producers	—
Autotrophic	Primary producers	Many uses
Autotrophic	Primary producers	Many uses

The first of these major groups, the *Monera,* consists of *prokaryotic* organisms. Prokaryotic organisms are characterized by lack of an organized nucleus and a lack of membrane-bound organelles in their cytoplasm. The blue-green algae (*Cyanophyta*), the various groups of bacteria (*Schizomycota*), and the viruses are the major divisions of the Monera.

BLUE-GREEN ALGAE (*CYANOPHYTA*)

The *blue-green algae* or Cyanophyta are predominantly planktonic members of the marine flora, but they do occur in lakes, ditches, and ponds worldwide. They contain chlorophyll, and as photosynthetic organisms, they are important as primary producers in oceanic food chains (see Chapter 8). They are best known for causing "algal blooms" and for fixing atmospheric nitrogen. Their rapid growth, sometimes resulting from nutrient pollution, can cause fish kills, poison seafood, and render drinking water unsafe for a time.

BACTERIA (*SCHIZOMYCOTA*)

The rather inclusive term *bacteria* includes several groups of organisms which, in deference to simplicity, will be treated as one. Although bacteria are most famous as pathogens (organisms that cause disease in plants and animals), the greater percentage of these organisms cause no harm to any group of organisms. Indeed, many bacteria are beneficial to both man and other species.

The nutritional habits of bacteria are quite various. Many are *saprophytic;* that is, they live on dead organic matter. Others are *autotrophic,* making their own food utilizing either light energy (photosynthetic) or energy from certain chemicals (chemosynthetic). Still others are *pathogens,* practicing a parasitic mode of nutrition.

In the world at large, the bacteria (along with fungi) are the true "ecologists." These organisms are the decomposers, the recyclers, because they break down dead organic matter into component forms that can be reused by other organisms. Economically, bacteria are important for a variety of processes, especially those dealing with fermentation products. The use of dairy products such as cheese, butter, yogurt, and cottage cheese is made possible by bacteria. The curing of tobacco and tea is accomplished by bacterial action. In the future, the role of bacteria as decomposers of human sewage will increase (see Chapter 20).

VIRUSES

Viruses have been called the simplest form of life. Although this is a matter of opinion, they are certainly the most reduced form of life.

Viruses consist of a nucleic acid core (DNA or RNA) and a proteinaceous sheath. These pathogens carry on no metabolism (being acellular) of their own, but they do invade cellular organisms and use the synthetic components of their cells to make new viruses. This is not a "simple" mode of life!

Viruses are included with the cellular prokaryotes in the Monera. This is acceptable, as current thought interprets viruses as being fragments of genetic material that became detached from bacterial chromosomes in times past. All viruses are parasites, causing serious or minor disease in their hosts. Some viruses, such as cucumber mosaic virus, tobacco mosaic virus, and aster yellows virus, are responsible for much crop loss each year.

Kingdom Fungi

The second segregate from the old plant kingdom is the kingdom *Fungi*. The fungi are characterized by having (1) eukaryotic cells with cell walls, (2) a lack of chlorophyll (and the resulting saprophytic or parasitic mode of nutrition), and (3) mycelial (or filamentous) growth and tissue differentiation without organized tissue systems (as in vascular plants). The major divisions within the fungi are the slime molds (*Myxomycota*), the algal-like fungi (*Phycomycota*), the sac fungi (*Ascomycota*), and the club fungi (*Basidiomycota*). The lichens (algal–fungal associations or symbionts) will be considered in this section as well.

SLIME MOLDS (*MYXOMYCOTA*)

The *slime molds* or Myxomycota are a group of fungus-like plants that have a vegetative stage which takes the form of a multinucleated mass of protoplasm (called a *plasmodium*). These organisms are decomposers and are often encountered in fields and woodlands.

ALGAL-LIKE FUNGI (*PHYCOMYCOTA*)

The *algal-like fungi*, or Phycomycota, are a rather heterogeneous group of fungi distinguished by a lack or paucity of cross walls (septa) in the hyphae of the vegetative body. Some of these fungi may be unicellular. Many of them are parasites on algae, other aquatic fungi, and vascular plants. A large number of species are found in fresh water, where they grow saprophytically. A lesser number of these algal-like fungi parasitize small fish, insect larvae and other aquatic animals. Research at the present time is aimed at trying to use certain of these organisms for biological control of mosquitoes. The most widely known member of the Phycomycota is probably *Rhizopus*, the bread mold. Other species act to decompose animal waste and other organic material in forest litter or in compost and manure piles.

SAC FUNGI (*ASCOMYCOTA*)

The *sac fungi*, or Ascomycota (Figures 2-1 and 2-2), are a large group of fungi with septate hyphae (with cross walls). Their name refers to the fact that they bear their spores in a saclike compartment, the *ascus*. Representative organisms of the Ascomycota include the yeasts; the brown, green, and pink molds; the powdery mildews; and the cup fungi.

The yeasts are important in anaerobic respiration (fermentation) in industry and in the home. These organisms are used in the leavening of breads and in the brewing of alcoholic beverages.

The brown, green, and pink molds are decomposers of organic matter. These organisms often cause molding of fruit in the home and degrade organic waste in the field. *Penicillium,* from which the antibiotic, penicillin, is obtained, is a member of this group.

The powdery mildew fungi are leaf parasites in a number of higher plants. These pathogens cause serious losses in grains, vegetables, fruits, and ornamental garden plants. The deadly poison of the ergot fungus,

FIGURE 2-1.
Gibberella fujikuroi, an ascomycete fungus from which the hormone gibberellin was first extracted. (Courtesy of David Bay.)

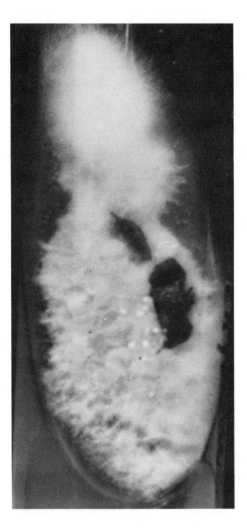

FIGURE 2-2.
Morchella, the edible morel, is an ascomycete. (Courtesy of Larry Mellichamp.)

Claviceps purpurea, which is related to powdery mildews, was responsible for thousands of deaths in the Old World. Ergot is also a natural source of lysergic acid, which is used in various forms as a medicine and an hallucinogen (see Chapter 4).

The cup fungi are fleshy forms, often colorful, and frequently encountered in the woods. These are saprophytic members of the forest's decomposers. The edible morel *Morchella* is also included in this group.

CLUB FUNGI (*BASIDIOMYCOTA*)

The last and most familiar group of fungi are the *club fungi,* the Basidiomycota (Figures 2-3 and 2-4). The characteristic feature of this

FIGURE 2-3.
*Coprinus, the shaggy mane,
is an edible basidiomycete.
(Courtesy of Van Baldwin.)*

group is the basidium, a specialized club-shaped hypha that bears the spores of the fungus on protruding stalks. The mycelium is always septate. Representative groups of club fungi are rusts, smuts, puffballs, mushrooms, and pore fungi.

The rusts and smuts are parasitic members of this group. They attack many grain crops and induce significant yield loss in wheat, oats, barley, rye, corn, and other cereal grains. The remainder of the group listed includes fleshy saprophytes, which play a major role in the decomposition of organic matter. Some of the members of this group are edible—and some are deadly poisonous. Some aid in nutrient uptake by orchids and forest trees by growing symbiotically with the roots of these plants as *mycorrhizal* fungi.

LICHENS

Lichens (Figure 2-5) are not considered to be a separate category of organism, for they are a symbiotic association between a fungus (usually as Ascomycete) and an alga. These are, then, aggregates of two dif-

FIGURE 2-4.
Polyporus, a bracket fungus, is abundant in wet woods. It is a basidiomycete. (Courtesy of Van Baldwin.)

ferent kingdoms of organisms. They occur worldwide on soil, rock, and tree bark. They serve valuable ecological roles as "pioneer plants" in breaking down rocks and in soil formation; they are a source of food for grazing animals. Some species (e.g., manna, or bread of heaven) are eaten by humans. They have been used for their medicinal properties. Also, dyeing of wool and other fibers is accomplished with the natural pigments found in lichens. Lichen classification is artificial (not assuming relationship) and refers to their gross external morphology. Crustose lichens cover trees, rocks, and some soils. Foliose lichens have leaflike protuberances arising on their vegetative parts. Fruticose ("shrubby") lichens have an erect form like that of a miniature shrub.

Kingdom Plantae

The kingdom *Plantae* includes the true algae (excluding the blue-green algae and certain protists) and the land plants. The land plants consist of the mosses and liverworts (Bryophyta) and the vascular plants (Tracheophyta).

ALGAE

The *algae* are a large and extremely diverse group of plants. They have eukaryotic cells bounded by a cell wall, and they reproduce asexually and/or sexually. The major groups of algae follow.

Red Algae (*Rhodophyta*)

The *red algae* (Rhodophyta) are predominantly marine forms containing the red pigment *phycoerythrin*. Other pigments give a variety of colors to members of this group. These algae exist as single cells or as branching or nonbranching filaments, or as sheets, generally less than 1 meter (3 feet) in length. Their most remarkable feature is the presence of sperm, which lack flagella, a rare phenomenon in living organisms.

Brown Algae (*Phaeophyta*)

The *brown algae* (Phaeophyta) are again predominantly members of the marine flora. The "seaweeds" are for the most part brown algae. The largest and most elaborate algae (the kelps) are found in this division. Some species of the kelp genus, *Macrocystis*, reach over 43 meters (140 feet) in length.

Green Algae (*Chlorophyta*)

The *green algae* (Chlorophyta) are mainly found in fresh water, although marine forms do exist. These algae exist as single cells, colonies, filamentous forms, and as fleshy, tissuelike forms. They have abundant distribution worldwide.

Stoneworts (*Charophyta*)

The *stoneworts* (Charophyta) are inhabitants of quiet freshwater areas. They are called stoneworts because of their coarse texture conferred by calcium carbonate deposits in their cells. They are a small group with only one extant family.

Other Groups

Other "algal-like" organisms, now transferred to the Protista, include the *euglenoids* (Euglenophyta), the *yellow-green algae,* the *golden brown algae,* and the *diatoms* (Chrysophyta). These are unicellular, colonial, or filamentous forms and are variously pigmented. They exist in aquatic habitats throughout the world.

The ecological importance of algae is impossible to overestimate. Their role in aquatic food chains is the foundation for all life on earth. They are primary producers, utilizing solar energy to make sugar. The life of land organisms is also dependent on algae. The oxygen balance in

FIGURE 2-5.
Usnea, old man's beard, is a lichen. This is a large genus of epiphytes. (Courtesy of Van Baldwin.)

the atmosphere is largely maintained by the oxygen evolved by these organisms as they grow.

In terms of human economics, algae have yet to come into their own. Although Oriental cultures have utilized seaweeds as foods and delicacies for ages, most other people have not yet turned attention to the marine algae as a source of food. As world food needs continue to increase, algae may grow to be a staple food on all the tables of the world.

BRYOPHYTES (*BRYOPHYTA*): LIVERWORTS, HORNWORTS, AND MOSSES

Liverworts (Hepaticae), *hornworts* (Anthocerotae), and *mosses* (Musci) are collectively called *bryophytes* (Bryophyta) (Figures 2-6 and 2-7). They are sometimes referred to as the "amphibia" of the Plantae. This is

FIGURE 2-6.
Marchantia, a common liverwort, produces its spores in the spring. It is common in wet areas and on stream banks. (Courtesy of Larry Mellichamp.)

because they do occupy an intermediate position between the aquatic algae and the dominant land plants, the Tracheophyta (vascular plants). The bryophytes not only share the terrestrial habitat with the vascular plants, but have another feature in common: an embryonic stage in their life cycle. In the Plantae only the bryophytes and vascular plants have this feature as an adaptation away from an aquatic environment.

These plants are familiar to most casual observers of nature, inhabiting predominantly wet and low places or the decaying bark of trees where ground water and dew are readily available. The main groups of Bryophyta differ in their morphology. The hornworts and liverworts are flat, thalloid, dorsi-ventral (with upper and lower surfaces) organisms with no differentiation into leaves, stems, or roots. The mosses exhibit more structural complexity, but also lack the same three organs.

Ecologically, these organisms can be pioneers on newly cleared ground or members of climax communities. They hold moisture close to the ground and thus improve water relations for themselves and neighboring organisms. Bryophytes are neither edible nor pathogenic. A few are of use to humans in commerce: *Sphagnum* or (peat moss), for example, is used as mulch by gardeners and has an antibiotic property that prevents fungal growth.

FIGURE 2-7.
*Sphagnum, or peat moss,
is a true moss. It forms the
basis of the peat industry.
(Courtesy of Larry Melli-
champ.)*

VASCULAR PLANTS (*TRACHEOPHYTA*)

The *vascular plants* (Tracheophyta) are by far the largest and most familiar of the plant groups. The vascular plants contain approximately 75 percent of all taxa at one time described as plants. The single salient feature of the Tracheophyta is the occurrence of xylem and phloem as conducting elements in the body of the plant. For simplicity's sake, we shall look at this great assemblage of plants in four groups: the lower vascular plants, the ferns, the primitive seed plants, and the flowering plants. It should be noted that this separates the Tracheophyta into groups by the level or morphological advancement (i.e., change) from the primitive fossil forms. This implies relation of the plants treated in a group—but the degree of relationship is different in each category, as explained below.

Lower Vascular Plants (*Lycopsida and Sphenopsida*)

The *lower vascular plants* (Figures 2-8 and 2-9) are a group of primitive plants that represent the remnants of two groups more common in times past. There are two major groups to consider here—*lycopsids* and *sphenopsids.* One, the Lycopsida, includes *club mosses* (*Lycopodium*), *spike* mosses (*Selaginella*), and *quillworts* (*Isoetes*). The other, the *sphenopsids* (Sphenopsida) contains the *horsetails* and *scouring rushes* (*Equisetum*).

The members of the Lycopsida have true stems, leaves, and roots, and they reproduce by spores in the typical manner of land plants (see "heteromorphic alternation of generations" in the glossary). The club mosses occur worldwide as ground herbs, bog plants, or, in the tropics, as epiphytes. The waxy spores of these plants have been used in various pyrotechnic devices (i.e., fireworks). The spike mosses are often confused with the true mosses (Musci), for *Selaginella* is about equal in stature, and their habitat is often the same. *Selaginella* occurs worldwide, with a concentration of epiphytic forms in the tropics. The spike mosses serve no real economic purpose. The quillworts are a small group of

FIGURE 2-8.
Lycopodium (center); this "club moss" is not a moss at all but a vascular plant. Cladonia (bottom right), or reindeer moss, is also not a moss but a lichen; it serves as winter food for grazing deer and other herbivores. (Courtesy of Larry Mellichamp.)

FIGURE 2-9.
Equisetum, or scouring rush, is a lower vascular plant that produces spores in early spring. (Courtesy of Van Baldwin.)

aquatic plants, seldom noticed by the layman, as they look like a rush or an aquatic grass. They are economically unimportant.

The horsetails and scouring rushes are inhabitants of wet areas and some dry sandy areas worldwide, excluding the Australian area. Their stems are jointed with distinct nodes and internodes. They grow for a few centimeters to $1\frac{1}{4}$ meters (4 feet) or so in height, one species (*Equisetum giganteum*) reaching over 12 meters (40 feet) in Central America. Some species are quite beautiful and are occasionally cultivated. The plant accumulates silica gel in its epidermal cells and has a coarse texture. This accounts for the use of these plants in earlier times to scour pots. One colloquial name for these plants is "Colonial Brillo."

FERNS (*Pteropsida*)

The *ferns* (Pteropsida; Figure 2-10) are a large assemblage of plants occurring as herbs, epiphytes, or small trees (in the tropics). A few species are aquatic. They reproduce by spores, as do the lower vascular

FIGURE 2-10.
Osmunda, the common fern, is a frequenter of wet areas. The young fiddle-heads are edible when sautéed, but reports vary as to their salubrious qualities. (Courtesy of Van Baldwin.)

plants, but are not related to the former group. Most species have a characteristic pattern of vascular tissue in their underground stems (or rhizomes) and a type of leaf that clearly separates them from the plants listed previously. This vascular pattern leaves a "gap," an area of parenchymatous cells in the vascular cylinder as a leaf trace branches off and enters a leaf. Some ferns have been used medicinally, but this use is of local importance. The young leaves (fiddleheads) of some species are eaten as a wild edible food (see Chapter 17). Their greater value to humans is their use as cultivars both inside and outside the home.

FIGURE 2-11.
Dioon, a cycad native to Mexico, is used as a starch source. (Courtesy of Van Baldwin.)

Primitive Seed Plants (*Gymnosperms*)

The *primitive seed plants* (the *gymnosperms* or naked seeded plants; Figures 2-11 to 2-13) again represent remnants of groups that have decreased with time. There are four groups of gymnosperms living today: the cycads (Cycadophyta), the conifers (Coniferophyta), the ginkgo (Ginkgophyta), and the gnetes (Gnetophyta).

The *cycads* are a small tropical group of palmlike plants. They bear no relation to palms, which are flowering plants. The cycads have thick, punky stems full of starch-filled parenchyma and little wood. They reproduce by seeds that are borne on modified leaves or in cones. By and large, they are of little value to the world's economy. Some cycads are used as ornamentals in gardens, greenhouses, and buildings. A number produce edible seeds, which are not widely marketed. Several species are cooked as a starchy foodstuff.

The *conifers* are, in general, pines, firs, junipers, hemlocks, larches, spruces, and their relatives. These are tall trees that reproduce by seeds borne in cones. These plants, in some places, are the dominant canopy species, or are at least significant contributors to the local flora. They are used extensively as Christmas trees and as ornamentals. Much of the paper and lumber we use comes from this group of plants. They contain

FIGURE 2-12.
Sequoia, the coastal red-wood of northern California, is one of the world's tallest trees. It is a coniferous gymnosperm often cut for lumber. (Courtesy of Van Baldwin.)

the oldest known living organisms, giant redwoods and twisted bristle-cone pines (see the Frontispiece) over 4000 years old.

One species, *Ginkgo biloba*, is the sole remnant of a once-widespread group of trees, the *ginkgos*. Now existing only in cultivation, this tree is widely used as a shade tree, as it is relatively resistant to air pollution. Only male trees are cherished, as the ripe seeds of *Ginkgo* smell so foul (due to butyric acid) as to cause nausea in the less hardy.

The *gnetes* (Gnetophyta) are a group of morphologically bizarre plants. One of these, *Welwitschia mirabilis*, grows in rainless Namib desert of southwestern Africa. The plant consists of a huge taproot, a crown, and

FIGURE 2-13.
Juniperus, or cedar, is a commonly encountered cultivated (and wild) shrub or tree. The fleshy cones (incorrectly referred to as berries) are used to flavor gin. Its scented wood is highly prized for furniture. (Courtesy of Van Baldwin.)

two large straplike leaves that trail over the surface of the desert. Another genus, *Ephedra,* is common in the southwestern United States. This is Mormon tea or joint fir, from which the alkaloid *ephedrine* was first obtained. The third genus, *Gnetum,* is a vinelike plant of tropical forests.

FLOWERING PLANTS (*Anthophyta*)

The *flowering plants* (Anthophyta; Figures 2-14 to 2-25) are the crowning peak of development, as we know it today, in the plant kingdom. They occur worldwide as trees, shrubs, herbs, epiphytes, parasites, and saprophytes. Some grow and bloom under water (even brackish water), and one species, an orchid, carries out its complete life cycle—including flowering—underground. This remarkable diversity of form is unparalleled in any other group of plants.

These are plants with a special reproductive structure, the flower, which contains the seed in an ovary (the carpel). This is a fundamental distinction from the naked seeded plants (gymnosperms) considered above. There are two major groups of flowering plants—*dicots* (Dicotyledoneae), having two seed leaves, and *monocots* (Monocotyledoneae), having one seed leaf. Dicots are the larger group and contain many trees and shrubs, as well as roses, daisies, geraniums, beans, and most vegetable crops. Monocots consist of only one-fourth the number of species represented by the dicots and contain lilies, orchids, sedges, rushes, aroids, bromeliads, palms, and the grasses.

Ecologically, flowering plants are the dominant life form of most of the terrestrial habitats today. They form canopy, subcanopy shrub, and herb layers in many forests. Economically, their importance is staggering. From the standpoint of food alone, one family of flowering plants (the grass family) provides most of the metabolic energy used by humans on earth. The grains—corn, oats, wheat, rice, barley, rye, millet, and others—are the staples of our diet. The grains and members of other families of plants, such as the legume family, and various fruits make up essentially all of our nutrition. Even the meat we consume is derived from plant material. Flowering plant products not only feed us, but cotton clothes us; we build with beech, oak, and poplar (as well as pine and other gymnospermous woods). Our homes and yards are decorated with flowering shrubs and tall trees, and we take our recreation in

FIGURE 2-14.
Scilla, a native to Europe and a cultivar in the United States, is another member of the lily family. (Courtesy of Van Baldwin.)

FIGURE 2-15.
Darlingtonia is an insec-
tivorous plant. It lures
insects into its nectar-
secreting, tubelike leaves,
where they are digested.
(Courtesy of Van Baldwin.)

herb-filled forests. This group of plants provides the most benefit directly to humans.

Summary

Nomenclature, the process of naming things, is fundamental to our understanding of nature. It is by names that we refer to the diverse units of life that we observe in the world. Classification, the process that is the complement of nomenclature, is also an integral concept in our ordering of nature. It is by classification that we can begin to understand the evolutionary categories which have established ecological interrelations

FIGURE 2-16.

Pedicularis, or elephant head, is a parasitic plant, taking water and mineral nutrition from the roots of another plant. Its leaves can be used for a tea, and the roots taste like a carrot. (Courtesy of Van Baldwin.)

through time. In our quest for sounder classification, we may help to understand the totality of interactions between organisms.

All groups of organisms benefit humans directly or indirectly. This is true, as all organisms are bound together in a number of interacting ecological cycles that depend on each other for stability. To infer that a particular group of organisms is of no importance because it is of no direct benefit to humans is both presumptuous and foolish.

Although all organisms are important from an ecological point of view, the degree to which each group benefits humans varies greatly. For a summary of the more important economic groups of plants, consult Table 2-1.

This has been a brief look at the plant kingdom. We have examined the rationale for classification and nomenclature of living organisms. We have examined the many different forms of plant life and noted how each goes about living. The importance of each of the groups of plants to man as a species and to the ecology of the earth at large has been surveyed. We hope that the information presented here will give some insight into the diversity, importance, and beauty of plant life on the earth.

FIGURE 2-17.
Mammilaria is one of many cacti native to North American arid regions. Here it is shown in flower. (Courtesy of Van Baldwin.)

FIGURE 2-18.
Digitalis, the foxglove, is the plant from which the heart stimulant digitalin is procured. A native of Europe, it is now established in the Pacific Northwest. (Courtesy of Van Baldwin.)

FIGURE 2-19.
Monarda, or bee balm, is one of the many members of the mint family (Lamiaceae-Labiatae) used in cooking and for making a tea. (Courtesy of Van Baldwin.)

FIGURE 2-20.
Magnolia is an example of a primitive dicot. This large tree, a native of southeastern swamp forests, is often cultivated for its huge flowers and evergreen foliage. (Courtesy of Van Baldwin.)

FIGURE 2-21.
Epilobium, or fireweed, is a member of the evening primrose family (Onagraceae). An inhabitant of disturbed areas, it often invades after fire has ravaged an area. (Courtesy of Van Baldwin.)

FIGURE 2-22.
Rhododendron, one of the many species of the heath family (Ericaceae). These large shrubs, which are native to many habitats, are cultivated for their evergreen leaves and showy clusters of flowers. (Courtesy of Van Baldwin.)

FIGURE 2-23.
*Ipomoea, or morning glory,
is a member of the sweet
potato family (Convol-
vulaceae). Many species of
this family provide a
starchy food source. (Courtesy of Van Baldwin.)*

FIGURE 2-24.
*Solanum, or nightshade,
is a member of the tomato
family (Solanaceae). This
species, however, is poisonous. The drug belladonna
comes from a related species.
(Courtesy of Van Baldwin.)*

FIGURE 2-25.
Gaillardia, a member of the sunflower family (Asteraceae-Compositae), is a native to prairie regions and to sandy areas and beach dunes in the Southeast. It is sometimes cultivated. (Courtesy of Van Baldwin.)

Selected References

ANDERSON, E. 1969. *Plants, Man, and Life.* University of California Press, Berkeley.

ARNETT, R. H., and D. C. BRAUNGART. 1970. *An Introduction to Plant Biology.* The C. V. Mosby Company, St. Louis, Mo.

BAILEY, L. H. 1975. *Manual of Cultivated Plants.* Macmillan Publishing Co., Inc., New York.

BOLD, H. C. 1973. *Morphology of Plants.* Harper & Row, Publishers, New York.

BOLD, H. C. 1977. *The Plant Kingdom.* Prentice-Hall, Inc., Englewood Cliffs, N.J.

FERNALD, M. L., and A. C. KINSEY (revised by R. C. Rollins). 1958. *Edible Wild Plants of Eastern North America.* Harper & Row, Publishers, New York.

GREULACH, V. A., and J. E. ADAMS. 1967. *Plants: An Introduction to Modern Botany.* John Wiley & Sons, Inc., New York.

JENSEN, W. A., and F. B. SALISBURY. 1972. *Botany: An Ecological Approach.* Wadsworth Publishing Co., Inc., Belmont, Calif.

KEETON, W. T. 1972. *Biological Science.* W. W. Norton & Company, Inc., New York.

RADFORD, A. E., W. C. DICKISON, J. R. MASSEY, and C. R. BELL. 1974. *Vascular Plant Systematics.* Harper & Row, Publishers, New York.

SHERY, R. W. 1972. *Plants for Man.* Prentice-Hall, Inc., Englewood Cliffs, N.J.

1. In what ways do air pollutants and herbicides deleteriously affect seed germination? Photosynthesis? Vegetative growth of plants?
2. What part of a grain seed contains the most protein? Starch? What is the nutritional significance of this as far as brown and white rice are concerned?
3. Why are certain species of plants such good indicators of water pollution? How can plants act as scavengers of heavy metal pollutants in our waterways and lakes?
4. What are some characteristics of green revolution crops, such as wheat and rice, that have resulted in increased yields? What are some results of increased use of fertilizers, pesticides, and irrigation water?
5. What are some of the practical ways by which we can induce plants to flower? To delay or prevent their flowering at certain times of year?
6. What is the best way to store fruits, seeds, and tubers to prolong their storage life?
7. What agents cause disease in plants? Can diseases be intensified by environmental factors?
8. When fruits such as grapes, currants, and apples are used to make wine, what basic fermentation reactions occur? How do you prevent wine from turning into vinegar?
9. How does the breakdown of compost to humus differ from the breakdown of plant material in a methane generator?

*Questions for
Consideration*

This chapter is concerned with five basic themes: seeds and seed germination, vegetative growth, the flowering process, fruit ripening, and decay of plants. All these topics are salient to our study of plants, people, and environment. They touch upon topics we take up in later chapters: air and water pollution, pesticides, green revolution crops, economic applications of botany, agricultural ecology, human uses of plants, recycling plants, and aquatic ecology.

Seeds are structures containing embryonic plants, varying amounts of nutritive tissue, and protective coverings. A typical grain seed, illustrating these parts, is shown in Figure 3-1.

*Seeds and Seed
Germination*

FIGURE 3-1.
Structure of the rice grain, which is typical for various cereals. (From D. F. Huston, ed., Rice Chemistry and Technology, *1972, p. 17. Used with permission of the American Association of Cereal Chemists, Inc., St. Paul, Minnesota, publishers.)*

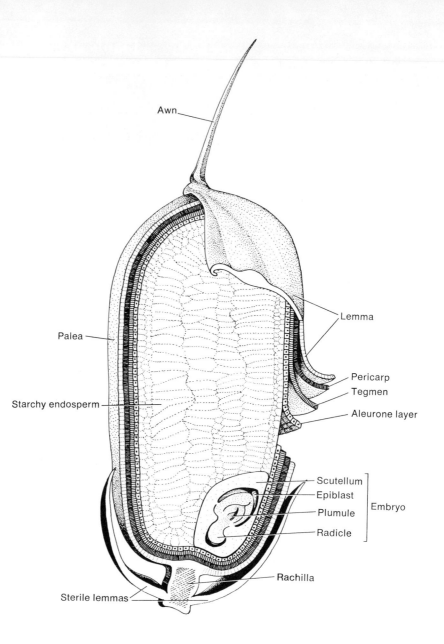

To germinate, seeds must take up water and oxygen and be able to release carbon dioxide from respiration. Many seeds have protective coverings that are hard and impervious to liquids (e.g., water) and gases (e.g., carbon dioxide and oxygen). Such seeds may lie in the soil for years before this hard seed coat is fractured by microbial action and weathering. This assures that, over a period of years, some seeds will germinate each year. The survival value of this type of behavior is obvious.

Seeds have other mechanisms to ensure that they do not all germinate in the same year. One involves the presence of chemical inhibitors in the seed. A hard rain will leach these inhibitors from the seeds and allow them to germinate when favorable weather conditions prevail. This is the case with many desert *annuals*. Following a heavy rainfall, the seeds of such plants will germinate, the plants develop, and a new crop of seeds will be set in a matter of a few weeks (Went, 1955). You can see the great survival value of this behavior.

The aleurone layer, illustrated in Figure 3-1, is a protein-storage "jacket" in the seeds of grasses (these include the grains we eat). This layer accounts for the color of brown rice. It contains most of the protein of rice seed, including some enzymes. One of these enzymes is amylase, which breaks down the starch in the endosperm tissue (Figures 3-2 and 3-3). In white rice, the aleurone layer is removed as rice polishings (Figure 3-4) and fed to livestock. People who eat rice as their major food rely very much on the aleurone layer; it is the major source of their protein. Other seeds, such as legumes, have most of their protein in the embryo and the cotyledons (seed leaves). In grains, the protein content is of the order of 8 to 10 percent of the seed's dry weight, whereas in legume seeds, protein content varies between 20 and 40 percent (Sinclair and deWit, 1975). Thus legume seeds are some of the most important sources of protein that we can obtain from plants.

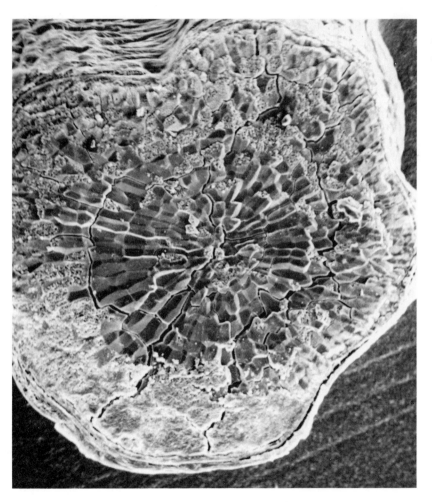

FIGURE 3-2.
Scanning electron micrograph of endosperm tissue in a rice grain. ×60. (Courtesy of P. Dayanandan.)

FIGURE 3-3.
*Scanning electron micro-
graph of starch grains in
the rice endosperm. ×2500.
(Courtesy of P.
Dayanandan.)*

The endosperm tissue of most seeds consists mainly of starch (Figures 3-2 and 3-3). As water is taken up by a seed and germination proceeds, the starchy endosperm breaks down as a result of the action of amylase, releasing soluble sugars utilized for the growth of the embryo. This release of sugar during seed germination is also utilized in synthesizing the alcohol desired by beer brewers. In such cases, barley seeds are germinated in an oxygen-poor atmosphere, which favors the alcoholic fermentation reactions involved in beer making.

Now that you are familiar with the basic structures in seeds and some of the causes of dormancy in them, let us look at a list of some of the methods successfully used to break seed dormancy. These include the following:

1. Scarification (abrasion) of the seed coat down to the endosperm or cotyledons (seed leaves) with a grindstone or file. This method is practical for seeds with hard seed coats, as in members of the legume family.

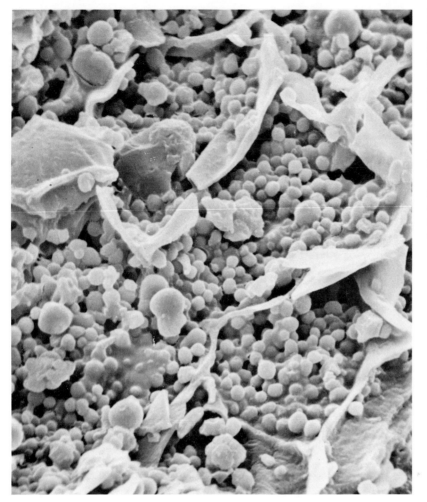

FIGURE 3-4.
Scanning electron micrograph of rice polish. The granules represent aleurone particles, which contain large amounts of storage protein. Polish is obtained in the milling of white rice, and it is often fed to hogs. As a result, the hogs get most of the protein from the rice seed, and humans get the bulk of the carbohydrate in the form of starch. ×1600. (Courtesy of P. Dayanandan.)

2. Another method for hard-coated species involves soaking the seeds in concentrated hydrochloric or sulfuric acid for 30 minutes, then washing the seeds with water.
3. Some hard-seeded species of plants, such as cultivated morning glory (*Ipomoea purpurea*), can be immersed in boiling water for about 30 seconds to induce the seeds to germinate.
4. For seeds that normally overwinter under snow, the best treatment is *stratification* or *after-ripening*. To do this, one simply puts the seeds between two layers of cheesecloth or muslin, places this "sandwich" in a mixture of moist sand and peat moss (1 : 1 mixture), and stores the seeds in this moist substratum in the cold around 0°C (32°F) for 4 to 6 weeks. This treatment usually is effective in causing the breakdown of chemical inhibitors in the seeds and otherwise releasing them from their dormant state.
5. Seeds with rudimentary embryos can be grown on nutrient agar supplemented with vitamins and sugar under sterile conditions. For orchids, green or brown seed pods are surface-sterilized in 10 percent

Clorox for 20 to 30 minutes, washed in sterile distilled water, the pods cut open with a sterile scalpel, and the seeds taken out and placed on a special orchid medium in glass bottles to germinate. Further directions for doing this kind of sterile culture work are given in works of Hartmann and Kester (1975), Withner (1975), and the American Orchid Society (1975).

6. Many seeds have a *light requirement* for germination, or have their germination greatly enhanced by light. In most seeds, the best way to fulfill this need is to illuminate the seeds with red light (either a ruby-red bulb or red cellophane over the seeds) after they have taken up water. The light treatment may be as short as 5 minutes (e.g., lettuce seeds); others may require as much as 60 minutes of exposure of light to germinate.

Concerning seeds, the question of how the storage life can best be prolonged is often raised. The answer is relatively simple for most seeds. They should be stored in a dry, cool place in recycled glass bottles with loosely fitting lids, or in plastic bags or envelopes. Dryness prevents the seeds from rotting, and a relatively cool temperature (about 4°C) will greatly slow the respiration processes that lead to "burning up" the stored food in the seeds. Some seeds, such as willows and maples, cannot be stored for any period of time and remain alive. They should be sown immediately after harvest. Seeds of many vegetables only remain viable (alive) for several years in storage so should be discarded after 2 to 3 years if not planted. Seeds of many woody plants can be stored for much longer periods, some up to 20 years (as in many pines and spruces). Seed storage is important to us because many seeds are very expensive and harvesting and storing your own seeds saves a tremendous amount of money.

Clearly, seeds and seed germination are of central importance to people and their environment. You will read in later chapters that seedlings can be very susceptible to herbicides and air pollutants; that many green revolution crops have yielded new varieties that have *less* protein and more carbohydrate in their seeds; that seeds can carry large quantities of pesticide residues and many microorganisms (e.g., fungi) that cause damping-off disease of seedlings after germination; and that germinating seeds have considerable commercial value (e.g., the seeds of barley and rice are the basis of brewing beer and saké, respectively).

Vegetative Growth of Plants

After seeds have germinated, and seedlings have emerged, plants begin to harvest light energy from the sun through the process of photosynthesis. Life on earth depends on photosynthesis; all our food is, and all our fossil fuels were, derived from products of photosynthesis. Table 3-1 shows how much carbon is fixed by plants in forests, the oceans, croplands, grasslands, and deserts. It is readily apparent that marine plants fix the largest amount of carbon, through photosynthesis, on the earth's surface. Pollution of the oceans by oil, garbage, industrial wastes, landfill material, heavy metals, and so on, can drastically reduce

TABLE 3-1 / Estimated World Production of Organic Carbon from Photosynthesis

Type of Area	Area $(km^3 \times 10^6)$	Organic Carbon per km^2 per Year (metric tons)[a]	Total Annual Production of Organic Carbon (metric tons $\times 10^9$)[a]
Forest	44	250	11.0
Cultivated land	27	160	4.3
Grassland	31	36	1.1
Desert	34	7	0.2
Total land	136	122	16.6
Ocean	371	340 ± 120	126 ± 57

[a] One metric ton = 1000 kilograms.

Source: Data compiled from Schroeder, *Naturwiss.* **7**, 27 (1919); Ritey, *Amer. Sci.* **32**, 132 (1974).

photosynthesis in marine plants, and in so doing, greatly disrupt food chains that operate in marine environments (see Chapter 8). Photosynthesis can also be greatly impaired in terrestrial environments by herbicides such as 2,4-D and 2,4,5-T, and by air pollutants such as smog, sulfur dioxide, nitrogen oxides, chlorine, and carbon monoxide. All of these agents can cause massive destruction of photosynthetic tissue by stimulating chlorophyll breakdown (see Chapter 9). In addition, herbicides cause plugging of plants' vascular systems (which shuts off transport of sugars, water, and nutrients in the plant) and sharp increases in respiration rates, "burning up" (oxidizing) needed photosynthetic products.

The vegetative growth of plants is dependent not only upon photosynthesis but also on water and on mineral nutrients in the soil and in water systems. Many of us have seen one of the effects of excess nutrients from agricultural lands and polluted water systems in our lakes and ponds—algal blooms. These blooms are particularly accentuated by high levels of phosphates from detergents in our water systems and by the vast quantities of nitrogen- and phosphorus-containing fertilizers needed to grow modern, improved varieties of soybeans, corn, wheat, and rice. These "green revolution" crops have higher yields, grow faster, and mature earlier than older varieties. In addition, the green revolution grains, wheat and rice, have less tendency to lodge (fall over due to action of rain or wind) and produce more lateral shoots (called tillers). These characteristics result in more food at harvest time, but at the expense of increased use of fertilizer.

The principle that fast-growing plants require large amounts of fertilizer (mineral salts) is also evident in the water hyacinth, *Eichhornia crassipes* (Figures 8-7 and 8-8). This plant, which occurs as a weed in southern United States waterways and ponds, has an enormous rate of vegetative growth. Recently, it was found that the water hyacinth is a scavenger for heavy-metal pollutants such as lead, zinc, and mercury. These plants literally take up huge quantities of mineral salts from the water, including the heavy-metal pollutants, to such an extent that they concentrate the minerals within their cells to levels far in excess of the concentrations found in the water. This ability of the water hyacinth is being utilized to depollute many water systems, and other aquatic plants, now being studied, will be utilized to remove water pollutants in the future.

Studies of mineral nutrition contributed to the development of the green revolution crops by providing plant breeders with an understanding of plant foods. Another spin-off from studies on mineral nutrition in plants is *hydroponics.* This method is used commercially to grow vegetables in greenhouses during the winter in countries of the north temperate zone. The method involves growing the plants in a liquid culture, where the liquid (water) contains salts of the major elements (nitrogen, phosphorus, potassium, sulfur, calcium, magnesium) and the minor elements (iron, molybdenum, manganese, boron, copper, cobalt, chlorine, zinc) essential for plant growth.

The Flowering Process

We all know that most seed plants come into flower after a period of vegetative development. But it is not quite as simple as this. There also are environmental cues for flowering in many plants, which include night length, temperature, or a combination. Let us look at a few examples.

Most plants have no specific day-length requirements for flowering; these are termed day-neutral in their light requirements for flowering. However, it was discovered by Garner and Allard in about 1920 that some varieties of tobacco and soybeans flowered only when given short days—less than 10 hours of daylight (these are called short-day plants). It was discovered later that other plants, such as spinach and barley, require long days to flower (long-day plants). It has been found that the length of the night period is the critical controlling factor in inducing short- or long-day plants to flower. Interrupting a long night (15 hours) with a 5-minute flash of dim light in the middle of the night has the effect of transforming the night into a "short" one. Short-day plants will not flower under such conditions, whereas long-day plants will.

You can now see why, when you grow poinsettias or chrysanthemums, both of which are short-day plants, you must subject the plants to a certain number of long nights for them to flower. Most people keep their poinsettias under long-night regimes until visible flowers appear. Commercial growers of chrysanthemums and poinsettias use overhead lights to lengthen the day, and black cloth to shorten the day. Under the long days, the plants are grown vegetatively until they reach an optimum market size; then they are switched to short days to induce flowering. With such combinations, it is now possible to produce flowering chrysanthemums at any time of the year. This timing of the onset of short days is especially critical with such plants as poinsettia and Christmas cactus, which are used at specific times of the year.

Temperature is also of major importance to the control of flowering. Many plants must experience a period of low temperature before they will flower. These include many bulbs and deciduous trees and practically all biennials (such as carrot and sugarbeet). The grains winter wheat and winter rye also flower only after they have overwintered, being exposed to chilling winter temperatures below freezing. In contrast, spring wheat and spring rye have no such requirement and are planted after the winter is over. The process of artificially chilling seedlings, such as those of winter wheat and winter rye, to bring about flowering is called *vernalization*.

The flowering process, as you are now beginning to see, has many important implications for people and their environment. We use low temperatures to force bulbs into flower in the middle of winter, thus providing spring flowers in the home out of season. We use short daylength to bring poinsettias into bloom for the Christmas season. Farmers rely on low winter temperatures to grow winter wheat and rye. Pineapple growers use synthetic hormones (similar to the native, auxin-type hormones found in plants) to induce pineapple to flower in the field. Greenhouse growers have found that the hormone, ethylene, can induce bromeliads (members of the pineapple family) to flower. In Chapter 10, you will learn that herbicides such as 2,4-D are used to prevent plants such as ragweed from growing in urban lots. Ragweed flowers (Figure 3-5) produce copious amounts of pollen (Figure 3-6), which causes "hay fever" (correctly called allergic rhinitis; see Chapter 9) in millions of people. Pollen from such plants is considered to be a serious air pollutant called an *aeroallergen* (see Chapter 19). Finally, you will learn that there are some very delicate ecological relationships between certain insects and other animals and flowering plants in the pollination process. Some of the most beautiful examples are to be found in the tropics; these are described in Chapter 7.

Fruit Ripening

In the flowering plants (the angiosperms), a fruit is a mature, ripened ovary or a group of ovaries. Some fruits are more complicated, by having included with the ovary or ovaries, adjacent parts which are fused to the ovary tissue. We need only be concerned here with simpler types of fruits. The structure of the apple is shown in Figure 3-7. It is the ovary tissue of the fruit that regulates its development, through hormonal control, and is most intimately involved in the ripening process.

Most fruits develop very quickly after pollination, their increase in size being controlled largely by hormones. This fact is utilized to increase fruit size by means of hormone sprays. Gibberellin-type hormones are used to increase berry size in grape, and auxin-type hormones to increase the size of figs. Hormones are also employed to increase *fruit set* (number of fruits that develop after pollination) in tomato, pear, apple, fig, and apricots (Weaver, 1972). In addition, the auxin-type hormones are sprayed on apple trees after pollination, as the young apples are first developing, in order to prevent fruit drop, which is a natural *thinning* process. Such treatments can result in significant increases in yields of apple fruits on the tree.

FIGURE 3-5.
Scanning electron micrograph of the male flower of a giant ragweed plant Ambrosia trifida. Ragweed pollen is a major cause of hay fever in the United States. Because of this, it is called an aeroallergen. ×50. (Courtesy of P. Dayanandan.)

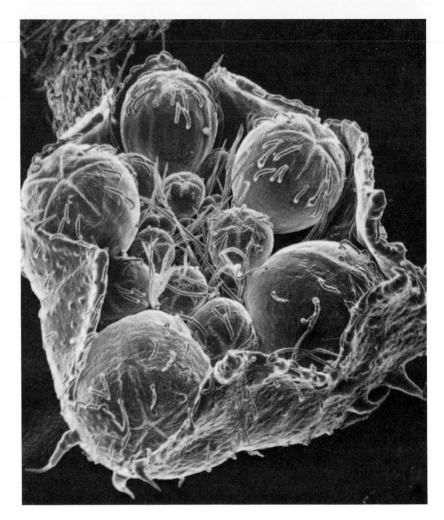

One of the most interesting of these treatments with hormones is that with figs. Normally, fig flowers are pollinated by a certain type of wasp. In regions such as California where the wasp may be absent or be present only in small numbers, the trees are sprayed with auxin-type hormones during flowering. This results in good fruit set. It also results in seedless figs, called *parthenocarpic fruits.* Some varieties of fruits are naturally seedless or parthenocarpic; such genetically selected varieties include Thompson's seedless grape, seedless oranges, and a watermelon selection from Japan.

Fruit maturation is also under hormonal control. The key ripening hormone is ethylene. This hormone is normally produced by all parts of the plant, but is especially copious in maturing fruits. Apple fruits give off so much ethylene in storage that they can cause other kinds of fruits, as well as vegetables, held in storage with the apples to overripen very quickly. To prevent this, it is best not to store apples in the same room with other foods.

Since apples give off so much ethylene while being stored, this in itself causes the apples to ripen in storage more quickly than if the ethylene

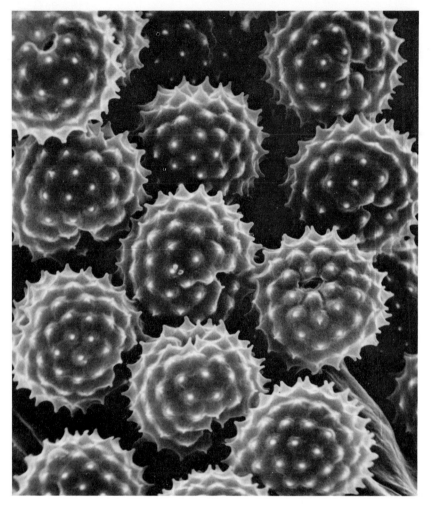

FIGURE 3-6.
Pollen grains from one of the anthers (pollen sacs) of the male flower of a ragweed plant. (Courtesy of P. Dayanandan.)

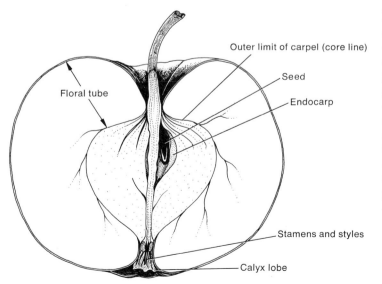

Outer limit of carpel (core line)

Seed

Endocarp

Floral tube

Stamens and styles

Calyx lobe

FIGURE 3-7.
Longitudinal section through an apple fruit, indicating the basic structural parts of this fruit. Apple fruits have five carpels. Each carpel contains one to several seeds. The seeds are mature ovules that lie within the endocarp tissue. It is the fleshy floral tube tissue that we eat in apple fruits. (Courtesy of Linda Kaufman.)

were removed. To prolong the storage life of the apples, the ethylene they give off can be removed with brominated activated charcoal. In addition, the ripening process can be slowed further by increasing the partial pressure of carbon dioxide and decreasing the partial pressure of oxygen in the storage room. This procedure slows down the rate of respiration—the rate of "burning up" (oxidation)—of the sugars and organic acids in the ripening apples. The carbon dioxide also inhibits the action of ethylene. This combination of treatments allows apple growers to store apples for periods up to 9 months in a good edible condition, instead of the 2 to 3 months encountered if no changes are made in the storage-room atmosphere.

Rather than slow down the rate of ripening of fruits, it is sometimes desirable to accelerate the process. This is necessary with green bananas so that they can be marketed in a ripe condition. Ethylene gas is used to accelerate the ripening of bananas. In the home, you can use an apple as the source of ethylene to ripen your bananas more quickly; simply place the bananas with an apple in a paper bag. This method can also be applied to other unripe fruits, such as pineapples or avocados.

Decay of Plants

The decay of plants is extremely important to people and their environment. It results in the formation of humus from forest floor litter, compost from a combination of soil and plant remains (e.g., grass clippings, cut tops of plant leaves, and straw), and methane from plant remains decomposed in a methane generator. These decay processes involve microorganisms that are able to convert the plant tissue to humus, compost, or methane (and other gas by-products). Anaerobic bacteria (growing in the absence of oxygen) in several genera are capable of producing methane. Aerobic bacteria (growing in the presence of oxygen) and many fungi carry out the breakdown of forest litter to humus and plant debris to compost. The result of all this decomposition activity is the recycling of both carbon and mineral nutrients for use by germinating seedlings and plants growing in the forest or garden. Composting and methane generators are taken up in more detail in Chapter 20.

The decay of plants is a problem to people as well as a help. The storage of apples, potatoes, and other foods is complicated by the presence of bacteria and fungi, which cause rot and surface blemishes. Many dollars are spent washing apples and peaches and treating them with fungicides to prevent the growth of such rot-causing organisms as *Penicillium* and *Botrytis*. Many dollars are lost when an organism does get the upper hand in a storage warehouse. Parasitic organisms also cause disease in growing plants, which again results in the expenditure of time and money in the control of disease, and the loss of money and crops when disease strikes.

Agents causing plant disease include bacteria, fungi, viruses, parasitic higher plants, such as mistletoe; nematodes, which are little worms; and abiotic factors, such as air pollution and improper application of fertilizers and herbicides. Under certain environmental conditions, diseases can be extremely severe. In 1970, a warm, wet fall provided ideal conditions for an outbreak of *Cylindrocladium* black rot in peanut crops in Virginia, resulting in financial disaster for farmers in the area.

Disease is an ever-present fact of life for all organisms. Contrary to the claims of some organic gardeners, all diseases of plants are not induced by modern agricultural practices. For instance, the American chestnut was exterminated as a forest tree by the chestnut blight, caused by a fungus, *Endothia parasitica,* imported from Asia. Before the blight, there were extensive, virgin stands composed largely of chestnut which had been untouched by man. The blight fungus was so virulent that it killed back to the ground a population of trees that had covered the equivalent of 9 million acres of land. This was a very organic process, which had nothing to do with the application of fertilizers or other man-made substances to the earth. The point of this is that it is sometimes necessary to use chemicals to control plant diseases. Otherwise, there simply would not be enough food to go around.

One of the most interesting aspects of decay involves fermentation reactions carried out by various strains of a yeast fungus, *Saccharomyces cerevisiae,* resulting in the formation of wine. The fermentation process is quite simple. Yeast converts the sugars from the wine ingredients, by means of a series of reactions, into alcohol and CO_2. This conversion occurs very rapidly in the first 3 to 5 days. This is why CO_2 bubbles are evident and why containers are not sealed right away. This is called the *primary fermentation* and produces an alcohol content of 12 to 13 percent. The *secondary fermentation* takes about 3 weeks and produces a final alcohol content of 14 percent for most wines. The yeast responsible for fermentation can only tolerate a certain amount of alcohol, about 14 percent. This naturally limits the strength of wine. Beverages of higher potency can only be made by distillation, and this is illegal in the United States without government license. In the secondary fermentation stage, it is desirable to exclude oxygen but to allow the CO_2 to escape. "Fermentation locks" can be made or purchased for this purpose. In making wine, baker's or brewer's yeast is generally used to carry out the alcoholic fermentation. Care must be taken to prevent the growth of vinegar bacteria, which carry out fermentation reactions that lead to the formation of acetic acid. This type of fermentation is usually prevented by the use of fermentation locks to keep the bacteria out, high acidity to kill the bacteria, or heat sterilization at every step. Acetic acid (vinegar) can also form from the nonbiotic oxidation of alcohol; this reaction can be prevented by the use of fermentation locks and the addition of sodium metabisulfite as an antioxidant. In Chapter 4, we shall take up a practical case of alcoholic fermentation—the making of dandelion wine. Beadle (1975) also gives you recipes for brewing your own wine.

Summary

You should now have a better understanding of the importance of seed germination, vegetative development, flowering, fruiting, and decay of plants to people and their environment. These processes touch on many environmental problems and human uses of plants: quality of protein in seeds, growth characteristics of green revolution crops, storage of seeds and fruits to prolong their storage life, production of methane from crop residues, composting, wine making, and induction of flowering by use of environmental cues or hormones with some of our crop plants and house plants. Each of these problems or applications is discussed in detail, and their relevance to people and their environment is elaborated on in depth. Our discussion of plants at work from seed germination to decay continues throughout the book.

Selected References

American Orchid Society. 1975. Meristem tissue culture. Selected articles from *American Orchid Society Bulletin*, 1–72.

BEADLE , LEIGH. 1975. *Brew It Yourself*, 3rd ed. Farrar, Straus & Giroux, Inc., New York.

GREULACH, V. A. 1973. *Plant Function and Structure*. Macmillan Publishing Co., Inc., New York.

HARTMANN, H. T., and D. E. KESTER. 1975. *Plant Propagation. Principles and Practices*, 3rd ed. Prentice-Hall, Inc., Englewood Cliffs, N. J.

RAY, P. M. 1972. *The Living Plant*, 2nd ed. Holt, Rinehart and Winston, New York.

RAYLE, D., and L. WEDBERG. 1975. *Botany: A Human Concern*. Houghton Mifflin Company, Boston.

SINCLAIR, T. R., and C. T. DEWIT. 1975. Photosynthate and nitrogen requirements for seed production by various crops. *Science* **189**, 565–567.

WEAVER, R. J. 1972. *Plant Growth Substances in Agriculture*. W. H. Freeman and Company, Publishers, San Francisco.

WENT, F. W. 1955. The ecology of desert plants. *Sci. Am.* **192**(4), 68–75.

WITHNER, C., ed. 1975. *The Orchids: Scientific Studies*. John Wiley & Sons, Inc. (Interscience Division), New York.

Economic Applications of Plants
Peter B. Kaufman and P. Dayanandan

4

Questions for Consideration

The human use of plants and plant products is as old as human beings themselves. From an early dependence on the plant kingdom for food and perhaps for protection, the use of plants has helped to create the very civilizations of human beings, including the modern one. The plant kingdom has provided people with food as cultivated and wild plants; drugs and medicines; wood for fuel, ships, railways and houses; plants for flowers and landscaping; fibers for rope and cloth; dyes for wool and fiber; feed for farm animals; substrate for beverages; fossil coal and gas; and a host of substances (such as latex, pectins, gums, resins, oils, and waxes) that make our lives comfortable and spicy. In this chapter, we shall look at some specific examples of interest. In Chapters 14 through 17 we shall probe even further into specific uses of plants by various peoples.

Medicinal, hallucinogenic, and narcotic drugs are closely interrelated, both because of their early use to cure various human ailments and because of their chemical similarities. Besides the microorganisms, which provide the important antibiotics, the higher plants yield several alkaloids, glycosides, resinoids, and essential oils that are used as drugs. A host of herbs, some well known to us, others little or unknown, are used all over the world for curing a number of minor ailments. In this category fall such herb teas as camomile (*Anthemis tinctoria*), mormon tea (*Ceanothus spp.*), mint teas (*Mentha spp.*), violets (*Viola spp.*), sassafras

Medicinal, Narcotic, and Hallucinogenic Plants

TABLE 4-1 / Examples of Plants That Have Medicinal Value

Plants	Drugs	Uses
Atropa belladonna	Atropine, hyoscyamine, and scopolamine	Heart stimulant, dilates pupils
Cinchona officinalis	Quinine	Antimalarial
Digitalis purpurea	Digitalin	Heart stimulant
Erythroxylon coca	Cocaine	Pain killer
Ephedra spp.	Ephedrine	Bronchial remedy
Papaver somniferum	Opium, morphine	Pain killer
Rauwolfia serpentina	Reserpine	Against hypertension

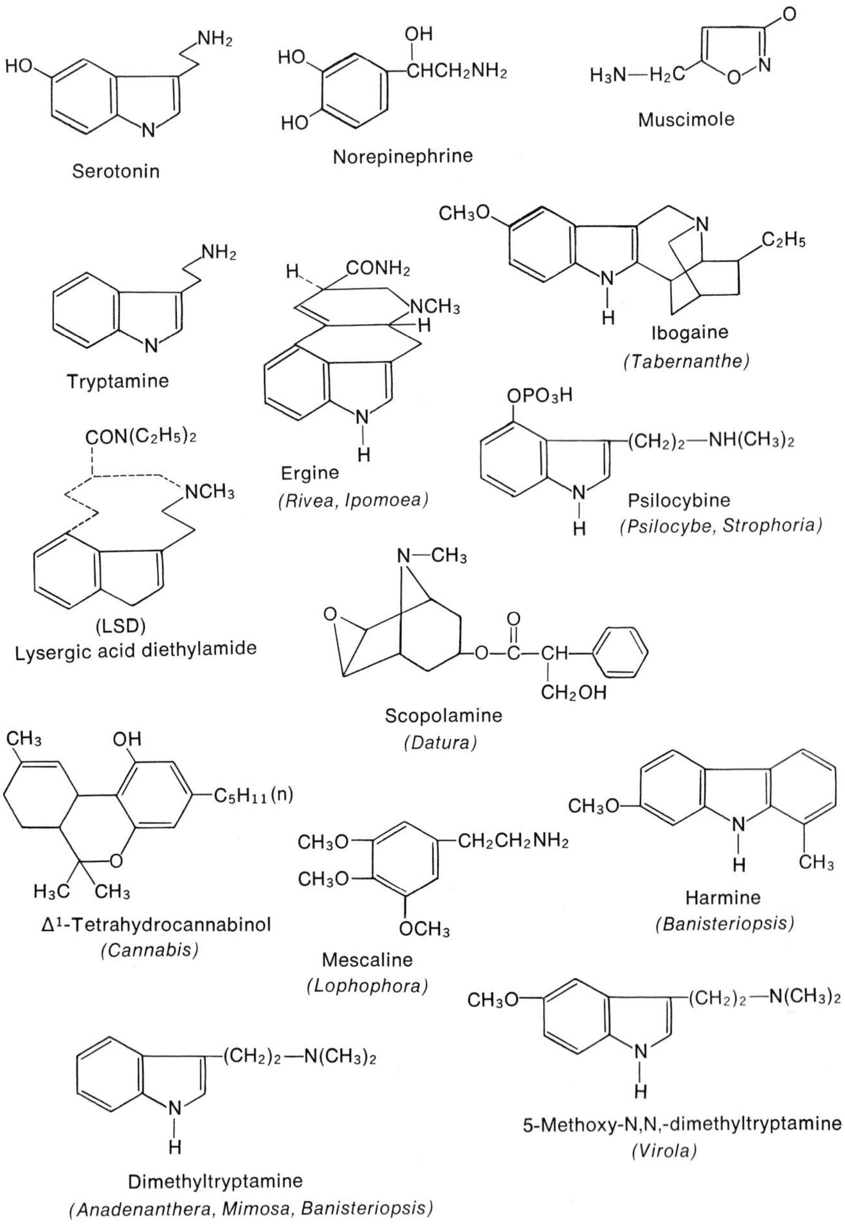

FIGURE 4-1.
Chemical nature of various hallucinogens.

(*Sassafras albidum*), and ginseng (*Panax quinquefolium*). In Table 4-1, we list a number of plants that are of medicinal value, the drugs obtained from them, and their uses.

In spite of its enormous popularity in recent years, narcotic and hallucinogenic use of plant products has been with us since antiquity. Among the nearly 350,000 species of plants, fewer than 100 are used for these purposes: almost all are flowering plants, although a few are fungi. Their uses range from the socially accepted coffee, tea, and tobacco, to such powerful hallucinogens as the peyote and psilocybe, to the very addicting cocaine and opium. The greatest concentration of hallucinogenic plants known is found in the New World. On a chemical basis, the hallucinogens can be divided into two groups. Most fall in the group that contains nitrogen as part of their chemical structure, and most of these are alkaloids. The nonnitrogenous group is exemplified by a well-known product of marijuana, tetrahydrocannabinol, and a compound from nutmeg (*Myristica fragrans*). Most of the alkaloids have an indole ring as a basic structure and resemble serotonin, a nerve hormone. Figure 4-1 indicates the chemical nature of some hallucinogens and their resemblance to serotonin and norepinephrine, another nerve hormone.

Although not as powerful as the major hallucinogens, other narcotic plants are, in fact, more important from a commercial point of view. These include such stimulants as coffee and tea. Table 4-2 lists the major narcotic plants in use.

The structure of hallucinogenic plants is now becoming better understood by botanists. We present below examples of the details that are to be found on a single plant of great importance, *Cannabis*. *Cannabis sativa* contains about 30 different cannabinoids, of which tetrahydrocannabinol is the major hallucinogenic component. These compounds are

TABLE 4-2 / Narcotic Plants

Plant	Drugs	Nature of Drug
Amanita muscaria	Muscimole	Hallucinogen
Claviceps purpurea	Ergoline	Hallucinogen
Psilocybe mexicana	Psilocybine	Hallucinogen
Lophophora williamsii	Mescaline	Hallucinogen
Cannabis sativa	Tetrahydrocannabinol	Hallucinogen
Rivea and *Ipomoea*	Ergine	Hallucinogen
Datura	Scopolamine	Hallucinogen
Banisteriopsis caapi	Harmine	Hallucinogen
Virola theiodora	Dimethyltryptamine	Hallucinogen
Anadenanthera peregrina	Dimethyltryptamine	Hallucinogen
Sophora secundiflora	Cytisine	Hallucinogen
Erythroxylon coca	Cocaine	Hallucinogen
Papaver somniferum	Morphine, opium	Sedative
Nicotiana tabacum	Nicotine	Stimulant
Coffea arabica	Caffeine	Stimulant
Thea sinensis	Caffeine	Stimulant
Theobroma cacao	Caffeine	Stimulant
Cola nitida	Caffeine	Stimulant

said to occur in glands over the surface of stem, leaves, and flowers. Figures 4-2 to 4-6 represent some of these glands as well as other details of the marijuana plant.

Dyestuff Plants

Table 4-3 lists a selected group of plants used for the preparation of natural dyes. Obviously, many other plants are sources of natural dyes. We simply present these to show the diversity of plants and plant parts that are used as sources of natural dyes.

The material for extracting dyes and for dyeing silk, wool, or other fibers, making the dyes fast or permanent with chrome, aluminum, tin, and iron mordants, are treated in detail in a work of the Brooklyn Botanic Garden (1973).

Beverage Plants

WINE MAKING

Give strong drink to him who is perishing and wine to those in distress; let them drink and forget their poverty, and remember their misery no more.

Proverbs

FIGURE 4-2.
Glandular and nonglandular hairs of marijuana as they occur at the leaf blade and petiole joint. In addition to the fully mature hairs, some hair initials are also seen. Some glandular hairs are in two- and four-celled stages. ×230. (Photograph by P. Dayanandan.)

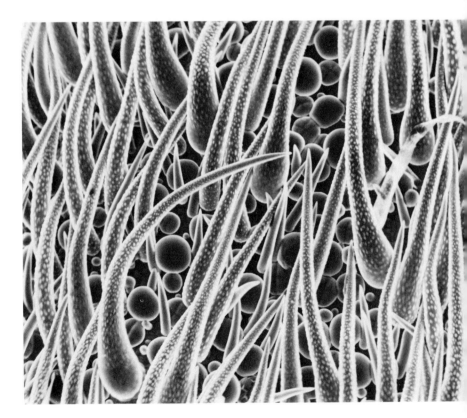

TABLE 4-3 / Plants Used for the Preparation of Natural Dyes

Plant	Parts of Plant Used	Color of Dye Prepared from Plant[a]
Tagetes (marigold)	Flowers	Gold, yellow, buff
Crocus sativus (crocus)	Anthers	Yellow (saffron)
Rhus typhina (staghorn sumac)	Roots, berries	Yellow, khaki brown
Solidago (goldenrod)	Flowers	Yellow-gold
Populus (aspen)	Leaves	Gold, green-gold
Rhododendron	Leaves	Gray
Allium cepa (onion)	Bulb scales or "skins"	Gold, orange, or red from onion skins
Rubus (blackberry)	Young shoots	Light gray
Dahlia	Flowers	Chartreuse, yellows, orange
Oak galls from *Quercus* (oak)	Galls	Browns, grays, blacks
Lichens	Thalli	Purples, greens, yellows, browns
Indigofera tinctoria (indigo)	Leaves	Blue

[a] These are typical colors. Colors will vary for a given dye source, depending on the mordant used.

The cultivation of grapes for wine making began in Armenia about 3000 years ago. The wonders of this beverage soon spread. The Greeks showed their admiration for wine by creating a god, Dionysus, in its honor. The Romans later spread their wine-making knowledge throughout their conquered European lands. When Leif Ericson came to America in A.D. 1000, he named the land "Vinland" because of the abundance of wild grape plants growing on the eastern coast of this new world. Twenty-six years after Columbus came to America, Cortez ordered wine to be made in what is now Mexico and southern California. The Franciscan friars there were vital in establishing this area as an important wine-producing center.

It is no longer necessary to rely on California, New York, or even Europe for your wine. It is possible for you to prepare delicious wines from plants that grow wild near your home.

Wine can be made from almost any part (fruit, flower, stem, leaves, roots, bark) of almost any plant. Generally, if the plant part smells and tastes good, it will make fine wine. Be creative, but *be careful!* Many plants are deadly toxic and must never be consumed (e.g., oleander, alamander, rhubarb leaves, foxglove, poison sumac, and castor beans).

For making good wine, here are a few very basic points to consider:
1. Use nontoxic plants that are in full bloom, picked early in the day.
2. Keep all wine-making equipment clean and aseptic. Wine is spoiled easily.

FIGURE 4-3.
Upper leaf surface of marijuana. In addition to the short and long hairs, three glands are also seen. No stomata are evident on this surface. ×250. (Photograph by P. Dayanandan.)

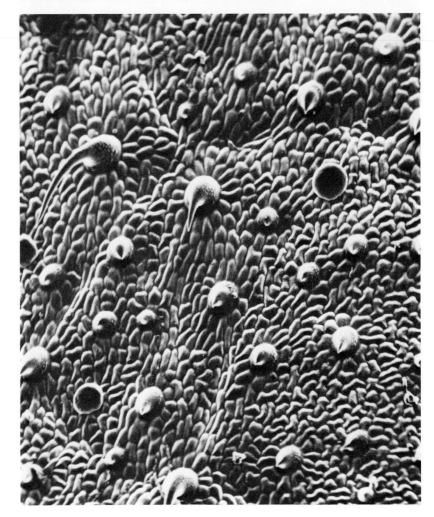

3. Do not use metal containers. Enamel is fine if it is not chipped.
4. Give your wine time to completely ferment. Do not rush the wine and drink it all before it is ready. Many flower wines reach their peak after a year or more. Be patient.

Following are two recipes for wine making. We hope that you will enjoy these drinks enough to encourage you to try your hand at making other beverages from plants that grow around you.

MAKING DANDELION WINE

1. Collect blossoms that are fully open (Figure 17-20). Remove only the heads; stems impart a bad flavor to the wine.
2. Place 5.6 liters (6 quarts) of dandelion blossoms in a crock or other suitable container.

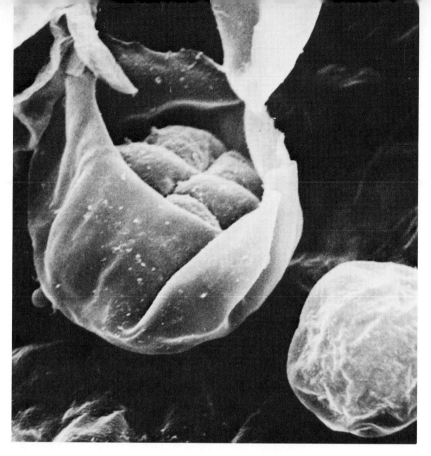

FIGURE 4-4.
Gland of marijuana whose outer envelope has been ruptured to show the cells inside. The resinous secretions of these cells accumulate inside the envelope until the latter gets over-filled and ruptures. The head is made up of eight cells. ×1125. (Photograph by P. Dayanandan.)

3. Add 3.7 liters (4 quarts) of boiling water and let stand for 4 days.
4. Remove the blossoms, squeezing a little.
5. Now add the following:
   1 cake of yeast
   2000 milliliters (8 cups) of sugar
   2 lemons—sliced with rind
   2 oranges—sliced with rind
6. Let stand 4 to 5 days. Then, strain and seal in jugs or bottles. Wire on corks on heavy bottles to prevent exploding.
7. Wine may be sampled any time after 6 to 8 weeks.

MAKING GRAPE WINE[1]

1. Crush grapes and dilute juice $\frac{1}{3}$ with water (to reduce final acidity of wine).
2. Bring specific gravity up to 1.0899 by adding sucrose (use a hydrometer).
3. Add yeast starter, and when fermentation begins, punch down and mix grape material and juice every 4 hours.

[1] Recipe furnished by Robert Lowry, Department of Botany, University of Michigan, Ann Arbor. Note that specific gravity is determined with a hydrometer, which can be obtained in most party stores or shops that have wine-making kits.

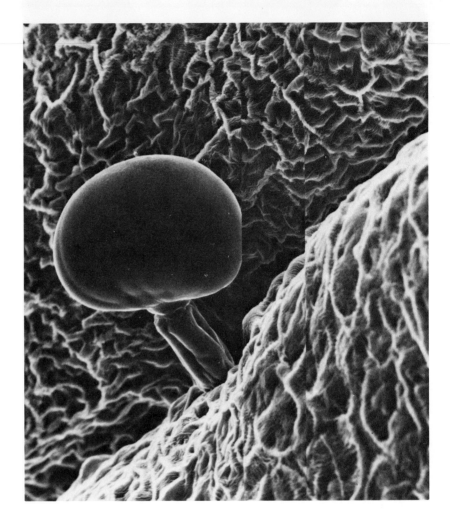

FIGURE 4-5.
Glands such as this occur on the surface of an anther of a male flower of marijuana. These glands have resin-filled heads and long stalks. ×250. (Photograph by P. Dayanandan.)

4. When the specific gravity reaches 1.0001, siphon the wine from under the cap of skins into a carboy 18.9-liter (5-gallon) glass jug and seal with a gas lock (fermentation lock).
5. Press the skins and add the press wine to the above.
6. When fermentation in the carboy is completed and the wine clears, rack it off the sediment into a clean carboy with a gas lock.

TEA PLANTS (*CAMELLIA SINENSIS*) FOR MAKING THE BEVERAGE, TEA

We select as our example the tea-growing region of the USSR in the Republic of Georgia. Tea plants are native to China. In Georgia, the plants are in horizontal rows in hedges on slopes approximating 15° on mountainsides where the rainfall averages 148 centimeters (58 inches) per year.

Over 96 percent of Russian tea is grown here. The tea leaves, buds, and

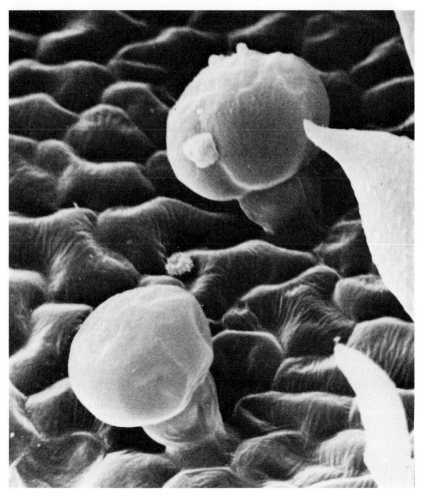

FIGURE 4-6.
Some glands of marijuana only possess a head made up of one to four cells. The nature of the chemical activity in these glands or in those with eight cells is not known. ×1000. (Photograph by P. Dayanandan.)

young shoots are harvested from May to October, with harvests made every fifth day. In the tropics, the harvesting is done throughout the year. This harvesting is mostly done by hand, but on large state farms, mechanical harvesters with "mechanical fingers" do it, replacing 40 to 50 workers for each machine. The young buds make the best tea. In the USSR, the greatest amount of growth occurs in May and June, coincident with a period of higher rainfall. The plantations must be irrigated in the summer, because of the much lower rainfall. With so many harvests, the plants require a great deal of fertilizer, especially nitrogen (N), phosphorus (P), potassium (K), and calcium (Ca). About 1 metric ton (over 1 ton) of fertilizer per hectare is applied per year, usually in the winter. Fungal pests are controlled by applying a nonpersistent pesticide in the winter, when the tea leaves and buds are not harvested, thus causing no danger to the consumer. In winter, the plants are also pruned.

Tea harvesting and processing for making black tea involves the following steps. The buds and leaves are harvested, and within 1.5 to 2 hours they are taken to the "tea factory." The harvested parts are withered in the shade (allowed to wilt), then dried. After this, the leaves are subjected to heat and high humidity, and concurrently are rolled and squeezed. The "squeezate" (sap), used to promote fermentation, and the pressed leaves are combined and allowed to ferment until the proper taste and aroma develop. The tea is finally dried and packed. For production of green tea, *no fermentation is used.* This kind of tea is preferred by many people in central Asia, Japan, and Mongolia. A green tea, called "pressed tea," is also made from coarse or more mature leaves (not the tips) and prunings; these parts are pressed, then used to make a soup in some of the oriental republics of the USSR. Some of the branches pruned from tea bushes are used in making caffeine for medicine.

Propagation of tea plants is by seeds, sown in the spring. In 3 years, the first crop is produced. The shrubs may produce for as long as 100 years. Pruning, fertilizing, and pest control provide work for laborers during winter months.

The laborers receive about 25 kopeks per kilogram of leaves and earn up to 200 to 300 rubles per month during harvest season. The average salary is 100 rubles per month, but the workers on the farms, as in the tea plantations, have houses with very low rent and can have cattle, vegetable garden plots, and fruit trees. These fringe benefits essentially make up the difference. Over 1600 workers are on each farm in the winter and some are added in the summer. Over 3.27 metric tons (3.6 tons) of dry tea are obtained per hectare, and over 1950 metric tons (2150 tons) of green, freshly harvested shoots and buds of tea are yielded per farm. The profit is approximately 1.25 million rubles per year.

Vegetative Propagation

Vegetative propagation of plants has been practiced for centuries. It basically involves taking a plant, or a plant part, and using this to produce another plant (s) of the same kind by asexual means. Asexual or vegetative plant propagation is used by horticulturists and gardeners to increase a particular variety or cultivar so that the individuals will be of the same type. This is often necessary where plants may be hybrid in origin. Such plants will not breed true when propagated sexually by seeds.

Let us now look at the various kinds of vegetative propagation.

PROPAGATION BY CUTTINGS

Cuttings include leaves, stems, roots, thalli, all usually having actively growing tissues present. The one exception is where one can use hardwood cuttings with buds that are ready to break dormancy.

PROPAGATION BY GRAFTING

Grafting means to "knit" two parts of a plant together. This involves taking a *scion,* to be used for the top or shoot part, and a *stock,* to be used for the base or root part, and grafting them together by various means.

Grafting is now being used in nurseries to produce more desirable nut and fruit trees, roses, and even conifers, such as the Douglas fir (*Pseudotsuga menziesii*), which through grafting produces cones earlier.

PROPAGATION BY LAYERING

Layering refers to propagation directly from the "mother plant" without cutting off a part during the propagation process. This may involve placing the tip of a lateral portion of a lower branch in moist soil and anchoring it so that it can take root, or using a mass of moist sphagnum moss in a plastic or paper cover directly on a branch so that roots may form (called Chinese or air layering).

PROPAGATION BY DIVISION

Division is one of the simplest means of propagating a plant asexually. It is used most frequently with herbaceous perennials. Basically, one simply divides the crown of a "mother plant" into two or more parts, each with a viable shoot and root system. It may simply involve a dormant bud or a root, or it may include a whole mass of shoots cut out of the original "crown." This method is employed when a plant becomes too large or more individuals are desired in a very short time.

PROPAGATION BY CELL OR TISSUE CULTURES

Culturing is one of the most recent means of plant propagation and has had wide application in cloning orchids to rid the plants of virus diseases and to save time. Sexual propagation of orchids with seed planted in soil takes 1 to 3 years longer, and many viruses are transmitted through seeds. The exciting possibility of using somatic cell hybridization and tissue culture to obtain new strains of lumber trees, such as the Douglas fir, and to attempt to get nitrogen fixation to occur in cereal plants, is now being explored. Figure 4-7 shows a scheme for propagating plants asexually by somatic cell hybridization.

Breeding of new plants by asexual crossing has been practiced for centuries to obtain new genetic recombinants in plants that would be useful to people. Such recombinants are obtained with a view to increasing the yield of a particular crop, to provide a new variety that is resistant to a disease-producing fungus or bacterium, or to provide for a crop or horticultural variety of garden plant that is drought-resistant. Mendel discovered the basic principles underlying plant variation and inheritance, and Darwin put the principle of natural selection on a firm basis of understanding. Now these principles are being put to work all over the world to produce new flower varieties, new and more desirable fruits and vegetables, and more satisfactory forest and garden plants.

Plant Breeding: Trees and Crop Plants

FIGURE 4-7.

Scheme of somatic hybrid-ization. (Courtesy of Oluf Gamborg, Prairie Regional Laboratory, National Research Council Canada, Saskatoon, Saskatchewan, Canada.)

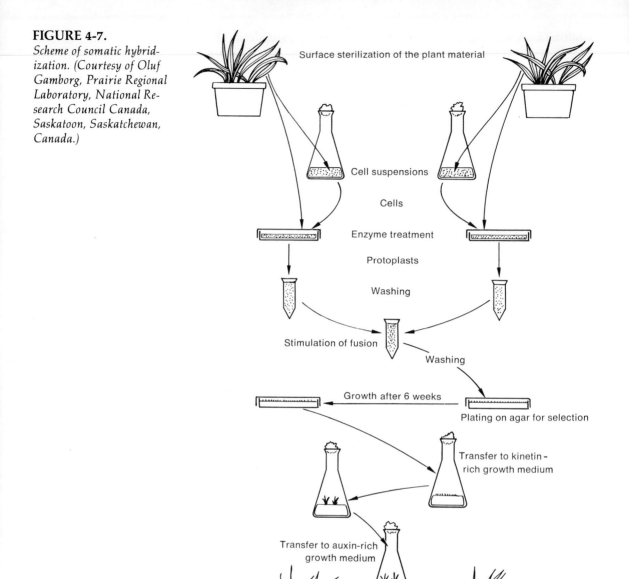

Surface sterilization of the plant material

Cell suspensions

Cells

Enzyme treatment

Protoplasts

Washing

Stimulation of fusion

Washing

Growth after 6 weeks

Plating on agar for selection

Transfer to kinetin-rich growth medium

Transfer to auxin-rich growth medium

Transfer to pots

The high-yielding corn we know so well is derived from crossing two inbred homozygous parental lines through a double-cross method. Such methods have also been used to produce the controversial new dwarf varieties of wheat and rice of the green revolution (see Chapter 11). The same methods are being used by agronomists to produce rust-resistant wheat varieties (rust is caused by a fungus). On the average, plant breeders require about 10 years to produce a new variety of wheat. After a time, the fungus may mutate and produce a new strain, which can then attack the new wheat variety. Because of this, the breeding program for disease resistance must be carried out continuously.

With forest trees, breeding has been a very slow and tedious process. One of the best hybrids developed for aspen (*Populus*), a pulpwood tree, is a cross between *Populus tremuloides* and *P. tremula*, the American and European aspens, respectively. It is a rapid grower, can produce good pulp-bearing trees in a short time, and is very hardy.

Finally, in tree breeding, we should mention the use of day length and of gibberellin hormones to induce male and female cone production in conifer *seedlings*. Normally, cones take years to be produced, instead of months, as with these treatments. You can see what a big advantage this would be to the tree breeder.

Uses of Wood

Wood, as a renewable resource, has been a staple item in our lives for centuries. In earliest times, it was used for house construction, manufacture of the wheel, and as a fuel for heat. Paper manufacture had its inception in China in about A.D. 105. Since those earliest times, the uses for wood have increased astronomically. Wood technology has become a major industrial field, as witnessed by the extremely diverse uses made of wood today: wood alcohol, house-building materials, fenceposts, wood for fuel, sawdust pressed into logs for fuel and into wall or flooring fiberboard, paper and cardboard, furniture, telephone poles, eating utensils, matches, picture frames, tongue depressors, toothpicks, dishes, toys, crates, chips for paths, sawdust for acidifying soil, airplane manufacture, cork, ship construction, veneer on doors and cabinets, musical instruments, tie clasps, and supports for light fixtures. Some of the common trees used for paper and paper products include aspen (*Populus tremuloides, P. tremula*), loblolly pine (*Pinus Taeda*), Norway spruce (*Picea Abies*), and Douglas fir (*Pseudotsuga menziesii*), teak, oak, mahogany, birch, walnut, basswood, cherry, maple, and pine are used for lumber and furniture.

It is seen from this list, which is by no means complete, that wood plays an essential role in almost every facet of our lives. This is why foresters and lumber companies are so active in trying to produce faster-growing trees—even resorting to the use of plant tissue cultures to produce new strains of trees that are high cellulose producers. In Sweden, a whole new technology has developed recently to use lignin wastes from paper manufacture to make new products. Since we know that our supplies of wood for making paper are finite, the practice of paper recycling has become of major importance in making magazine paper stock and even letterhead. You will read about this in Chapter 20.

Control of Pests: Weeds, Insects, Viruses, Bacteria, and Fungi

Pests have plagued the efforts of people to grow crops for centuries. By pests, we refer to weeds, insects, fungi, bacteria, and viruses, which interfere or compete with our efforts to produce healthy plants.

The most primitive methods employed by people to control pests involve roguing out diseased or insect-infested plants, or in the case of weeds, simply pulling them out. This is a lot of work. So, mechanical methods were developed, especially in the case of weeds. These vary from a hand hoe to rototillers to field cultivators. With insect and disease-causing organisms, we have resorted to other methods. Disease-resistant varieties are continually being bred. Plants that give off volatile compounds may ward off insects. Cultural methods involving crop rotation and the use of highly competitive crops or garden plants are used for weed control. We also use gravel, plastic, and straw or other mulches to control weeds in the garden and the field.

One of the most controversial methods of pest control, of course, involves the use of pesticides (see Chapter 10). In the 1950s and 1960s, worldwide production and use of herbicides (to control weeds), insecticides (to control insects), and fungicides (to control fungal and some bacterial pathogens) reached their highest levels. Ecologists then began to bring forth new evidence of alarming damage being wrought in the entire ecosystem, including ourselves, by the flow and persistence of pesticides, particularly DDT, throughout the entire food chain. This flow has been so serious that many species of birds have almost disappeared (e.g., falcons and eagles). The damage to the mangrove swamps brought about by the use of herbicides (e.g., 2,4,5-T, cacodylic acid) in Vietnam wiped out vast areas of wildlife habitat (e.g., mangrove swamps), which ultimately will grow back, but only after a couple of centuries.

What has been done about all this? The use of DDT is now banned in many countries, and production of hard pesticides of all kinds is being greatly restricted or banned altogether. Some insecticides that are not so persistent are now being used. We shall discuss all of this in greater detail in Chapter 10. The main point is that *we must rely on cultural practices whenever possible to control pests* and use the least-damaging pesticides judiciously and where they will not pollute our ecosystem.

Revegetation of Denuded Areas

One of the most ambitious schemes for revegetation of cutover areas is found in Europe, where thousands of hectares have been replanted into forest. The planting of trees is done in rows, as in a nursery, to make large blocks of different species of forest trees. Such blocks are managed meticulously in that trees are harvested periodically for firewood and lumber, and dead limbs are removed for burning in houses. This procedure is carried out in still another way in the Ukrainian steppes of the USSR near Donetsk and Voroshilovgrad, where the prairielike steppes, once 47 percent covered with forests, were nearly denuded to make way for planting crops and for pastureland. Today, large tracts of land are being replanted to species of poplar and aspen (*Populus* spp.), English oak (*Quercus robur*), mountain ash (*Sorbus aucuparia*), and elm (*Ulmus laevis, U. scabra*).

In Western Germany, the Rhenish brown coal area is one of the best examples in the world of revegetation of a strip-mined area. First, after the coal has been surface-mined, the holes are filled by large earth-moving machines. Where the soils are better, the land is used for agricultural purposes. In this case, loose soil from the northern region (over 2 meters thick) is transported in water in pipes to the south, where it is poured into diked areas, called polders, to a depth of 1 meter. After 1 to 3 years, crops (e.g., wheat) can be planted. In the mined regions, where less favorable, sandy-gravely soils prevail, the land is leveled and planted to lupine (*Lupinus*), a nitrogen-fixing leguminous plant, in combination with desirable deciduous trees such as beech (*Fagus*), hornbeam (*Carpinus*), maple (*Acer*), and oak (*Quercus*), as well as coniferous trees such as pine (*Pinus*) and spruce (*Picea*). These tree seedlings get nitrogen and shade from the lupine and start growing immediately. Figures 4-8 to 4-15 illustrate a strip-mined coal area and agricultural and reforested areas of once barren strip-mined regions in the Rhenish brown coal areas of Western Germany.

HARVESTING FOREST TREES

Forest trees harvested for lumber, pulp (used in paper manufacture), or opening land for crops are either cut off completely (*clear-cutting*) or partially cut (*selective cutting*). Many arguments have been made against the practice of clear-cutting, because of the dangers of soil erosion and intrusion of other species of plants that are not representative of the original forest. Soil erosion is especially serious in the tropics on mountainsides where slash-and-burn agriculture is practiced. Today, many farmers are practicing selective cutting. The farmers leave a number of "seed trees," so that new trees of the original forest can revegetate the farmed area once it is abandoned, after 1 to 2 years of use. Selective cutting is also essential in areas where regeneration of the forest is slow; this is especially true in more arid regions, as in the front range of the Rocky Mountains or east of the Cascade Mountains of the North American continent.

Clear-cutting, according to informed foresters, is not always a deleterious practice. In regions where seed can be dispersed by airplane, or where trees can be planted by hand, and *where regeneration* of a desired tree species is rapid, as occurs in more moist temperate climates, clear-cutting has a distinct advantage. The age class of the succeeding generation of trees is more uniform, and harvesting the trees is easier and less expensive. Thus the species of trees and the environmental conditions dictate the practice one should use—clear-cutting, or selective cutting. Clear-cutting has been very bad for tropical ecosystems where "swidden agriculture" (slash and burn) is used. The soils in such clear-cut regions are rapidly lost by erosion, and, being relatively nutrient-poor initially, they become impoverished relatively rapidly, especially after intensive cropping with corn or other annuals. This situation has been an *ecological disaster* in many tropical regions, but now, with enlightened practices, it is being brought to a halt in many areas through the use of selective cutting methods and other cultural practices.

BURNING TREES IN FORESTS

Burning is used in forest clearing after tropical forest trees are removed by selective cutting. The seed trees are usually left at the top of a hill or mountain clearing, and the area to be cultivated is burned to

FIGURE 4-8.
Brown coal in the Rhenish district between Köln and Aachen in Germany is obtained by open mining of large flat areas. Shown here is the world's largest open coal mine at Fortuna-Garsdorf of the Rhenish Brown Coal Works, Inc., Köln, which extracts over 100 million tons of brown coal annually. The most effective machines excavate with a daily working performance, since 1976, of 200,000 cubic meters per day. (Courtesy of E. H. Erwin Gartner, Rhenish Brown Coal Works, Inc., Köln, Germany.)

FIGURE 4-9.
This machine in the Rhenish brown coal district has a working capacity of 100,000 cubic meters per day. It weighs 7400 tons and is 70 meters high and 200 meters long. (Courtesy of E. H. Erwin Gartner, Rhenish Brown Coal Works, Inc., Köln, Germany.)

remove cut debris, stumps, and so on, before the land is planted to crops. Burning may also occur in a natural way. In some dry forest habitats, fires occur naturally, started by lightning. This burning may not kill all the trees, and has the effect of starting a new succession. Closed, resinous pine cones, especially, open and release their seeds under the high temperature conditions that prevail when they are subjected to the heat of a forest fire. The seeds that are released can germinate in an open seed bed. Thus, in many federal and state forests, such burns are allowed to occur unchecked. This is a relatively recent policy based on burning as a natural part of the forest ecosystem and the process of succession in it. Finally, burning is used artificially to provide special habitats for wildlife, one example involving the jack pine warbler in Michigan. These birds will only nest in young stands of jack pine, not in jack pine that has reached the climax level of succession (see Chapter 5).

DEFOLIATION AS A FOREST PRACTICE

Defoliation of forest trees with herbicides such as 2,4-D; 2,4,5-T; picloram; and cacodylic acid is well documented by their destructive use in Vietnam (Westing, 1972). Some of these defoliants are very soluble

FIGURE 4-10.
As high as a three-story apartment (16 meters), this paddlewheel of the excavator can scoop in 1 day's full-time operation 100,000 cubic meters of material. (Courtesy of E. H. Erwin Gartner, Rhenish Brown Coal Works, Inc., Köln, Germany.)

FIGURE 4-11.
Polder in the southern part of the Rhenish brown coal basin at the site where coal mining operation once occurred. In this land-reclamation process, loess soil is mixed with water and pumped into the polder. After the water dries, the fertile loess soil reaches a thickness of 1.5 meters. (Courtesy of E. H. Erwin Gartner, Rhenish Brown Coal Works, Inc., Köln, Germany.)

FIGURE 4-12.
The heath mining lake by the Bruhl, formerly a brown coal surface mine of the Rhenish Brown Coal Works, Inc., Köln, Germany. The need for a recreation area near the adjoining city was thus realized in an exemplary way in the Rhenish brown coal district. The heath mining lake is one of the most outstanding scenes here, which is becoming established together with a harmonious land-reclamation project involving extensive forests. (Courtesy of E. H. Erwin Gartner, Rhenish Brown Coal Works, Inc., Köln, Germany.)

FIGURE 4-13.

In the Rhenish brown coal district, agricultural reclamation is occurring directly next to the surface coal mine. To favor the agricultural recultivation, local working farms take over cultivation of the new land. After 5 years, the farming can be given over to agricultural settlements. (Courtesy of E. H. Erwin Gartner, Rhenish Brown Coal Works, Inc., Köln, Germany.)

and persist in the ecosystem. Some foresters use defoliants in deciduous vegetation to make the harvesting of trees easier. The herbicide 2,4-D is quickly metabolized and is thus a much better one to use than picloram. Another herbicide, 2,4,5-T, is now banned because it contains the impurity "dioxin," which has been shown to cause birth defects in mice and possibly in humans. One of the real dangers in using forest tree defoliants is that they also destroy the habitats of various beneficial insects, mammals, and birds. As a result, these species then begin to disappear. Thus, the use of defoliants in forests as a forest practice is looked upon with much disfavor by many environmentalists and ecologists. Perhaps the best way to achieve defoliation of deciduous trees is to wait for nature to do it in the fall with low temperatures, or in the dry season, and then to harvest the timber.

Plant Microfossils and Oil Exploration

Plant fossils have been used as an aid in locating particular layers of rock that may contain good grades of coal and oil. Oil companies, especially, have hired palynologists (pollen and spore experts) to

FIGURE 4-14.
Woods and lakes are becoming a reality in the recultivated southern part of the Rhenish brown coal region agricultural design. The receding forest along the lake affords a splendid view of one of the many lakes. (Courtesy of E. H. Erwin Gartner, Rhenish Brown Coal Works, Inc., Köln, Germany.)

examine cores taken during drilling, to correlate assemblages of pollen and spores (palyomorphs) with particular rock strata known to lie above and below "oil traps." By this means, palynologists can also locate old marine coastlines known to be important oil reservoirs. Finally, plant fossils have given us clues as to how major groups of plants have evolved through time, and have given us insights into the types of climates that existed in past geologic history. Such a record has contributed in an important way to geology, introductory biology, botany, and engineering. It has, indeed, given us some clues as to how life originated on our planet and what kinds of plants first appeared here.

Summary

You have seen how utterly dependent we are upon plants for food and beverages and for many of the products that we make. Plants are also vital in our efforts to "reclothe" the landscape after destruction by fire, lumbering, mining, and other pursuits. Living with nature involves using plants constructively. Some must be left to reseed areas scarred by slash-and-burn agriculture. Some must be propagated to make new forests, to establish new croplands and cover strip-mined areas, and to have for our own enjoyment. Others must be used for food and drink. Essentially all of these constructive uses of plants involve "recycling" them. Plants are doing this continually in nature—reclothing the land-

FIGURE 4-15.
Through the demolition law, the recultivation of the Rhenish brown coal region has resulted in new, attractive recreational areas. More than 62 million trees and shrubs have been planted in this area since 1948. (Courtesy of E. H. Erwin Gartner, Rhenish Brown Coal Works, Inc., Köln, Germany.)

scape, propagating themselves. Living with nature involves our understanding of how plants live in associations, how they propagate, which ones thrive best on disturbed areas, learning how they reproduce, and learning not to destroy plants so that only a few species are left. We hope you have gained some insights into these matters. Having done this, you should be ready, in Section II, to pick up the "ecological bug."

Brooklyn Botanic Garden. 1973. Natural plant dyeing. *Plants and Gardens* **29**(2), 1–65.

HARTMANN, H. T., and D. E. KESTER. 1975. *Plant Propagation. Principles and Practices,* 3rd ed. Prentice-Hall, Inc., Englewood Cliffs, N.J.

HEISER, C. B., JR. 1973. *Seed to Civilization.* W. H. Freeman and Company, Publishers, San Francisco.

JANICK, J. 1972. *Horticultural Science.* W. H. Freeman and Company, Publishers, San Francisco.

JENSEN, W. A., and F. B. SALISBURY. 1972. *Botany: An Ecological Approach.* Wadsworth Publishing Co., Inc., Belmont, Calif.

LaRUE, J. I. 1977. Natural dyeing with plants in Michigan. *Mich. Botanist* **16**(1), 3–14.

LEWIS, W. H., and M. P. F. ELVIN-LEWIS. 1977. *Medical Botany. Plants Affecting Man's Health.* John Wiley & Sons, Inc., New York.

SCHERY, R. W. 1972. *Plants for Man.* Prentice-Hall, Inc., Englewood Cliffs, N.J.

SCHULTES, R. E., and A. HOFFMAN. 1973. *The Botany and Chemistry of Hallucinogens.* Charles C Thomas, Publisher, Springfield, Ill.

WEAVER, R. J. 1972. *Plant Growth Substances in Agriculture.* W. H. Freeman and Company, Publishers, San Francisco.

WESTING, A. H. 1972. Herbicides in war: Current status and future doubt. *Biol. Conserv.* **4**(5), 322–327.

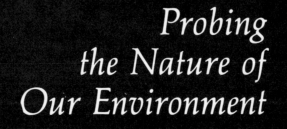

Probing
the Nature of
Our Environment

In Section I, you learned basic information regarding the various types of plants, their way of life, and how they are used for our economic benefit. In this second section, we shall examine the many roles that plants play in the ecology of our environment: fundamental ecological processes at work in terrestrial, tropical, and aquatic environments and in our atmosphere; agricultural ecology and the influence on it of the green revolution; and finally, urban ecology, the population problem, and the deleterious effects of overuse of the environment in a highly populated area of the United States. Plants are the dominant theme in this section. We must thoroughly understand their ecology before we can make good, rational judgments on solutions to the serious problems of misuse of pesticides, air and water pollution, strip mining, overgrazing, soil erosion, urban sprawl, rapid disappearance of timber resources, thermal and nuclear pollution, and destruction of natural areas caused by all of us. This "ecology" section will lay the basic groundwork and framework for Section III, which deals with solutions to our environmental problems.

The Ecology of Plant Communities in Temperate Regions
James R. Wells

5

Questions for Consideration

Plant communities are dynamic; plants grow and die, the climate is dynamic, soils develop, water relationships change as topography is modified by forces of erosion, and bodies of water may become filled in while islands are eventually worn down and returned to the sea. Rocks are subjected to forces of weathering, and new rocks are continually being formed. Thus, not only are the plants in a given area undergoing change, but their habitats are changing also.

Plant communities possess many unique attributes. Among these is aspect, or physiognomy. The aspect may be different at different seasons. A vegetational unit may be analyzed on the basis of its major parts; for example, a deciduous forest is composed of layers or strata. These include the principal tree species that constitute the canopy, an understory of smaller tree species, a shrub layer, a herbaceous layer, and ground cover.

There have been many efforts to characterize vegetation. One system is based essentially upon the water requirements (Figure 5-1). These categories include *hydrophytes,* plants that grow in water; *mesophytes,* plants that occur in moderately moist sites; and *xerophytes*, plants that grow in very dry places. Another system of classification takes into consideration the location of overwintering buds relative to the soil surface. The buds may be below the soil surface, just above the soil surface, or more than 25 centimeters above the soil surface. Quantitative parameters are commonly used to characterize plant communities. For example, *density*, the number of plants per unit area, can be the standard of comparison; or *frequency* can be useful, as it refers to the regularity with which a given species is encountered.

A large bare boulder may become covered with lichens, which, in time, would be followed by communities of mosses, herbs, and grasses; *plant succession* is occurring. Species of woody plants may eventually become

FIGURE 5-1.

*Representation of the
interrelation of climatic
types and plant formations.
From* The Study of
Plant Communities: An
Introduction to Plant
Ecology, *Second Edition,
by Henry J. Oosting. W. H.
Freeman and Company,
San Francisco. Copyright
© 1956.*

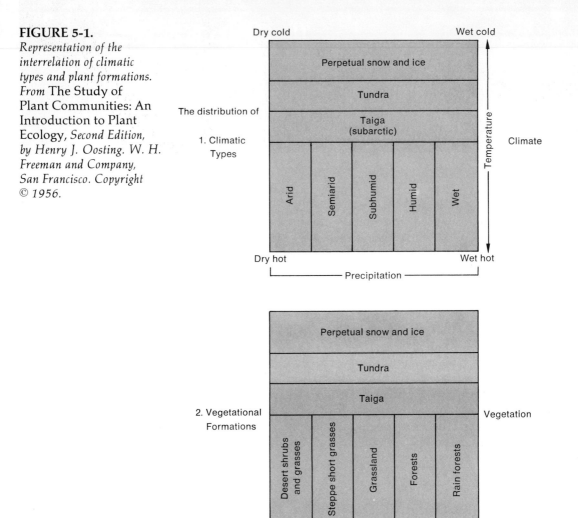

established upon the rock. Organic acids produced by these plants help bring about the disintegration of the rock into soil particles.

As one moves from the middle of a sphagnum bog lake toward the nearby forest, distinct zones of vegetation are encountered. The first zone might include plants with floating leaves, such as water lilies, followed by a floating mat composed of sphagnum moss, grasses, and sedges. Behind this mat lie broad-leaved plants, including shrubs and ferns. Still further inland we notice a zone of trees, which includes larch, black spruce, and white cedar. Finally, a mature coniferous forest is found. This bog lake will finally become filled in and covered by forest. The gradual aging process of lakes can be observed as each zone of vegetation grows toward the water and occupies the space formerly held by the community adjacent thereto. A model of this process is depicted in Figure 5-2. A more formal definition of *plant succession,* then, is that it is the process of orderly change in the composition of plant communities in a given area. An abandoned cornfield in the eastern United States undergoes a series of successional changes, the first being invasion by,

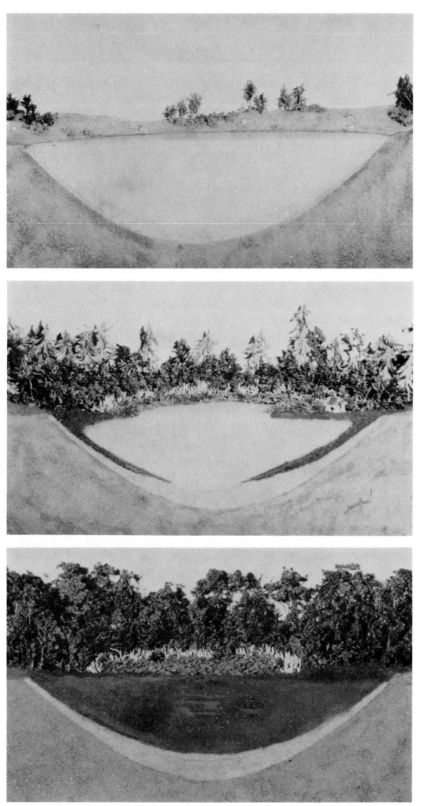

FIGURE 5-2.
*Model of the aging process
of lakes. (Courtesy of the
Cranbrook Institute of
Science Museum.)*

and establishment of, grasses and herbs. In time, this community will become inhabited by shrubs and will eventually develop into a forest community. The "final" grouping of plants in a successional series is referred to as the *climax.*

There is a tendency among plant ecologists to characterize the major floristic regions on the basis of their dominant plants as well as by major climatic regimes. These regions, called *formations,* are shown for North America in Figure 5-3. Some of these formations are discussed in the following sections.

Tundra

As one approaches the North Pole, the limit of trees is reached; beyond lie the vast stretches of the Arctic tundra. Tree limits may also be reached on sufficiently high mountains. The tundra, circumpolar in distribution, comprises about 20 percent of the earth's land surface. Most simply envisioned, the tundra can be thought of as a cold desert. At present, it is relatively undisturbed by the forces of civilization and, together with the Antarctic region, represents the "last frontier" on earth. The aspect of the tundra (Figure 5-4) is that of a rather drab yellow-brown, treeless, rolling plain.

There are essentially two seasons in the tundra: a cool, growing season 6 to 8 weeks in duration, followed by winter. In summer, periods of daylight may be 24 hours long, with winter nights about as long. About 0.3 to 1.3 meters below the soil surface, the ground is perennially frozen. In places, this permafrost extends down many meters.

The tundra climate is characterized by severity. The winters are long and cold, and precipitation averages about 10 centimeters per year (e.g., at Point Barrow, Alaska). The mean annual snowfall may be no more than 65 to 70 centimeters. The scant snow is often shifted by the wind and much of the ground may be overlain with but a thin blanket of snow all winter. The climatic summary shown in Figure 5-5 pertains to Point Barrow, Alaska, but may be typical for much of North American tundra.

The alternate freezing and thawing of tundra soils aids in the formation of hummocks and pingos. Hummocks are raised clumps of grasses and sedges. Pingos are small volcano-shaped domes caused by forces of expansion of freezing water situated between the frozen surface of the ground and above the permafrost.

Floristic diversity in the tundra is perhaps the least varied of any plant formation in North America. Regions covering hundreds of square kilometers may contain fewer than 350 species of plants, and most of these may be representatives of fewer than a dozen plant families. The vegetation is shallowly rooted because of the permafrost. Owing to the continual low temperatures, decomposition is slow.

Although tundra plants do not include large tree forms, one finds dwarfed species of birch, alder, and willow as common constituents. Stems of these plants may be less than 1 centimeter in diameter even though the plants may be 50 to 100 years old. Some characteristics of tundra plants include the following: small number of annuals; reproduction by vegetative means, although few species occur with tubers or bulbs; the tufted and trailing habit of growth; xeromorphic characteris-

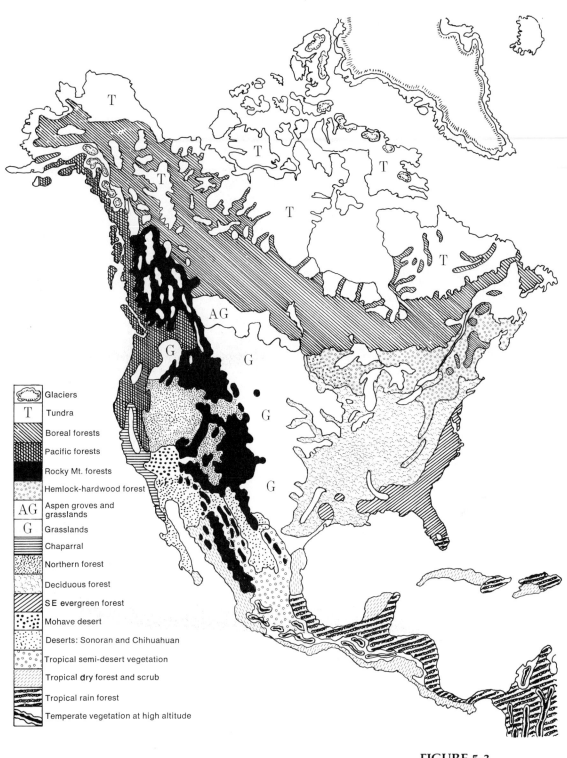

Glaciers

T Tundra

Boreal forests

Pacific forests

Rocky Mt. forests

Hemlock-hardwood forest

AG Aspen groves and grasslands

G Grasslands

Chaparral

Northern forest

Deciduous forest

S E evergreen forest

Mohave desert

Deserts: Sonoran and Chihuahuan

Tropical semi-desert vegetation

Tropical dry forest and scrub

Tropical rain forest

Temperate vegetation at high altitude

FIGURE 5-3.
Map of major plant formations of North America. (Map prepared by E. N. Transeau in 1948; courtesy of Miss Annette F. Braun.)

FIGURE 5-4.

Aspect of Arctic tundra, Greenland. (Courtesy of the U.S. Geological Survey; photograph by William S. Benninghoff.)

tics such as leathery or waxy inrolled leaves; absence of poisonous, climbing, parasitic, or thorny plants; and root systems that are shallow, yet constituting a substantially greater biomass than the above ground parts. Sexually reproducing plants must be able to flower, undergo pollination, fertilization, and maturation of seed all within the short growing season of 6 to 8 weeks.

On high mountain peaks in North America is found tundra of another kind, alpine tundra. In the eastern United States it may be seen on Mt. Washington in New Hampshire, and western examples include many high peaks in the Rockies and Sierras. As one proceeds up high mountains toward timberline (Figure 5-6), the trees become dwarfed, twisted, and prostrate. This growth form is termed *krumholz*, and it marks the lower limits of alpine tundra. Shrub height is, in general, proportional to the average snow accumulation. Growth beyond observed heights is eliminated by windblown ice and sand. Alpine tundra is reached at increasingly lower altitudes as one proceeds farther north. It is found at about 3000 meters in the Rocky Mountains and at only 900 meters in Alaska. Above alpine tundra lie perpetual snow and ice.

The alpine tundra environment is notably different from its arctic counterpart. The rarified air at high altitudes permits more light in the short wavelengths to reach the ground than that obtaining in arctic tundra. High levels of ultraviolet radiation are implicated in polyploidy, a characteristic of many tundra plants. More total hours of sunlight occur in alpine tundra than in arctic tundra.

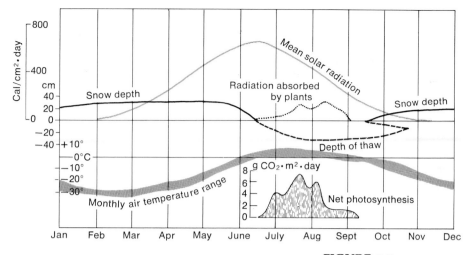

FIGURE 5-5.
Climatic summary of a typical tundra locality, Point Barrow, Alaska. [Reprinted from Volume 5, Number 1 of Mosaic, the magazine of the National Science Foundation.]

On mountain slopes there is a tendency for substrate movement down the slope. Freezing action forces rocks and soil particles outward from the soil surface. Upon thawing, these rock and soil constituents do not return to their original position because the forces of gravity pull them downward. The resultant movement is a continual, but gradual, downhill movement. Arctic alpine plants are adapted to survive such soil disturbance.

Boreal Forest

Like the tundra, the boreal forest is circumpolar in distribution. Viewed from a distance, the boreal forest presents an aspect of many tall spires, the dominant conifers. This formation is the youngest in North America. Only about 12,500 years ago, Wisconsin glacial ice covered much of North America while excluding areas both north and south of the boreal forest. Thus the area has only recently been open to colonization by plants. The soil and flora has had little time in which to develop compared with other regions. Other names for the boreal forest are taiga, northern coniferous forest, and spruce–fir forest.

This is the first forest type encountered as one moves south from the Arctic tundra. The bulk of the land mass in Canada is covered by boreal forest. To the south it is bordered by the hemlock–hardwood forest, but in the western reaches its southern border is occupied by aspen. Although this formation extends across the North American continent, its best development probably occurs just east of Hudson Bay. Westward and northward the forest becomes more open. Proceeding southward it extends along the Appalachian Mountain range, where it is found at 1525 meters (5000 feet) in the Smoky Mountains but occurs at 900 meters (3000 feet) in the Adirondacks and at only 150 meters (500 feet) on Mt. Kathadin, Maine.

The two species that best characterize the boreal forest are white spruce (*Picea glauca*) and balsam fir (*Abies balsamea*). In the southern Appalachians these species are absent, but Fraser's fir (*Abies fraseri*) and

FIGURE 5-6.
Timberline at approximately 3500 meters near Fall River Pass, Rocky Mountain National Park, Colorado. (Photograph by James Wells.)

red spruce (*Picea rubens*) occupy corresponding niches. Associated species in the taiga are tamarack (*Larix laricina*), black spruce (*Picea mariana*), paper birch (*Betula papyrifera*), jack pine (*Pinus banksiana*), mountain ash (*Sorbus* spp.), and arborvitae (*Thuja occidentalis*).

Boreal forests typically occur in regions where highly acidic mineralized (podzolic) soils are found. This soil is characterized by a superficial layer of slowly decomposing coniferous litter.

The growing season ranges from about 2 to 4 months. The coldest temperatures are between −30 and −35°C. Precipitation varies widely. In the east, annual precipitation may be up to 1 meter and drops off westward to about 38 centimeters. Tree size decreases in the west. Snowfall may be more than 2 meters, and an early snowfall may effectively insulate the soil and prevent its freezing throughout the winter months.

Factors limiting boreal forest extension northward are permafrost and a shorter growing season. Also, a lack of snow cover results in colder soil temperatures. Along its northern border, the taiga may occur only along stream banks. To the southeast, the forest is bordered by hemlock–hardwood forest. Shorter winters here preclude the total amount of low temperature necessary for breaking of dormancy of buds. Also, the low light intensity under broad-leaved trees prevents boreal species from reproducing. In the west, the boreal forest is limited by the rain shadow created by the Rocky Mountains. The plains, with their high winds, low precipitation, and heavy sod development, act as additional barriers to its spreading.

Factors favoring boreal forest development include high effective rainfall due to low temperatures and high humidity. The acidic podzol soils are conducive to growth of the dominant conifers and associated species.

Hemlock–Hardwood Formation

According to some ecologists, including Braun (1950) and Oosting (1956), the hemlock–hardwood region does not qualify as a separate formation but is merely a northern extension of the deciduous forest formation. Some synonyms for this hardwood formation are transition forest, lake forest, white pine–hemlock, northern hardwood forest, south Canadian forest, and hemlock–white pine–northern hardwoods forest.

This formation has probably been subjected to the activities of people more than any other forested region in North America. Early settlers removed choice white pines for use as masts of sailing vessels. Many small battles and skirmishes were fought over choice stands of this species of tree. More recently, harvesting has been carried out for the hardwood timber contained therein. Some of the prime examples of this forest type still remaining are in Cook Forest, Pennsylvania, and in the upper peninsula of Michigan, including the Porcupine Mountains (see Figures 1-1 to 1-3). Two centers of the hemlock–white pine–hardwood forest are the Great Lakes region and the St. Lawrence River valley region. In Figure 5-7 is shown a spring aspect of this forest in upper Michigan, with hemlock and sugar maple. In such regions, hemlock and white pine may be more than 30 meters tall, more than 1 meter in diameter, and over 500 years old. Heights in excess of 50 meters, diameters of 2 meters, and ages over 600 years are possible for white pine. One reason for considering this to be a distinct formation is its location in a snow belt where it possesses a distinct macro-climate and a distinct vegetation. Some species, including red pine (*Pinus resinosa*), are limited to this formation.

The dominant coniferous species are hemlock (*Tsuga canadensis*) and white pine (*Pinus strobus*). Associated hardwood species include sugar maple (*Acer saccharum*), red maple (*Acer rubrum*), yellow birch (*Betula alleghaniensis*), beech (*Fagus grandifolia*), northern red oak (*Quercus borealis*), basswood (*Tilia americana*), and black ash (*Fraxinus nigra*). White pine, although a successional species, persists into climax stages.

If successional stages in this formation begin with a sandy habitat, they might proceed as follows: bare sand → grasses and broad-leaved herbs → jack pine → red pine → white pine → hemlock–white pine–hardwoods region. A spruce–fir forest might prevail in poorly drained soils, but the hemlock–white pine–hardwood forest would be found on drier sites.

FIGURE 5-7.
*Virgin hemlock—white
pine—northern hardwoods
forest in northern Michigan
in the spring. The large
tree in the center is a sugar
maple (Acer saccharum).
(Photograph by James Wells.)*

Deciduous Forest Formation

The deciduous forest formation, the most complex of all, possesses a summer green aspect and a showy fall coloration. The fall color, followed by leaf drop, is altogether absent or is less evident in other formations. Deciduous forests occur mostly in the northern hemisphere, including central Europe, eastern Asia, and North America. The deciduous forest of eastern North America is centered in the Cumberland Mountains region (Braun, 1950). As one proceeds outward from its center, there appear distinct regions, termed *associations* by Braun, each characterized by a unique flora. Why is this formation so complex? Some of the reasons are: (1) its antiquity—there have been up to 80 million years with relatively little major disturbance; (2) great diversity of habitats as a result of differences in soil types, exposure, and microclimatic variability; (3) great contrast in soils (and parent materials); (4) large number of species, up to 40 tree species as dominants in some localities; (5) southward migration of northern species as glacial ice advanced; (6) xeric periods following glaciation, which allowed eastward invasion of prairie species; and (7) persistence of coastal plains plants in mountain valleys as far north as Kentucky.

Although now restricted to the eastern United States, it is believed that the deciduous forest once extended across the North American continent. Deciduous forest fossils are known from as far west as Oregon. The uplift of the Rocky Mountains some 65 million years ago created a rain shadow, which eventually led to the arid conditions prevailing in the central plains region.

The climate of the deciduous forest formation is predictably quite varied because of the large area encompassed. Annual rainfall along the western edge averages about 76 centimeters, while on the southern boundary there are approximately 127 centimeters of rainfall. Evaporation is 51 centimeters in the north and 102 centimeters in the south and west, and this factor is taken into account in ascertaining the rainfall effectiveness of a region. The frost-free season varies from about 175 days in the north to 230 days in the south. The average temperature in the north is 10°C and in the south is 18°C.

The *mixed mesophytic* forest region is considered to be the center of the deciduous forest. From it may have radiated the other forest types. Located on unglaciated land of the Appalachian plateau, the mixed mesophytic is a remnant of a forest some 80 million years old. Much time has thus been available during which a unique and varied flora could evolve as compared to only 13,000 postglacial years available for spruce–fir forest development. In the mixed mesophytic there are optimal conditions of drainage, light, and soil development. It is toward this type of association that all other deciduous forest associations tend to evolve and would likely develop except for the limitations imposed by edaphic or climatic factors. In this situation we find the largest size and number of tree species per unit area of any region. Up to 40 species of trees make up the dominants. These species include tulip tree (*Liriodendron tulipifera*), beech (*Fagus grandifolia*), cucumber magnolia (*Magnolia acuminata*), yellow buckeye (*Aesculus octandra*), white oak (*Quercus alba*), red oak (*Q. borealis*), red maple (*Acer rubrum*), sugar maple (*A. saccharum*), basswood (*Tilia americana*), wild black cherry (*Prunus serotina*), white ash (*Fraxinus americana*), shagbark hickory (*Carya ovata*), sour gum (*Nyssa sylvatica*), and silverbell (*Halesia carolina*).

In the drier western part of the deciduous forest is found the oak–hickory association, containing several species of oak and of hickory. Although occurring from Canada to Texas, it is centered in the Ozark and Ouachita Mountain regions. Its principal range of distribution includes parts of Illinois, Arkansas, Missouri, and Oklahoma. Oak–hickory forest types may be found in other deciduous forest regions where rainfall is less effective, such as on sandy ridge tops within the beech–maple association. The oak–hickory association is found on the southern slopes of the western mesophytic or within portions of the oak–pine region. As one proceeds westward from the oak–hickory association, the trees become more scattered, reminiscent of that found in a savanna. Farther westward, the final vestiges of the oak–hickory forest are restricted to stream banks. Beyond this lie the grasslands.

The *western mesophytic* association, extending from northern Alabama and Mississippi to Wisconsin, is considered to be a region of transition. Although many climax vegetation types exist, the principal type of transition is from mixed mesophytic communities in the east to oak–hickory communities in the west. The variation in forest types is due in part to soil differences, reflecting a variety of parent materials, and to the youthful age of the region, which continues to carry with it results of past influences upon the vegetation. The forest communities of the western mesophytic are not so extensive as those of the mixed mesophytic.

The *beech–maple* association is located principally in Ohio, Michigan, and Indiana. The climate is characteristically cool and humid. Located almost entirely within the glacial boundary, the soils are typically heavy and slightly drained. Although as many as 15 species may occur as forest canopy constituents, beech (*Fagus grandifolia*) and sugar maple (*Acer saccharum*) may occupy as much as three-fourths of the stand. Outliers of beech–maple forest type may be found within the hemlock–hardwood region as well as on north-facing slopes of the mixed mesophytic forest region.

The *maple–basswood* association is typical of much of Wisconsin and Minnesota. It is the smallest association of the deciduous forest. Sugar maple (*Acer saccharum*) and basswood (*Tilia americana*) are the principal tree species and often comprise the great majority of tree species cover. Located within parts of Wisconsin, Minnesota, and Iowa is a region known as the "driftless area." This area, presumably unglaciated by Wisconsin glacial ice, is included in the maple–basswood association.

The *oak–chestnut* association is today perhaps not the best terminology to describe the dominant trees herein. Just after the turn of the century, the American chestnut (*Castanea dentata*) (Figure 5-8) was all but eliminated due as a result of chestnut blight (Figure 5-9). This left many large gaps in the forest canopy, which presently are being filled by tulip tree and red oak, among other species. Well-developed oak–chestnut communities were found on dry noncalcareous southerly slopes within the mixed mesophytic association. It is in a similarly dry situation that one looks today for the remnants of this forest, which, with the chestnut, existed from northern Virginia into New England. A well-developed shrub layer, including many heaths (*Ericaceae*), occurs in this association.

Along the piedmont plateau from lower New Jersey southward into the Gulf states is a forest region known as the *oak–pine* association. This is a region of transition between the pine forests to the south and the deciduous forests to the north. White oak is the most abundant species, with an admixture of other oaks and hickories. Loblolly pine occurs throughout and longleaf pine is found in the Gulf Slope section. Pines may be considered transitional and are ultimately replaced by deciduous species. In the piedmont region of North Carolina are found excellent examples of oak–pine habitats.

Some of the factors limiting the deciduous forest to the west are extensive drought; low light intensity near the ground among tall grasses; heavy sod, which precludes seed germination; high winds; and fire. Eastward boundaries are ultimately reached due to the higher temperatures associated with the southeastern evergreen forest. In the

FIGURE 5-8.
View of the trunk of an American chestnut tree (Castanea dentata), with Philip Clampitt holding chestnut fruits from this tree. Highland Recreation Area, Oakland County, Michigan. (Courtesy of Larry Mellichamp.)

north, the acidic podzol soils and shorter frost-free period are limiting factors. In the south, it is limited due to the effects of high rainfall, high temperature, and too few cold days needed to cause buds to break dormancy.

The southeastern evergreen forest, also referred to as the southeastern pine forest, extends from the James River in Virginia to near the limit of forested lands in Texas. Remnants of beach terraces remain. The soil is generally unconsolidated sterile sand and gravel. The climate is mild, with an average temperature of about 10°C, and annual rainfall averages 114 to 152 centimeters. Leaching and percolation are common, and intense rainfall adds to their intensity.

The principal species of pine are slash pine (*Pinus elliottii*), longleaf (*P. palustris*), loblolly (*P. Taeda*), shortleaf (*P. echinata*), and Virginia pine (*P. virginiana*). Not all evergreens in this formation are conifers. Broadleaved

Southeastern Evergreen Forest

FIGURE 5-9.
The two fallen trunks shown are those of the American chestnut (Castanea dentata). The trees died as a consequence of the chestnut blight disease caused by the chestnut blight fungus, Endothia parasitica. American chestnut trees once dominated this forest, but only a few trees not attacked by the fungus remain, as shown in Figure 5-8. Highland Recreation Area, Oakland County, Michigan. (Photograph by James Wells.)

evergreens include the evergreen magnolia (*Magnolia grandiflora*) and live oak (*Quercus virginiana*).

According to some plant ecologists, the southeastern evergreen forest is a subclimax vegetation type perpetuated by recurrent fires. That it really is a segment of the deciduous forest is suggested by the fact that many local forests are comprised predominantly of deciduous species, while in other forest situations, the understory is mainly deciduous. Alternatively, other workers point out that the region is floristically unique in North America and that many species, including certain trees found here, do not occur elsewhere.

Stands of longleaf pine often occur on sandy ridges and elevated plateaus and are characteristically open, with little understory development (Figure 5-10). Scrub oaks may occur in the understory and become dominant following fire.

Poorly drained areas in this region harbor distinctive swamp forest types (Oosting, 1956). Some examples are black gum—bald cypress (*Nyssa sylvatica—Taxodium distichum*), and southern white cedar (*Chamaecyparis thyoides*). The white cedar may become established in boggy areas following fire. If black gum—cypress swamps are burned, *pocosins* may develop. A pocosin contains relatively few herbs and is basically a dense entanglement of evergreen shrubs. *Savannas* are favored by fire, which often results in the complete loss of organic matter down to the mineral soil layer. A savanna is a fire-maintained community of grasses, sedges, and herbs in which trees are widely interspersed.

Fires play a very important role in the maintenance of the southeastern evergreen forest. There are relatively few forest stands where fires have

FIGURE 5-10.
Longleaf pine (Pinus palustris) stand near Baxley, Georgia. Some trees have been "boxed" for resin collection. The resin is used to make turpentine. (Photograph by James Wells.)

been prevented over a long period, but presumably oak–hickory forests would develop in many sites, given sufficient time. For this reason, the southeastern evergreen forest, although a fire-maintained subclimax forest type, may for all practical purposes be considered a separate formation, since fires seem to be a natural part of the ecosystem.

Factors limiting the distribution of the southeastern evergreen forest northward are temperatures below freezing over a longer period, while to the west less sandy soils and reduced rainfall favor oak–pine and oak–hickory.

Desert

Large regions of desert occur in tropical eastern Africa, Chile, Australia, and China. The four subdivisions of the North American desert are outlined in Figure 5-11. The Chihuahuan desert extends into southern New Mexico and southwestern Texas, but is situated primarily in north-central Mexico. The smallest segment, the Mojave, is in the southern parts of California and Nevada. The Sonoran desert is located in northwestern Mexico, southern California, and Arizona, as well as in Baja Peninsula. The Great Basin desert, by far the largest, conforms essentially to the physiography of the Great Basin region.

North American desert regions have several common traits: low rainfall, evaporation exceeding rainfall, high percentage of sunny days, great fluctuation in daily temperatures, high soil-surface temperatures, and high concentration of salts in the soil. A desert is not merely vast expanses of sand devoid of plant and animal life, is not always hot, and is not topographically modified by wind as much as by water.

All four deserts have an average annual rainfall of about 40 centimeters, usually less. A hardpan mineral layer, called *caliche*, frequently occurs at or just below the soil surface. This layer is an effective barrier to root penetration and, consequently, to plant development. High evaporative rates result in salt accumulation at the soil surface.

Considerable bare soil is often visible between plants in desert environments. This factor helps to obscure plant associations which in

FIGURE 5-11.
*Deserts of North America.
(Courtesy of the U.S.
Department of the Interior.)*

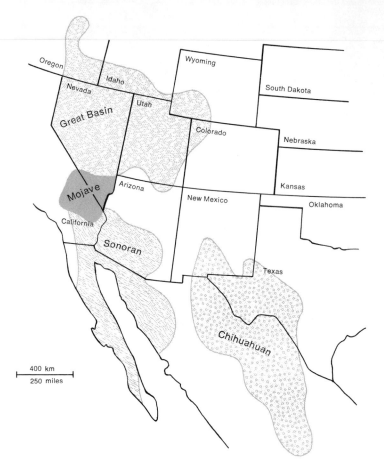

the desert are simple or nonexistent. Plant succession in the usual sense is unknown in the desert. If plants are removed from an area, many seedlings appear at first, but the stand will ultimately be essentially as it was before the disturbance.

Desert annuals are often able to develop through the stages seed → seedling → mature flowering plant within the short period of 6 weeks. Some seeds of desert plants contain chemicals that inhibit germination (see also Chapter 3). These substances may be removed from the seed by a combination of leaching and abrasion against soil particles: for example, during periods of intense rainfall, accompanied by rapid flow of water and erosion of soil. Chemical plant-growth inhibition is also operative in subsurface soil environments. Roots of desert plants producing growth inhibitors may cause neighboring plants to "keep their distance" (termed *allelopathy*). This chemical inhibition, together with shallow, wide-spreading root systems, helps to account for the open space between desert plants.

A number of features of desert plants have adaptive significance, leading to an increased ability to accumulate and retain water. Some of these adaptations are roots that accumulate water; deep root systems; leaves shaped like spines or thorns, which in clusters reduce air movement and create shade; hairy or white surfaces, which reflect light and heat; waxy covering (cuticle), which reduces water loss; stomata limited to lower leaf surface or sunken in hair-lined cavities; photosynthetic stems; stems that accumulate water; leaves absent or falling during critical drought; cell walls thickened; and hard seed coats.

Combinations of these characteristics enable plants to survive periods of drought, that is, any period in which transpiration exceeds water intake. In general, there are four categories of desert plants, based upon their water relationships: (1) *drought-evading*—growth rates are reduced especially during drought or water usage may be reduced by leaves falling off (e.g., ocotillo, *Fouquieria splendens*); (2) *drought-escaping*— plants grow and mature quickly when soil moisture conditions are most favorable (e.g., desert annuals that survive in seed form); (3) *drought-enduring*—includes most desert shrubs (e.g., creosote bush, *Larrea tridentata*); and (4) *drought-resisting*—plants that accumulate water following rains and retain it in fleshy tissue (e.g., cacti and other succulents).

The Great Basin desert has been termed a "cold desert." As the name "basin" implies, the Great Basin is functionally a sink having no seaward drainage. Mineral-laden water from surrounding high areas flows downhill, evaporates, and leaves the salts as residue. An extreme example of such a "nonhabitat" is the great Bonneville Salt Flats near Salt Lake City, Utah. Here the salt accumulation may be as much as 3 to 4 meters thick and over 45 kilometers long. Sagebrush (*Artemisia tridentata*) is the most typical plant species of the cold desert. Rainfall in this desert may be up to 30 centimeters annually.

Of the three remaining "warm deserts," the Mojave is the smallest. Death Valley is included in this desert near its northern boundary. The Mojave is more arid than the Great Basin and is about 70 percent covered with creosote bush (*Larrea divaricata*). Also found is the characteristic Joshua tree (*Yucca brevifolia*), which attains heights up to 9 or 10 meters (Figure 5-12). Rainfall in the Mojave may be as much as 12 centimeters annually but is usually less.

Located mainly in north-central Mexico, the Chihuahuan desert has about 8 to 40 centimeters of rainfall annually. Almost three-fourths of this occurs from June through September. Temperatures in the Chihuahuan desert average somewhat lower than in the Sonoran desert. Species occurring in both deserts include ocotillo (*Fouquieria splendens*) and creosote bush.

In the Sonoran desert are localities in which are found the extensive sand dunes popularized by many Hollywood movies. Low rainfall (5 to 10 centimeters per year), together with high average temperatures, makes this one of the largest extreme environments in North America. Although a number of genera are common to this desert and to the Chihuahuan, perhaps the best-known plants of the Sonoran are the saguaro cactus (*Carnegiea gigantea*) (Figure 5-13) and the organ-pipe cactus (*Lamairocereus thurberi*).

Many people advocate farming the deserts utilizing irrigation techniques. Those most knowledgeable about desert soils would probably agree that in many instances, eventual high salt concentrations in the upper soil caused by evaporation would preclude extensive agricultural activity.

FIGURE 5-12.
Joshua tree (Yucca brevifolia) in the Mojave Desert, California. (Photograph by James Wells.)

Grasslands

If trees are removed from an area, grasses usually follow. If the conditions are unfavorable for reseeding of tree species, grasses may persist. Large areas composed predominantly of grassland vegetation occur in both hemispheres. Much of the area occupied by grasslands in North America formerly contained extensive forests. The uplift of the Rocky Mountains created a rain shadow that led to relatively arid conditions more conducive to grasslands than to forests. Grasslands are possible because the species possess a unique growth form, a long dormant period, and water requirements that coincide with adequate rainfall during spring and early summer. The grasslands of North America have been called the most extensive and varied of all climax vegetation types. These grasslands extend from the eastern range of the Rocky Mountains to North Dakota and Illinois, and from Alberta south to the Gulf Coast of Texas. Outlying regions of grasslands occur in Indiana, Ohio, Michigan, Kentucky, and Alabama.

Many plant geographers and ecologists recognize two subdivisions of the grasslands: the short- and tall-grass prairie regions. Others recognize a third subdivision, the midgrass region. The latter may be viewed as a region of transition (ecotone) between the short- and tall-grass divisions.

FIGURE 5-13.
Saguaro cactus (Carnegiea gigantea) in Saguaro National Monument, Arizona. (Photograph by James Wells.)

The tall-grass prairie is bordered all along its eastern boundary by the oak–hickory forest. It extends westward to about the 100th meridian. In the transition region between grassland/oak–hickory forest and the tall-grass prairie, there is a gradual thinning of trees, resulting in an open, parklike appearance in the forest. The forest gradually disappears as the trees become limited to stream banks or other locally moist areas. The conditions that favor grassland over forest include: annual rainfall less than about 60 centimeters; precipitation not evenly distributed throughout the year, but occurring at critical periods of maximum growth of grassland species; and a precipitation/evaporation ratio of less than 0.8 (Shimek, 1911).

No major success has yet been achieved in preserving original grassland vegetation in North America. As a result, very little virgin tall-grass prairie remains today. In Figure 5-14 is shown part of one

FIGURE 5-14.
*Virgin prairie tract
(Sheeder Prairie) near
Guthrie Center, Iowa.
(Photograph by James Wells.)*

small tract of virgin prairie of about 8 square hectometers (20 acres), the Sheeder Prairie, located near Guthrie Center, Iowa. Its former owner, Willis Sheeder, aged 82 years, informed the author that this area was left in its original (unplowed) condition to provide "natural hay" for his livestock and was not preserved for reasons of prairie conservation.

Dominant plants of the tall-grass prairie include big bluestem (*Andropogon gerardi*), little bluestem (*A. scoparius*), Indian grass (*Sorghastrum nutans*), smooth dropseed (*Sporobolus heterolepis*), tall panic-grass (*Panicum virgatum*), slough-grass (*Spartina michauxiana*), and porcupine grass (*Stipa spartea*). Of these, big bluestem is by far the most abundant species, constituting more than two-thirds of the original vegetation in some areas. Indian grass is almost always an associate. During years of optimal rainfall, plants may reach heights in excess of 3 meters.

The western boundary of the tall-grass prairie, like the eastern boundary, is not distinct but gradually intergrades into shorter forms of the same species, as well as taller forms of species characteristic of the short-grass prairie. The ecotone constituting the midgrass region of the grassland formation covers more area than either the tall- or short-grass regions. Typical midgrass prairie is found in eastern and central Montana, central and western North and South Dakota, eastern Wyoming, and central Kansas and Nebraska. The midgrass prairie is limited on the east by more rainfall, which favors taller grasses, which cast shade over the shorter species, thus eliminating them. To the west the lower rainfall favors the short-grass species. Construction of sod houses was possible because of the compacted root systems of prairie grasses.

The short-grass region, also referred to as "plains" or "steppe," is located, in general, west of the 100th meridian. Plants here attain heights up to 20 centimeters and frequently occur scattered, with open ground visible. The dominant species of the short-grass prairie are buffalo grass (*Buchloë dactyloides*), blue grama (*Bouteloua gracilis*), and hairy grass (*B. hirsuta*).

The climate in the plains is more severe than in the tall-grass prairie. Annual precipitation averages less than 50 centimeters and summer droughts are often severe. The soil is seldom wet to any appreciable depth because of the occurrence of the carbonate hardpan 20 to 60 centimeters below the surface. Although the tall-grass and midgrass regions are quite suited to growing cereal crops, the more arid short-grass prairie would not support these crops. Consequently, the plains are greatly valued by the vast livestock-growing industry.

Western Forests

Extending westward from the Rocky Mountains to the Pacific, northward into British Columbia and Alaska, and southward into southern California and Arizona is a vast mosaic of forest types designated here, for the sake of simplicity, as "western forests." The geologic, topographic, climatic, and floristic diversity of this vast region is so varied as to make a brief, adequate description of western forest types very difficult. The beginning student contemplating the forests of western North America will probably likely envision the giant redwood forests of California. However, the redwoods occupy a small, rather narrow region less than 720 kilometers (450 miles) long. The coast redwood (*Sequoia sempervirens*), shown in Figure 5-15, does not attain the size of the more inland giant sequoia (*Sequoiadendron giganteum*). By the

FIGURE 5-15.
Virgin stand of coast redwood (Sequoia semper-virens) near Crescent City, California. (Photograph by James Wells.)

time both species have achieved ages of 200 years, seed production is abundant. Seedlings often grow best in burned-over or otherwise disturbed areas. A factor that helps to account for the great antiquity of redwoods is the fire-resistant bark, which in some cases may exceed 1 meter in thickness.

Just below timberline in the Rocky Mountains occurs the subalpine spruce–fir forest. Engelmann spruce (*Picea engelmannii*) and alpine fir (*Abies lasiocarpa*) are the dominant species. Farther west in the Sierra Nevada the composition of the corresponding zone varies and includes red fir (*Abies magnifica*) as the dominant, with western white pine (*Pinus monticola*) an associate.

The Douglas fir (*Pseudotsuga menziesii*) zone in the central Rockies occurs between about 2000 and 3000 meters. White fir (*Abies concolor*) and blue spruce (*Picea pungens*) are frequent associates. In the Sierra, Douglas fir may be found at altitudes between 1500 and 1800 meters. Lodgepole pine (*Pinus contorta*) often becomes established in this region following fire.

The widely distributed stands of western yellow pine (*Pinus ponderosa*), which become dominant at a level below Douglas fir, are relatively open, with widely spaced trees (Figure 5-16). This forest type is best developed on plateaus between 1000 and 2500 meters.

Extensive coverage obtains for other types of plant communities characterized by small trees and shrubs: piñon–juniper climax, the oak–mountain mahogany region, and chaparral. The piñon–juniper, because of the low stature of the trees (10 meters or less), is referred to as woodland with widely spaced individuals. The ecotone between needle-leaved forest and nonforested regions is called oak–mountain mahogany. Both deciduous and evergreen species are found in this zone of the southern Rockies. Chaparral, characteristic of the southern California foothills, is essentially scrub vegetation consisting of broad-leaved evergreen trees and shrubs. Chaparral is subject to frequent fires, but the same species reappear in essentially the same relative abundance.

Summary

Species of plants with similar or complementary adaptations occur together in communities. Assemblages of communities constitute physiognomic formations. Those formations that occur in temperate regions tend to have the greatest pressure of exploitation by various activities of people. Tundra and desert regions are becoming increasingly modified (by oil pipelines, subdivisions, overgrazing, and the like) as never before. Wise use of the world's natural resources should include plans to preserve in perpetuity large tracts of land within each land formation. In this way, these areas can serve as base-line data sources for the measurement of ecological changes in each of these formations, to assess pollution and the effects of physical disturbance in them, to preserve wild and endangered species of plants and animals in them, and for nature education and enjoyment.

FIGURE 5-16.
*Stand of ponderosa pine
(Pinus ponderosa) near Bend,
Oregon. (Photograph by
James Wells.)*

BRAUN, E. L. 1950. *The Deciduous Forests of Eastern North America.*
McGraw-Hill Book Company, New York.
DITTMER, H. J. 1937. A quantitative study of the roots and root hairs of a
winter rye plant (*Secale cereale*). *Am. J. Bot.* **24,** 417–420.
JENSEN, W. A., and F. B. SALISBURY. 1972. *Botany: An Ecological Approach.*
Wadsworth Publishing Co., Inc., Belmont, Calif.
KREBS, C. J. 1972. *Ecology.* Harper & Row, Publishers, New York.

Selected References

ODUM, E. P. 1971. *Fundamentals of Ecology*, 3rd ed. W. B. Saunders Company, Philadelphia.

OOSTING, H. J. 1956. *The Study of Plant Communities.* W. H. Freeman and Company, Publishers, San Francisco.

SHIMEK, B. 1911. The prairies. *Bull. Lab. Nat. Hist., State Univ. Iowa* **6**, 169–240.

WALTER, H. 1973. *Vegetation of the Earth.* Springer-Verlag New York, Inc., New York.

WHITAKER, R. H. 1975. *Communities and Ecosystems,* 2nd ed. Macmillan Publishing Co., Inc., New York.

Overuse of Our Environment: Case History in New Hampshire

Timothy Mahoney

6

1. Why are New Hampshire's White Mountains so vulnerable to overuse by campers and backpackers?
2. How rapidly do overused areas (as from overuse by campers) recover from such damage?
3. What solutions are available to prevent overuse of the White Mountains and similar fragile wilderness areas by backpackers, horseback riders, and campers?
4. How much "development" should be allowed in fragile wilderness areas like the White Mountains?
5. How effective do you think the Forest Plan for management of the White Mountains is in terms of saving this wilderness area from overuse by people? By lumber interests? For ORV use?

Questions for Consideration

Much of the attention of the environmental movement has been given, quite justifiably, to the establishment of public recreation areas, parks, greenbelts, wildlife preserves, and, perhaps most important, to wilderness areas. Far less attention has been paid to the effects of the increased recreational use that often accompanies their establishment. Increased use produces its greatest changes in those areas which can least afford it: backcountry areas and wilderness. Wilderness areas that are easily accessible to large population centers are especially vulnerable to overuse. And where there is little backcountry available, use becomes more concentrated in those few "green spots" that a visitor can find on a road map.

In the late 1960s and the early part of the 1970s, this new backcountry population crunch became apparent in several western national parks, particularly those accessible to large cities. But overuse is particularly prevalent in the East, because the population is so dense and the backcountry is so limited. The Smokies, Virginia's Shenandoah Valley, the Adirondack's High Peaks area, and New Hampshire's White Mountains have been particularly hard-hit.

The White Mountains encompass about 1 million acres of northern forest, lakes, streams, and bogs, three-fourths of which are in public lands. They include 46 peaks over 1219.2 meters (4000 feet), topped by Mt. Washington, 1898.2 meters (6228 feet) high, the highest point in the northeast. The Appalachian Trail winds through the region for 209.1 kilometers (130 miles), including a 19.3-kilometer (12-mile) stretch above timberline in the Presidential Range, the only such alpine area in the east. Both Great Gulf Wilderness on the northeastern side of the Presidentials and the newly created Dry River Wilderness on the south side have been set aside by the federal government, but there are also nine "scenic areas" set aside with similar provisions for preservation and over 1610 kilometers (1000 miles) of mountain and forest trails.

There are probably only about 50,000 people living in this region of the state, and, despite rapid growth in the last decade, there are only about 2 million people in all of northern New England. However, within one driving day of the region live 60 million people, equivalent to almost *one-third* of the entire U.S. population. Boston, New York, and Montreal are the nearest large cities, but in addition, a host of medium-size cities in southern New England constitute one of the densest populations in the country. These people are close enough to visit the region any weekend, and millions do year-round. In the summer, moreover, vacationers from all over the East descend on the mountains for extensive stays (Figure 6–1).

Of course, all visitors to the White Mountains do not use, or even see, the backcountry. But their numbers exert pressures on the region that set off chain reactions. The most spectacular controversy in recent years has been that of highway development.

The five interstate highways of northern New York and New England are all north–south routes that connect the region to the megalopolis to the south. The tourist agencies in all the states have pushed for rapid completion of more and faster highways into the region to spur development. Figures from the New Hampshire Department of Public Works and Highways show that automobile traffic in the White

FIGURE 6-1.
Traffic growth on Route U.S. 3 in Lincoln, New Hampshire, as a result of highway construction in southern New Hampshire. Figures after 1969 are projections. (Prepared by the Planning and Economics Division of the New Hampshire Department of Public Works and Highways, February 1970.)

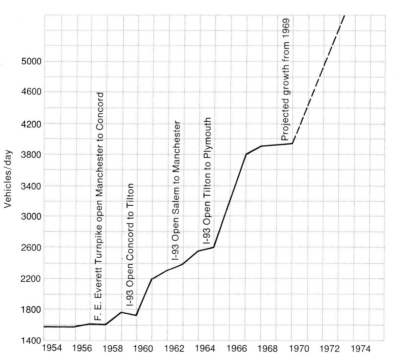

Mountains increased 150 percent between 1958 and 1970 as successive links of Interstate 93 were completed north from Boston (Figure 6-1).

Unfortunately, from the earliest times of settlement, most traffic in the White Mountains has had to funnel through three major notches, which were formed by Pleistocene glaciation from north to south. The westernmost of the three, Franconia Notch, is also a state park, home of the Old Man of the Mountain, a giant rock profile formed by glacial action, which juts out from the cliffs of Cannon Mountain 365.7 meters (1200 feet) above the floor of the notch. The Old Man, which served as the inspiration for Hawthorne's story, "The Great Stone Face," is New Hampshire's most famous tourist attraction, and has become a symbol of the state. The profile had been saved from destruction recently when frost action weakened the many slabs of granite that form it. The state patched it then and now monitors it for any movement in the stones. Now, plans call for Interstate 93 to run through Franconia Notch (Figure 6-2), and it has been completed to within a few miles of it on either side.

Conservation groups, notably the Appalachian Mountain Club and the Society for the Protection of New Hampshire Forests, fought the original plan to run a four-lane tunnel through the Notch, claiming that the construction might well damage the Old Man as well as disrupt the forest land within the narrow U-shaped valley. In 1970, the Secretary of Transportation, John Volpe, acceded to their demands and postponed further construction indefinitely.

But the sentiment of local political leaders and many townspeople living north of the Notch was very much opposed to the Secretary's decision. The New Hampshire north country, the most economically

FIGURE 6-2.
View of Franconia Notch from the Franconia Ridge. The Old Man of the Mountain protrudes from the cliffs on Cannon Mountain. Vermont is on the horizon. (Photograph by Timothy Mahoney.)

depressed area of the state, had not been sharing the population and economic growth that overtook the southern part of the state during the 1960s. The I-93 highway was to be their link with the cities to the south. Many felt that if this one segment was not completed, or if an alternative route were planned to avoid the mountains, that portion of the state would, in the words of Senator Norris Cotton, "become a desert."

Secretary Volpe was flown over the area by the governor and the state highway commissioner and shown the bottleneck of traffic where the two-lane highway entered the Notch. A compromise was worked out whereby I-93 would be completed up to the beginning of the Notch and a new "improved" two-lane highway would be constructed above ground through the Notch.

Senator Cotton's amendment to the 1973 Federal Aid Highway Act then provided federal funds for the completion of I-93 as a special "parkway" if the environment was protected and the parklike nature of the Notch preserved. The new Secretary of Transportation, Claude Brinegar, authorized the state to undertake an environmental impact statement on the proposed section of highway and to consider certain local alternative routes.

While that study proceeded, however, the conservation organizations went to court to halt construction of another, more northerly segment between Littleton, N.H., and St. Johnsbury, Vt. Their argument was that if this segment was completed, alternative routes would not be seriously considered, and the highway would be a *fait accompli*. Hence, in August 1974, Judge Hugh Bownes of the U.S. District Court granted the injunction, stating: "It is evident that Franconia Notch stands squarely in the path of I-93 and from an engineering and construction view, its location invites the final link of I-93 through it." The injunction would grant time for the consultants to study all the alternatives. In his decision, Judge Bownes touched on the broader aspects of the problem. "The effect of interstate highways in attracting traffic is by now well established. It seems that such highways not only attract automobiles but breed them."

The injunction, which could not be appealed, left the matter unresolved. State of New Hampshire officials were incensed and vowed to go to Congress to fight for new laws. A newly formed group, Save the Old Man, Inc., charged that the region is already adequately served by two other interstates, I–89 and I–91. Since the state has not empowered the consultants to consider this nonlocal alternative, the "no-build" option is not receiving just consideration. The Governor of Vermont has since stated that I–93 will not be completed into this state, so the entire argument may be moot. Meanwhile, the consultants continued their environmental impact study. Now, at last, the environmentalists' arguments have been vindicated. Only a two-lane portion of I–93 will be allowed through Franconia Notch.

While dramatic, the Franconia Notch controversy is not the first example of these arguments for more highways and more development. To paraphrase the judge, development breeds new development. While those who drive through the area may not harm the backcountry directly, all types of users appear to increase proportionally as the total number of visitors increases.

The White Mountain National Forest (WMNF) records use in terms of "visitor days," that is, a 12-hour period spent by one person in the area. Three people spending 4 hours each would equal one visitor day. Likewise, one person spending 2 days in the woods equals four visitor days. In 1973, there were 2.5 million visitor days in 5.9 million visits to the national forest, a 45 percent increase since 1966. The Forest Service has estimated that by the year 2000, the number of visitor days will double. In the entire New England region, 20 percent of the nation's recreational visits will occur on 1.5 percent of the nation's supply of recreation lands.

The damages resulting from highway construction, vacation home developments, new ski areas, and increased numbers of snowmobiles and other ORV's is evident. Less obvious are the effects of the quiet users, hikers, climbers, and backpackers. They are among the most well-meaning users and probably among those most aware of environmental damage. Figures are not readily available, but it appears that they only constitute about one-seventh of the total visitor days in the White Mountain National Forest. Yet in the most remote areas of the backcountry, they more than any other user, are responsible for recent changes. What have they done?

Throughout their nearly 1610 kilometers (thousand miles) of trails, the White Mountains are dotted by a system of manned huts and shelters maintained by the Appalachian Mountain Club (AMC), the WMNF, and some smaller organizations. On an average August night in 1970, 19.4 people slept at the Liberty Springs shelter on the Appalachian Trail near Franconia Notch. By 1972, there were 31.2 persons using this area nightly, and the shelter had been removed because of overcrowding and vandalism. A series of tent platforms were established in its place. But if campers were discouraged from using the area because they now had to carry in their own tents, it did not show in the use figures.

Actually, an average nightly figure is misleading. Most damage is done on those nights when campers far exceed average. Weekend and holiday use is especially heavy. Nightmarish reports are heard from backpackers who have found themselves in the middle of the mountains on Labor Day weekend with 75 others at a 12-man shelter. At such times misuse is inevitable. But misuse aside, overuse causes enough problems itself.

In 1971, the AMC was forced to close a shelter because the small mountain pond which was its water supply had become polluted. Shallow soils were not able to leach out all the human waste, and the water supply had become dangerous for drinking. The next year another shelter was abandoned because heavy use exceeded its small and unpredictable water supply.

At Tuckerman Ravine on the side of Mt. Washington, skiers regularly hike 3.2 kilometers (2 miles) in spring and early summer to find the last snow in the east. Their increasing numbers, along with those of year-round hikers and climbers, have resulted in more shelters, so that now, 100 persons can be accommodated. During ski season these spaces are given out on a first come—first serve basis each morning. Most weekends in June, they are gone by 7 A.M. for the following night. Much of the area around the shelters is now bare of vegetation.

Trail erosion (Figure 6-3) is also a horrible problem, particularly on ridges and above the treeline, where the fragile alpine tundra is not able to recover from meandering hikers. And increased numbers of users means more first-time users. Perhaps one-half of the parties in the mountains in the summer are there for the first time. Many of these users are not aware of the problems of overuse, and many are not equipped properly for the mountains. With no stoves or tents, they may be in danger themselves. But the mountains cannot stand for them to clear areas, cut branches from trees, or even gather firewood. There are few animals to be seen along busy trails. And what value can be placed on the loss of solitude for one who seeks it in the mountains?

If the problems of overuse disappeared tomorrow, their effects would still be felt. Two or three years after their closing, the land at the abandoned shelters is still found to be largely bare of vegetation. Since many of these sites are on slopes, the thin layer of mountain topsoil has been eroded, leaving exposed rock for an eternity. Observations made at lowland logging camp sites, which were active early in the century, indicate that succession is so slow in this cool climate that recovery is not complete after 60 years. At higher elevations, where the climate is more severe and there are steeper slopes, recovery will be many times slower. Experiments in replanting and fertilization, which have been made at Tuckerman Ravine, show that no recovery can be expected until people can be excluded from the replanted areas.

The organizations responsible for the management of the mountain land have not been idle. In 1972, the Forest Service declared the shelters to be concentrated use areas where new, stricter rules were in force, prohibiting the clearing of brush, using open fires, leaving any garbage or trash, disrupting the soil, or disposing of human wastes away from the privies. Camping elsewhere in the National Forest was forbidden lest these abuses spread.

At several locations at high elevation, the AMC converted pit toilets to sealed systems. Wastes were then flown out periodically to a nearby firm which made compost. But this is expensive, and at other shelters,

FIGURE 6-3.
Trail erosion is particularly disturbing on the shallow soils of alpine tundra. This is part of the Appalachian Trail on the Franconia Ridge; Mt. Washington is on the horizon. (Photograph by Timothy Mahoney.)

experimental composting toilets have been installed so as to treat the sewage on the site. Caretakers were sent in to several of the busiest sites. They served two purposes. First, they could manage the use at the site, juggle parties, and reduce misuse. Second, they could educate users as to the reasons behind these new rules. Capacities at the most heavily used areas were increased slightly through the addition of tent platforms. These were usually placed on sites that had previously been cleared by overflow crowds. The presence of the permanent platforms encourages campers to use them rather than to clear new areas. Furthermore, because the platform is raised, there is less soil compaction and erosion on the site.

One important action was the long-overdue development of a forest use plan for the White Mountain National Forest. The Forest Service is an unusual caretaker for such popular recreational lands, for, unlike the Park Service, they are responsible for developing timber, water, and mining resources, as well as taking care of recreation, wildlife, and wilderness. The *Guide for Managing the National Forests in New England,* published in 1973, tried to balance these demands and setoff areas where certain types of activities would be permissible (Figure 6-4). Naturally, no interest group felt that they had been given enough consideration, but the overall plan was important as a tool for shaping future policy rather than resorting to the current practice of taking care of crises as they arise.

For the WMNF, the *Forest Plan,* published in 1974, established four management areas:

1. Much of the low elevation forest (54 percent of the total) would continue to be managed for *timber production.* Recreation pressures are not high in these areas, but uses such as hiking and snowmobiling would be allowed.
2. Another area would be developed for *high-concentration activities,* such as downhill skiing, roadside camping, scenic drives, and visitor education. This area includes such land as already exists and much of the public land along the major highways of the region.
3. Lower mountains and low-elevation backcountry would emphasize *"dispersed recreation opportunities"* for a "near natural visitor experience." Limited timber cutting and snowmobiling are allowed, but there will be little development. People interested in hiking, back-packing, snowshoeing, and cross-country skiing would be expected to coexist with these other users.
4. The highest peaks, much of the Appalachian Trail area, and certain remote lower mountains would emphasize a *"natural recreation experience with a high degree of solitude."* There will be no timber cutting, no vehicles, and a minimum of site modification. This is for the backpackers.

In addition, there exist the aforementioned wildernesses and scenic areas which are already under a management plan for their preservation.

Necessary as it has been to take action, and laudable as these actions have been, the problems remain while the symptoms are being treated. Fortunately, the rate of user increase has slowed somewhat in recent

FIGURE 6-4.

Former campsite at Garfield Pond, since abandoned because of fouled drinking supply. Area is still bare of vegetation a year after closing, but soil has been retained. Grids are part of a survey of vegetation. (Photograph by Timothy Mahoney.)

summers. This may have resulted from a rainy summer followed by gasoline shortages and an economic downturn. As such, it will probably not be permanent. Moreover, it is uncertain whether many areas in the mountains can survive as they were under current use pressures. A solution to the problem would have to involve a *limitation on user numbers*, possibly organized as a reservation system or perhaps even as user permits, like hunting licenses. Painful as these ideas might be to one whose visits to the mountains are expressing independence and freedom, they may be quite necessary to preserve the solitude necessary for that independence, indeed to preserve the mountains themselves.

Summary

So mixed within the conflict of development and conservation in a region, and increasingly important because it offers values missing in an urban society, is the perplexing problem of the backcountry user. The first problem could be solved through wise regional planning and cautious development, which will maintain the character of the area, which attracts people in the first place. But in the backcountry itself, we run straight into the heart of a dilemma that may ultimately become apparent even in our most isolated areas. That is, in order to preserve the land for future generations, it is necessary for people to *know the land, to experience the wilderness*. But by their use, we may endanger the very land we seek to protect. It appears that once land has been set aside, it does not mean that it has been preserved.

Ecology of Tropical Regions
Dan Janzen

7

Questions for Consideration

1. Are the things you believe about the tropics really true?
2. How does a tropical tree deal with the vicissitudes of tropical nature?
3. How do tropical roots feed, seeds move, vines climb, and epiphytes survive?

The first thing to remember about tropical terrestrial habitats is that there are (were) many kinds, perhaps even more kinds than at higher latitudes. The second message is that this tropical diversity is being homogenized so rapidly by attempts to breed more people and, by some persons, to line their own pockets that most readers will never have the opportunity to see even a small fraction of this diversity. In fact, much of what we say here is derived from old reports and by reconstruction from tiny and scattered fragments of once-very-large tropical habitats.

The richness of kinds of tropical habitats is both appealing and frustrating. It is appealing because it illustrates so well what a multitude of arrays of interacting plants and animals can be produced when a single dominating and omnipresent constraint, the northern winter, is removed and the soil, plants, animals, rainfall, and so on, are allowed to interact on somewhat equal terms. It is frustrating because it renders almost any generality about tropical ecology valid only for certain closely circumscribed habitats and circumstances, rather than valid for all that land sitting between the Tropic of Cancer and the Tropic of Capricorn.

By way of introduction to the tropics, let us examine briefly a few standard platitudes about tropical habitats.

"There Are More Species in the Tropics"

Certainly, there are some very species-rich habitats in the tropics. A square mile of lowland evergreen forest on lateritic soil in western Africa may contain several hundred species of large trees; and in Malaya or Brazil, on similar soil, there may be as many as 500 tree species in 13 square kilometers (5 square miles). On the other hand, only a few meters from such forests may be climax forests where 90 percent of the canopy is made up of only one species of tree, such as the pure stands of *Shorea albida* in the peat swamps of Borneo, the *Eperua* forests on white sand soils in Guyana, and the *Parkinsonia* forests in deltaic swamps of northwestern Costa Rica.

As soon as one moves away from those tropical lowlands that have almost year-round rains, the species richness of plants begins to drop dramatically. The author knows of no tropic desert with a richness of annual plants (and their attendant bees) that even begins to match that of the southwestern United States and north Mexican desert flora. In the deciduous (seasonal) forests of Kenya, Costa Rica, and Venezuela, it is easy to locate undisturbed hillsides with not more than 5 to 15 species of large trees, a number that would not seem exceptional in many mid-latitude habitats. On poor soils in tropical deciduous forests, nearly mono-specific stands of trees are commonplace, such as the *Quercus oleoides* forests on lateritic hills in lowland Veracruz/Oaxaca, Mexico, and the *Shorea robusta* forests that used to cover major portions of India. At elevations above about 2800 meters (9200 feet) on mainland tropical mountains, it is commonplace for nearly pure stands of large trees to cover thousands of square miles, such as the *Podocarpus* forests of Mt. Kenya and the *Quercus* forests of Costa Rica.

The large trees are not the only plants that display very strong interhabitat variation in species richness in the tropics. Above 4000 meters (13,000 feet) in the Venezuelan Andes we encounter many square kilometers in which 90 percent of the biomass of the vegetation is a single species of *Espeletia* (Figure 7-1), a woody and shrubby composite. There are thousands of square kilometers of Venezuelan and East African grasslands where five or fewer species of grass make up 90 percent or better of the vegetation biomass. While several square miles of Costa Rican lowland Caribbean rain forest may contain several hundred species of epiphytic orchids and bromeliads, an equal-size patch of deciduous forest on the Pacific coast less than 160.9 kilometers (100 miles) away may contain only four or five orchid (e.g., *Catasetum, Brassavola, Laelia*) and bromeliad (e.g., *Tillandsia, Aechmea*) species. Prior to human destruction of Indian and Burmese tropical habitats, mono-specific stands of bamboo or *Strobilanthes* often covered many square miles. The understory shrubs and small trees of the *Shorea* and *Eperua* forests mentioned previously are also very poor in species.

Perhaps the most glaring exceptions of all to tropical plant species richness are the mangrove forests that fringe most tropical land masses and line river and estuary mouths throughout the tropics. These forests have existed for millions of years (although they may move in location a bit as bays fill and storms carve out shorelines) and may contain as few as one or two species of large trees and almost never more than 20 species of woody plants. Furthermore, they almost entirely lack an understory vegetation (except their own seedlings and saplings on occasion). Independently derived from a number of plant families (Verbenaceae, Rhizophoraceae, Combretaceae, Theaceae), mangrove trees not only have a highly convergent life form, which makes for the extreme monotony of mangrove forest, but are highly convergent in their chemical defenses. Mangrove forests the world over have served as a commercial source of vegetable tannins and of termite- and rot-resistant posts. This is not too surprising, however, when one considers that they have no winter, no dry season (the ocean is always wet), and no spatial heterogeneity to protect them from herbivores, and thus it is easy to

imagine that the only plant that can evolve into the mangrove community is one that also evolves the most impenetrable chemical defense. Leaves and bark with as high as 45 percent dry weight tannin content appear to be *the* answer to this requirement.

"The Tropics Are More Predictable"

No one can deny the perpetual hothouse climate of Singapore, the monotonous "winter at night, summer in the day" temperature regime of high tropical mountaintops, or the steady rainlessness of the central Sahara. However, predictability is very much in the eye of the beholder; the meteorologists (or the biometeorologists) have not really set out to document just how predictable the physical world is for wild tropical plants. And forget not that the more generally predictable the environment actually is to a plant, the more severely unpredictable will be the unexpected weather event.

Every spring in northern latitudes there is the chance of unpredictable frosts, ice storms, hot spells, and the like. However, at the beginning of every tropical rainy season there is the chance of a false start, such as in 1971 in the deciduous lowland forests of northwest Costa Rica. That year the first week of rains were followed not by the usual 20 to 40 centimeters (8 to 16 inches) of rain, but by 5 weeks with no rain. Trees better than 100 years old were killed by this drought, and many species aborted not only their leaf crop (with an attendant loss of food for their host-specific herbivores) but their flowers or fruits (a loss to the flower-, seed-, and fruit-eating animals of the habitat). In deciduous forests between 5 and 10° north and south of the equator, in the middle of the rainy season there is often a short dry season of highly unpredictable length and intensity in the months when the sun has passed overhead and moved the thermal equator up to one of the Tropics. It is commonplace for herbaceous plants to respond to this short dry season by wilting, dropping their flowers, aborting fruits, and greatly slowing the rate of production of new foliage.

Even in the wettest rain forests there are dry spells when no rain falls and the leaves become dry underfoot for a few days to as long as a month. In rain forest sites with the most continuous rain, such as in the lowlands of Borneo, such dry spells are not correlated with the time of year; the tree suddenly has to deal with the unfamiliar and potentially damaging dry air. It is often hard to see gross responses, but the subtle ones are there. For example, trees in the dominant family Dipterocarpaceae use particularly intense dry spells to synchronize their fruiting with other members of their species, genus, and family. Other species of trees wilt their growing shoot tips, and yet others change cambial wood increment rates, resulting in irregularly spaced rings in the wood. Speaking of using such unpredictable events for timing of synchronized reproduction, the pigeon orchid of the Malaysian forests (*Dendrobium crumenatum*) uses exceptionally cold nights (caused by the absence of cloud cover, which in turn leads to high net radiation from the surface) to determine when to flower synchronously.

Although hurricanes and cyclones occur in some extratropical habitats, it is clear that many lowland tropical evergreen forests are subject to devastating winds at highly irregular intervals. The author has seen hundreds of square miles of British Honduranian rain forest that had been flattened as though a giant steamroller had passed over it. There are Pacific island forests that appear to have a devastating hurricane often enough that the forest is in a continual state of succession. At the very slow growth rates of plants on the tops of tropical mountains, devastation by either volcanic or wind action may need to occur only every several hundred years to keep the vegetation in a permanently disturbed state. You may think that perturbations occurring at intervals of tens of years might not be of importance, but to a tree or an epiphyte with a life span of several hundred years, such unpredictable events are far more common than are cold spring frosts to a northern annual.

And cold is no stranger in the tropics. In 1964 at Christmas the lowland forests of Veracruz were thoroughly brown, with leaves killed by a winter cold air mass driven south by hurricanes in the Gulf of Mexico. At 3000 meters (9800 feet) elevation in the Costa Rican mountains, the weather is so predictable that snow has never been recorded, yet it may reach 0°C almost any night of the year. Yet in this habitat, once or twice a year, the air movements are just right, such that by dawn there is a dense layer of frost on the foliage of all the plants on north-facing slopes. Withstanding such an event is not something that the plant can gradually harden into, since these peculiarly cold days are not associated with any particular wet or dry season or other cue, but rather with just the right movement of clouds off the mountains.

But let us decide that at some levels the physical environment of the tropics is more predictable than outside the tropics. What is then generally forgotten is the question of how predictable biotic interactions are in the tropics. There are whole classes of biological events that are far more unpredictable than in extratropical habitats. If we examine the trees growing in a species-rich tropical rain forest, and ask what vertebrate and insect species will encounter its fruits, flowers, or leaves in a given day, week, or year, a far less certain answer can be given than in most extratropical habitats. Of course, if we move into some of the other less species-rich habitats mentioned previously, this generality no longer holds.

In the more species-rich tropical habitats, the large number of herbivorous insect species generate a very interesting kind of biological unpredictability. Many of these insects are highly host-specific folio-vores or seed predators, with their specificity centered around getting past the specific chemical, morphological, and behavioral defenses of their host. These insects probably produce new mutants and recombin-ants with novel abilities to circumvent host defenses at a fairly constant rate per species. Very simply, the more species there are in the habitat, the more frequently occurs the totally unpredictable (in time and character) event of the plant being confronted with a new herbivorous mutant. Since herbivores may pass many hundreds of generations in the lifetime of a single tree, this problem can be a very great one in the evolution of a tree, providing strong selection pressures for outcrossing (so as to be sure to pick up mutant genes for resistance from conspecific trees with maximum haste).

Seed dispersal is also subject to biotic unpredictability. As many as 40 species of frugivorous birds may visit one fruiting tree. That *a* frugivorous bird will arrive at a given tree is highly predictable (but one wonders if really more so than at a mountain ash in the French Alps), but which frugivorous birds in which proportions is quite another story. And which birds arrive makes a great difference, since some will be from climax forest, some will live primarily in swamps, others will live along river edges, and so on. If you consider a tree whose seeds have almost no chance of attaining adulthood except in some relatively specific subset of the total habitats available in the area, those seeds that end up in a bird that defecates or regurgitates them in the wrong habitat are as dead as if they had never been swallowed. One here expects selection for fruit traits that maximize the chance that the seeds are swallowed only by those bird species which provide only the very most appropriate transport. However, the tree is confronted with the problem that no one species of bird is likely to be *that* predictable in its presence at the fruiting tree.

Leaving aside the interactions with animals, plants in a northern community are probably coevolved for morphological and behavioral traits of value in competition with species with similar life forms. In short, beech-tree competitive traits have probably been influenced by the regular presence of maples as competitors (as well as by the presence of beeches themselves). Beech and maple saplings are probably competitively coevolved with understory dogwoods and cherry. Early succession dogwoods and sumac have almost undoubtedly influenced each other's evolutionary histories. In the more species-rich tropical habitats, the chances of such fine tuning are greatly reduced. In a forest with 500 tree species, the chances that a seedling or sapling will have the same other species for a major competitor in two successive generations appears to be very close to zero. In short, there is almost no way to predict with any certainty in some lowland evergreen forests what species will be competing with a given tree sapling. On the other hand, it should be noted that this is not the case in the species-poor tropical habitats. Furthermore, in some Malaysian evergreen forests with hundreds of species, many of the species have highly clumped distributions, and thus a sapling is quite likely to be competing with only a very small subset of the total number of tree species in the forest.

"The Tropics Have More Constant Environments"

This gross misconception probably stems from the general impression that the tropics are generally hot and moist, an impression easily gained from living in most tropical lowland cities. Seasonality is a conspicuous part of the lives of virtually all tropical plants, and quite frankly, we do not know if it is more or less important to tropical plants than to extratropical plants. The central problem may be exemplified by noting that if a tree lives in a tropical habitat receiving 30 centimeters (12 inches) of rain per month for 10 months of the year and 10 centimeters (4 inches) of rain per month the other 2 months, the dry season may just as effectively stop shoot-tip production as does 5 months of winter weather in Georgia or Kansas. In short, seasonality is in the eye of the beholder, and the sharpness of vision is directly related to the sorts of background against which the seasonal change is set. Costa Ricans recognize eight distinct seasons in their "constant" tropical habitat.

In tropical deciduous and semideciduous forests the world over, and they occur in all major tropical areas, it is commonplace for no rain at all to fall for 4 to 6 months, and then to receive 1 to 3 meters (40 to 120 inches) of rain in the remaining months. Associated with this lack of rain in the dry season, the number of hours of unobstructed sunlight per day may change from as little as 3 to as much as 12, and winds up to 32.2 kph (20 mph) may become a daily fact of life. That these winds are of very great importance is illustrated by the fact that in coastal Nigerian habitats there may be a 4-month dry season when the 40-m-tall trees stand evergreen, their semiwilted leaves drooping motionless throughout the day, while in the Pacific lowlands of Costa Rica, with the same rainfall regime but a steady offshore trade wind, the trees stand leafless and rarely exceed 30 meters (100 feet) in height. During the tropical dry season it gets hotter during the day and colder at night than during either the rainy season or in adjacent areas with a less severe dry season (less severe, i.e., as recorded by the meteorologist). This is because on the cloudless days there is more direct insolation, and the cloudless nights result in more rapid reradiation of heat.

Accompanying the tropical dry season is the ever-present threat of fires, although people have so badly altered tropical habitats that it is impossible to know the natural frequency of fire. People set fires so frequently that one cannot know how and with what frequency other fires are started. Second, their frequent firing has opened up forests, rendering them much more fire-susceptible than they would have been with natural firing frequency. In some areas we might as well speak of the fire season as of the dry season. Intact tropical deciduous forests were probably only very rarely, if ever, burned. At present, when a fire is burning in a nearby pasture, it only very rarely invades the forest more than a few meters and that only at ground level. If cinders alight on a dead tree within the forest, the tree usually burns to a tree-shaped pile of ash on the forest floor without spreading even a few centimeters (but note, the resultant fire-sterilized micro-habitat is quite a different site for early successional plants to occupy than is an ordinary tree fall). If tropical deciduous forests are attacked repeatedly by fire (and axe), they gradually give way to grasslands (Figure 7-2) as the seed sources for the woody plants are eliminated, although a few fire- and drought-hardy species such as *Curatella* and *Byrsonima* remain (to be misnamed "savanna trees").

Yet the dry season is not always the inimical season. For many deciduous forest trees, it is the time of year when vegetative activity comes to a halt and energy-demanding processes such as flowering and

FIGURE 7-2.
Tropical grasslands at Maturin, Venezuela. (Photograph by Dan Janzen.)

fruiting can proceed without strategically jeopardizing the plant through exhaustion of reserves at a time when they may be (unpredictably) needed to replace a major branch or deal with defoliation by a herbivore. To wild bees and other flower visitors, to say nothing of seed and fruit eaters, the dry season may be a time of plenty as compared to the barren weeks of the early rainy season when all the plant's resources are being channeled into leaf and branch production. For example, the author knows a commercial apiary in the Pacific lowlands of Costa Rica that can make two major honey extractions during the dry season and then has to turn around and feed large quantities of sugar to the bees during the last half of the rainy season, as they have exhausted their reserves during the first half of the rainy season. This perhaps tells us why wild honeybees (escaped colonies) are such a rarity in neotropical deciduous forests, and why wild social bee species richness is only 5 to 10 species in these forests, as compared to the evergreen forests only a few miles away where there may be 25 to 35 species of wild social bees in an area of several square kilometers.

In evergreen rain forests, the dry season months may be the time of maximum growth (especially for the understory plants). During these months, the increased average numbers of hours of unobstructed sunlight per day (say, from 2 to 6 hours) increases both the heating of the vegetation and the penetrance of light to the lower levels of the canopy as diffuse light and as sunflecks. This should increase photosynthetic rates and, at least in some cases, increase the rate of foliage replacement as removed by herbivores. It is of interest in this context that the overall biomass of insects (and to a lesser degree, species richness) in the earlier stages of succession and in the forest understory in lowland Costa Rican rain forest is highest during the dry season.

"The Tropics Have the Highest Productivity"

This generalization seems to stem in part from the failure to realize that standing crop (biomass) of vegetation is not a measure of productivity of the habitat. The generalization must be further dissected with the observation that there are two kinds of productivity to be considered. First, there is the actual amount of material fixed by the plants minus the respiration costs of that fixation. It is obvious that any calculation based on this kind of primary net productivity must always be accompanied by a statement about the time scale—per hour, per season, per year, and so on. Second, there is the amount of material generated that can be harvested whoever is interested in it. In statements about this kind of productivity, it is imperative that we state for whom this is harvestable productivity, and realize that its quantity is not a direct function of how much sunlight, water, and so on, there is, but rather, directly depends on the (highly capricious) strategic demands of a particular species of animal or plant.

Bearing these qualifications in mind, it is obvious that the plants in a tropical rain forest on lateritic or better soils have a higher net primary productivity per year than does arctic tundra, if for no other reason than that they can photosynthesize year round. However, if we try to pick habitats more similar in gross structure, one in an extratropical habitat and one within the tropics, the comparison becomes much more feeble. A mixed hardwood forest on an Ohio raised floodplain may very well have a higher net primary productivity than a mixed evergreen forest growing in a Sarawak peat swamp (bedded on white sand soils) at sea level on the equator and receiving 4 meters (160 inches) of rainfall evenly distributed throughout the year.

There is a very large technological problem in the measuring of net primary productivity. One cannot march out into a tropical forest and measure primary production by measuring anything that is standing there. We may infer that primary production is very high owing to year-round growing conditions, but even this assumption is vulnerable to the criticism that on poor soils (which most tropical soils are) the plant can no more make use of this constantly available heat, light, and water than a desert plant can make use of the light and minerals when water is wanting. Thus, we are forced to use empirical measures if we wish to know how productive tropical habitats are from the plant's viewpoint, but these measures require large masses of complex machinery lugged about in the forest, an event only rarely attempted by physiologists.

But even if we could determine how much photosynthate a tropical plant can make, and how much it expends to make it, we still do not know how much net primary productivity there will be for the plant in a tropical habitat as opposed to an extratropical one. It costs the plant a lot to live. Warm tropical nights burn up photosynthate that an extratropical plant may not have to expend. Even when leafless during a tropical dry season, the deciduous tropical tree is steadily eating up its photosynthate reserves, a drain that is certainly not nearly so great per dormant day for extratropical trees that are essentially in cold storage for the winter.

Once we know how much the plant has left over after its basic metabolic needs have been met, the organism harvesting the plant's actual material production may get less than it would in an extratropical habitat, simply because the plant makes things with its remaining photosynthate that are imperceptible, inedible, or otherwise undesirable to animals. Many tropical trees, for example, make large quantities of secondary compounds as defenses against herbivores (insects, mammals, fungi, and bacteria). Some of these are familiar—rubber, cacao, coffee, tea, tannin-rich mangrove bark, acacia gums, resins, chinchona, pyrethrum, derris, strychnine, rotenone, opium, chicle, and so on. When standing-crop or other biomass units are measured to determine the primary production, these expensive parts of the energy budget are generally ignored. Many tropical trees wait a number of years, storing their reserves until they suddenly make a large seed crop. If harvestable productivity is calculated in the seeding year, these trees seem to be very productive, while in other years they appear amazingly unproductive. Finally, the reader should note that biomass says absolutely nothing about net or harvestable productivity. Very large trees and forests can accumulate very slowly on a site, and a site with a very small standing crop may have a very high turnover of that standing crop, thus having a very high productivity of harvestable material if that plant is what you are bent on harvesting. Small wonder that studies of southeastern Asian rain forests have shown the same biomass and even production of standing crop as Japanese extratropical forests with only one-third of the net primary production.

With these qualifiers in mind, a few examples of tropical rates of production of standing crop are appropriate. On fairly recent volcanic soils from sea level to about 1500 meters (4900 feet) elevation, with a 3-month dry season, forest growth rates in the tropics can be phenomenal. Starting with bare ground, a forest with a canopy at 45 meters (150 feet) height can be generated in 20 to 30 years. This assumes, of course, that there is a seed source for the forest trees, the site is not repeatedly burned, and cattle are not allowed to graze as succession proceeds. Individual trees may gain height much faster. In such a site, one of the middle-successional species of tree (e.g., a Costa Rican *Bursera simarouba*) may attain a height of 40 meters (130 feet) and a diameter 1.2 meters (4 feet) off the ground of 0.6 to 0.9 meters (2 to 3 feet) in 10 years. Balsa trees may do it even faster. However, the author's reaction as a biologist is "What difference does it make?" What we have really said is that some trees put much of their photosynthate into size rather than sexual reproduction or hard trunks (and therefore smaller trunks). Such other trees in the succession appear "to grow more slowly," while in fact they are likely to simply be putting their photosynthate into quite different kinds of output than mere size (which, we know from balsa trees, may involve a very small wood investment indeed).

But there are many tropical circumstances where *all* the plants in the habitat grow more slowly by any measure. For example, on Bornean, Camerounian, or Brazilian white sand soils, a clearing may require as long as 30 years to regenerate secondary succession only 3 to 9.1 meters (10 to 30 feet) tall. Plants on these soils, which are about as nutrient-poor as ground glass, grow so slowly that throughout the tropics there is almost no native agriculture on them. The vegetation on such poor soils (derived, incidentally, from old beaches or weathering of sandstone mountains) has other odd traits. Vines are very rare in the various successional stages; presumably, a plant that depends on very rapid growth to overtop other plants has little chance in such a site. The species richness of early to middle successional plants is very low (even though nearby mature forests may have high species richness). The standing crop of flowers and seeds is very low indeed, as though what is produced in one year on a good soil site had been spread out over 20 years on a bad soil site. The replacement rates of plant parts of all the various types eaten by vertebrates and insects is very, very low, with a concomitant very low standing crop of these animals. (*Note:* Animals do not have as free an option of growing very slowly and therefore gradually accumulating a large biomass as do the plants in a low-productivity site.)

As one moves up a tropical mountain, again a lowered production of plant parts becomes evident. Above about 3000 meters (9800 feet) on Costa Rican and Andean mountains, when land is cleared by fire or other catastrophic means, it may remain bare of vegetation for as long as 5 years, even when there is a nearby seed source. A 1.8-meter (6-foot)-tall *ericaceous* or *rosaceous* shrub may be as much as 30 years old, a height-increment rate expected in England rather than in the tropics.

A Case History

We have been speaking here in gross generalities, and by now, you should be getting impatient with their inability to help you understand particular tropical systems. Yet we must strongly resist the urge to try to generate narrow generalizations about the tropics, or even about a single tropical habitat, at this state of our knowledge of the tropics. Such generalizations can be made *solely* as hypotheses to guide our construction of experiments and observations, rather than as definitive statements. Perhaps the next trail to take is a very narrow look at a specific tropical tree species, realizing all the time that this is not intended to be a "typical" or "representative" tropical tree. No one tree can be that. With respect to leaf phenology it may be in the left-hand tail of the frequency distribution for the habitat, with respect to dispersal agents it may be in the right-hand tail of the frequency distribution, and with respect to vegetation defenses against herbivores, it may be right on the mean. So which of these traits will we choose as the label for this tree in deciding if it is typical or representative? The answer is "None." In short, a single tree species is too much a composite to ever be representative of anything as a whole. However, a detailed examination of one species may give some feeling for the sorts of things that one does notice when thinking about a tropical tree. *Hymenaea courbaril* has been chosen for an example, mostly because the author knows more about that tree than any other in Central America. As we consider its biology, ask yourself how each of these traits compares with the analogous traits of an extratropical tree species with which you are very familiar.

Hymenaea courbaril (Figure 7-3) is a legume, a giant bean plant if you will. Its common name in Central America is "guapinol" ("stinking toe" in Jamaica and Trinidad, "algarrobo" in Puerto Rico), and you may know it indirectly, since its fossilized resin is the source of most Central American amber. This amber is Oligocene in age, which means that the guapinol has been with us a long time. The tree ranges from western lowland Mexico (south of Mazatlan) down the Pacific coastal plain and foothills of Central America, with occasional excursions over to the Gulf and Caribbean lowlands (Mexico, Nicaragua) and thence out onto the larger Caribbean Islands. In short, it is a widely distributed tree and the only one of its genus in Central America. It extends on down through much of lowland and foothill South America, where in places it is sympatric with a number of other species of the genus. In most of its habitats the forest ranges from fully deciduous (with *H. courbaril* primarily growing along creeks and on north-facing slopes) to evergreen with a 2- to 3-month rather distinct dry season (e.g., the Osa Peninsula, southwestern Costa Rica), where it is found on dry ridges. As the forest becomes semideciduous as we move toward the center of these two extremes, guapinol are found on most kinds of topography. In ecological summary, guapinol shows site preferences, just as do other trees. It should be noted that the author has never found one growing on a

limestone ridge or even on the soils derived from such parental material. In this sense it is also normal in having certain edaphic "preferences." If the seeds are transplanted, guapinol can be grown in rain forest (e.g., IICA-Turrialba and near Puerto Viejo, Costa Rica).

In areas with a 4- to 5-month severe dry season, such as the lowland coastal plain of northwestern Costa Rica (Guanacaste Province), guapinol trees reach their maximum height (40 to 45 meters) (130 to 150 feet) at about 100 years of age but can live at the least another 200. The largest tree I have ever found is 109.3 centimeters (47 inches) in diameter at chest height, but most reproductive adults are 50.8 to 88.9 centimeters (20 to 35 inches) in diameter. Trees grown in a forest have straight clean boles and are very valuable as lumber trees. Their wood is very hard (special sets to the saws are required in the mill to work it) and makes high-class furniture. The trees are often left as "cash in the bank" when the remaining forest is cleared; they can be cut and sold at any time.

Hymenaea courbaril would be classed as "evergreen" by most phenology schemes. In fact, it is the fully leafed green crown standing out in contrast with its leafless neighbors during the dry season that leads to such names as "semideciduous forest." However, this is an illusion. The guapinol tree holds its leaves until the first month or so of the dry season, and then abruptly sheds the entire crop and immediately grows another crop. It is thus truly leafless for only a week or two. The tree takes a substantial risk by producing its new leaves at a time when soil moisture is waning. In the driest parts of its range, the author has seen numerous cases where the soil (or air?) was too dry for the leaves to expand fully and the tree has ended up with only a partially filled crown during the rest of the year, owing to the leaves being small and less numerous than usual. The leaf itself is as one would expect of an "evergreen" leaf—thick, leathery, tough, stiff, waxy, and rich in resins (and perhaps other antiherbivore devices). Also, as one might expect, there is almost no sign of herbivore damage, although sucking insects (leafhoppers) may take a very heavy toll of the new foliage as it is expanding (presumably, at this time the leaf resins are poorly developed). In short, guapinol has an evergreen leaf lasting less than a year, with synchronized leaf drop.

The entire branch and trunk system is invested with a fine network of resin ducts and pockets just below the bark. When the tree is wounded, a copious flow of this resin results, sealing off the wound and in some cases directly killing the offending insect with volatile toxic terpenes or phenols contained in the resin. The detailed chemistry of this resin varies from tree population to population, among the parts of the tree, and with the age and health of the tree. This is not surprising, because different chemicals and mixtures will have different drying and flowing properties, toxicity, costs, and autotoxicity. The resins in the developing green fruits (pods) are quite different from those in the tree, and it is noteworthy that they serve two quite different defensive functions at different times in the fruit's ontogeny. When the fruit is green, the resin needs to be liquid so as to flow out of a wound; when the fruit is mature, the resin needs to be very hard to protect the seeds from boring insects and certain vertebrates.

In the lowlands of Costa Rica, *H. courbaril* has a most complicated pattern of sexual reproduction, one that becomes evident only after keeping individual records on hundreds of trees for many years. In most years, 60 to 80 percent of the adult trees bear flowers (Figure 7-4). Adult is here defined as large enough to have borne flowers at least once—this size is attained with trees of quite different ages and depends on the site, competition, tree health, and so on. On a given tree, flowers open for about a month, but the population as a whole bears flowers from late February through mid-March, with a few individuals flowering as late as June. In these sites, late February is about 2 to 3 months into the dry season, and the rains come in May. The flowers are bat-pollinated and open at night; introduced honeybees and other insects scavenge nectar from them at dawn. The majority of these flowering trees will then abort all flowers, but on 1 to 5 percent of the trees in any given year, about one fruit will be set for every 100 to 200 flowers produced. A large tree might have as many as 100,000 flowers. The large number evidently has an adaptive significance from two viewpoints. First, if the tree is scheduled to only be a pollen donor that year (effectively being a male), it is producing a very large amount of pollen on the chance that some of it will be carried to one of the few trees that is scheduled to produce a pod crop that year. Concomitantly, it is producing a very large amount of nectar to attract the bats that will make this transfer. Second, if the tree is scheduled to make a pod crop that year (be effectively hermaphrodite, or even female as viewed by the dispersal agent or seed predator), it cannot

merely make a number of flowers equal to the number of pods for which it has food reserves. Such a tiny flower crop would be quite unlikely to be attractive to the bats, unless it produces phenomonal amounts of nectar (or very high quality nectar), the cost of which could easily equal that of simply making a lot of flowers.

The question that comes immediately to mind is why many of the trees in a given year reproduce only through the production of pollen grains, rather than seeds as well. The most proximate answer appears to be that the three classes of flowering but nonseeding trees do not have enough reserves to make a large seed crop. The smaller trees, which are just entering reproductive status, have small reserves just because they are small. The full-grown trees that flower but do not seed are weakened by disease, major structural damage, root erosion, fire, and so on, or are in the third category. Trees in this category have exhausted most of their reserves with a previous seed crop and are now accumulating enough reserves for another seed crop years later.

The ultimate question should now be uppermost in your mind. What selective pressure could have led to a tree phenotype that makes a large seed crop every n years rather than $1/n$ as many seeds every year? It should be noted, as a relevant background to this question, that the guapinol trees in a habitat are not synchronized with each other with respect to the year in which they will bear a large seed crop. The author thinks that the primary selective agents are two weevils in the genus

Rhinochenus. These weevils kill up to 50 to 300 of the seeds in a guapinol pod crop. A large pod crop for a tree growing in a forest (see below for a qualifier with respect to competition effects) is 100 to 500 pods. It seems that the same number of pods are killed in a given tree's pod crop irrespective of how many more pods there are than that number. That is to say, if there are going to be enough weevils to kill 167 pods in a given tree's crop, they will do that irrespective of whether the tree produces 200 or 500 pods.

With that background we are in a position to ask why some original mutant that skipped a year between crops was favored. First, it forced the weevils to leave the tree (assuming that they cannot survive at the tree for 2 years without seeds). This meant that the only weevils to find the mutant's seed crop were those that immigrated to the tree that year. This number is almost certain to be lower than the number that would be there if there were a local population of weevils associated with each guapinol tree. But note that our new mutant tree is now going to have a bigger seed crop each time it seeds, assuming that it can store its unused reserves. If it waits 3 years between crops, it can have an even bigger seed crop, and so on. Of course, the balancing force to this is the fact that every year longer that it waits between seed crops is one more year in which it is reproducing only by pollen donation rather than by pollen donation and seed production. The increase in the number of years between seed crops rises and should eventually cease when it gets to the point where the costs about balance the gain. In the deciduous forest lowlands of Costa Rica, it appears that this is about 4 years in intact forest.

It might be of interest to the reader to note that in Puerto Rico, where the weevil has apparently never occurred but the guapinol tree is native, the adult trees bear seeds every year. Also, in Puerto Rico, the trees begin producing seed crops when they are as little as 5 to 10 years old (as contrasted with about 50 to 100 years of age in Costa Rica), and as they grow, their seed crops gradually increase in size. Finally, the Puerto Rican pods have very little resin in the pod walls; in Costa Rica, the walls of the green pods are rich in resin, and this appears to be responsible for the fact that the weevil can only enter the pod very late in the pod's development time, and thus there is time for only one generation of weevils per seed crop.

It is instructive to ask what happens to this system when the surrounding forest is cut down, leaving the guapinol adults as isolated pasture and fencerow trees. Suddenly provided with sunlight from all directions, and probably with greater water and mineral availability, these trees often produce not only a flower crop every year but in many cases make a large pod crop every year. Some of the crops may be huge, with as many as 2000 pods in extreme cases. The number of seeds killed by the weevils stays the same, however. There appear to be two causes. First, the weevils are genetically programmed to leave the tree after a crop, and thus there appears to be no buildup of a local weevil population on the annually produced seed crops of a given pasture tree. Second, with the removal of the forest, the vertebrate dispersal agents that used to open the pods are removed. This extinguishes one of the species of weevil; it depended on a dispersal agent chewing open the pod to escape from the pod. The other species can chew its way out of the pod, but by itself, it is not as effective as the two were together at killing seeds. Finally, we may add that the seeds that do escape weevils on these parent trees bereft of forest do not produce seedlings, because there are no animals to free them from the woody indehiscent pod. By the time that the pod has rotted open, it is too late in the rainy season for enough growth to occur for the seedling to make it through the dry season. Even if there were enough time, the seeds generally rot in the rotting pod rather than germinate there.

A Complex Interaction

In the previous paragraphs we have dealt with the autecology of a tree, with a few cross references to its interaction with a couple of beetles. The complexity of this system is about at the level experienced by many tropical plants, but there are some with what appear to be substantially greater interspecific complexity. A neotropical rain forest vine in the genus *Anguria* (the Curcurbitaceae or squash family) provides an example, although again I hasten to add that this should not be taken as an exemplar or average system.

Anguria vines occur throughout the Caribbean lowland rain forests of Costa Rica, Panama, Trinidad, and so on. At the base they are as thick as one's arm or leg and woody, and at the top they reach nearly to the canopy, from which they drape themselves out over tree branches. On the pendant branches are clusters of tiny bright-orange flowers. The plants are dioecious, with male vines outnumbering female vines. A given vine may produce flowers 365 days of the year, with each inflorescence producing a new flower every 2 to 3 days. Our story begins well before dawn, when a female *Heliconius* butterfly (Figure 7-5) comes fluttering up to a *Gourania* inflorescence hanging 30 meters (100 feet) up in the rain forest canopy. It is nearly dark and there is only one tiny open flower. But she comes, as she has for many months of mornings in the past, because she knows the exact location of that plant and its inflorescences. She knows because she has learned the location long ago, and this plant is one of a series of this species that she will visit that morning. Her feeding route may include plants spread over several square kilometers.

Most of the plants she will visit will be male, and they will be producing not nectar but pollen at this hour. By brushing her tongue (with its long recurved hairs) repeatedly past the newly opened anthers, the butterfly collects pollen on her tongue until she has a large ball of it glued onto the coiled tongue. She then regurgitates nectar from her gut, nectar collected the day before from *Anguria* and other plants, and soaks the pollen with it. Amino acids leach out of the pollen into the nectar, which she then drinks as a way of obtaining scarce building blocks for the proteins she needs for the few large eggs she will lay in her long lifetime.

As the butterfly moves from plant to plant, she occasionally visits a female plant and achieves pollination. As dawn comes on and the sun

FIGURE 7-5.
*Female Heliconius
butterfly, Guanacaste,
Costa Rica. (Photograph
by Dan Janzen.)*

rises, the plants begin to produce nectar and the butterfly may again run its "trapline." The plant cannot set seed without the butterfly, and the butterfly probably would have a much smaller clutch size were it not for the amino acids gathered from the pollen. But the complexity does not stop there. If we examine the nectar, we will find that it also contains amino acids, secreted independently by the plant. Further, this female also gains amino acids by mating with many males and then internally "digesting" the sperm.

But our story is not complete, since more than just the *Anguria* and the *Heliconius* adult are required for the system to operate. The female *Heliconius* lays her eggs on a particular species of vine in the genus *Passiflora* (passion flower in the family Passifloraceae). Each of the several species of *Heliconius* in a habitat feed on only a small subset of the *Passiflora* species present. Each female spends much of her working day running a different trapline from the *Anguria* flowers, going from *Passiflora* vine to *Passiflora* vine, checking to see if there is a new shoot tip that does not yet have some other *Heliconius* egg already on it. She will then lay only one egg per shoot tip, and even that on the very tip of a tendril. By doing this, the young caterpillar can eat its way down the tendril, and by the time it reaches the main stem, it is perhaps large enough to deal with the large ants that patrol and hunt on the *Passiflora* vine because of its large extrafloral nectaries. If there is more than one caterpillar per new growing shoot, there is not enough food for both of them to grow to full size, so it behooves the female to be very careful about not putting an egg down where there is already another.

Now, *Passiflora* is a very poisonous plant, being rich in cyanide-producing and perhaps other noxious compounds. Perhaps associated with this, *Heliconius* butterflies are highly distasteful. Being "models" in mimicry systems, or at the least warningly colored, it is not surprising to find that they live a very long time as adults. Thus, it is not surprising to find that during their 6- to 9-month or longer life span, a female *Heliconius* may well have several dozen *Anguria* plants in her trapline. It is hardly necessary to say that the plant's small daily expenditure of energy, which yet gets it very reliable pollination services, could only work with a faithful and diligent pollinator such as this.

Now we might expect the *Anguria-Heliconius—Passiflora* system to be rather self-contained in some areas. However, in many mainland Central American rain forests, there is a second genus of cucurbitaceous vine in the genus *Gourania*. Not surprisingly, *Gourania* flowers are quite similar in color and behavior to *Anguria* flowers, and *Heliconius* butterflies visit them, too. In many areas the density of both of these plants would probably not be sufficient to support much of a population of *Heliconius* butterflies, and there are probably other flower species in the same area about which similar stories could be told.

What Are the Major Guilds?

There are major sets of organisms in any habitat that do about the same thing ecologically, even though they may be rather unrelated taxonomically. In recent ecological literature such sets have been called *guilds*. There are some conspicuous guilds in tropical habitats, especially in the more species-rich lowland rain forest habitats, which can be roughly characterized. Some examples that come quickly to mind are those fungi that form mycorrhizal associations, those animals that act as seed-dispersal agents, the vines, and the epiphytes.

MYCORRHIZAE

The problem is a straightforward one. We may visualize a tropical rain forest as a thick mantle of large trees perched superficially on a thick layer of clay or sandy soil into which the roots penetrate only a few inches. As leaves, branches, dead insects, and so on, fall from above, they land on the thin layer of roots in the upper 2.5 to 12.7 centimeters (1 to 5 inches) of the soil, and their contained nutrients are rapidly extracted and taken up by the roots. This is in strong contrast to the humus-rich upper soil of deciduous or upper elevation tropical habitats, or in many extratropical forests, where the nutrients from decomposing litter go into a soil nutrient pool from which they are later extracted by the plant roots. In the tropical rain forest case, any mineral ion that gets out of the litter and is free in the soil rather than taken up by a root is very likely to be leached out of the soil by the frequent rain. Here, then, we expect strong selection for specialization among tropical rain forest trees with respect to the ability to quickly and thoroughly pick up mineral ions as or before they are released from the litter or rainwater. Possible areas for specialization involve such things as patterns of root placement, annual patterns of root growth, and physiology of ion uptake. To the best of the author's knowledge, these aspects of *tropical* tree biology are totally in the dark. However, there is another area of specialization that has been at least discussed and is now the subject of experimentation. In short, it appears that certain species of fungi form mutualistic associations with tree roots. It appears that the tree "feeds" carbohydrates, vitamins, and amino acids to the fungus (these items being relatively cheap), and in return the fungus "feeds" the tree mineral ions. Apparently, the fungus is a specialist at trapping and extracting mineral ions (especially phosphorus) from newly fallen litter, and then transports them into the plant roots. Associated with this, tropical rain forest trees often have roots almost completely lacking root hairs. The tree roots with their associated fungal hyphae are called *mycorrhizae,* and the interaction is often termed a *mycorrhizal association.*

Now, with that background in mind, a number of predictions have recently been proposed and are being tested by David Janos in a Costa Rican rain forest. We would expect, for example, that the plant species which are first to colonize a newly exposed piece of rain forest soil would not have much, if any, interaction with mycorrhizal fungi. They could not depend on the right species of fungus to be reliably present at the same place at an unpredictable and ephemeral resource such as a tree fall in the forest or a newly exposed gravel bar in a river. We might also predict that the less easily a plant can obtain the carbohydrates to feed the fungus, the less likely it is to take on the association. For example, if there are forest tree seedlings of a particular species that sometimes grow in the shade and sometimes in the sun, we might expect them to have a mycorrhizal associate less frequently in the shady than the sunny site. If fungi are more resistant to tannins in the litter than are tree root hairs, for example, we may find mycorrhizal associations to be more abundant under tannin-rich vegetation. On the other hand, we would expect seedlings growing from large seeds to develop mycorrhizal associations faster than those from small seeds. On the other hand, there is a balance provided by the value of the mycorrhizal association to the plant. A seedling from a large seed may be able to survive much longer without a needed mycorrhizal association than can a seedling from a small seed. Not only is Janos finding this to be the case, but he is also finding that seedlings with mycorrhizal associations can withstand much greater levels of herbivory than can conspecifics lacking the fungal association. On the other side of the coin, we may note that there are two circumstances in which it is maximally advantageous to the fungus to form the association. First, we may expect it where carbohydrates are in exceptionally short supply in the litter. Second, we may expect it where minerals are in a form or concentration easy for the fungus to obtain but difficult to obtain by the root hairs of the tree.

It is tempting at this point to try to list those tropical habitats in which mycorrhizal associations play a very major role in the structure of the community, but our ignorance is too great to do this effectively. What can be said is that the poorer the soil, the more omnipresent appear to be tree species with mycorrhizal associations. Likewise, it appears that in general the plants of early stages of succession often lack mycorrhizal associations. Going in the other direction, it seems likely that tropical deciduous forests and desert habitats will have no more mycorrhizal associations than do most extratropical habitats. It is of interest here that perhaps one of the stronger long-term negative effects created by clearing large tracts of tropical rain forest may be the extinction of the mycorrhizal guild, with the result that even if there is a seed source nearby, succession back to forest may be ridiculously slow, since not only the seed but the fungal associate has to immigrate into the cleared area.

SEED-DISPERSAL AGENTS

In practically all tropical habitats, a large proportion of the seeds of woody plants are moved about by animals that are either after the fruit for food, the seed for food, or both. We hasten to add that there are numerous species of tropical woody plants whose seeds are dispersed by wind, water, and explosive capsules, but they tend to be in the drier areas and in the earlier stages of succession. The study of seed dispersal by animals in the tropics is just now in its first stages of interesting growth, although as an anecdotal area of endeavor, it has been with us a hundred years or more. However, certain tentative generalizations are now possible and productive in understanding tropical ecology.

1. Any tropical tree that produces a fruit of high attractiveness to one animal (perhaps the optimal dispersal agent) will be potentially attractive to very many other species of mammals and birds. Since only a small fraction of these animals will (a) not kill the seeds and (b) will defecate or spit them in an optimal place for seed/seedling survival and growth, there should be strong selection for flavor, hardness, timing, minor element content, and so on, that will minimize the attractiveness to the wrong animals. Indeed, tropical fruits are extraordinarily varied in these traits, and the diversity of dispersal agents (and nondispersal-agent frugivores) is most likely responsible. Wild tropical fruits, ranging from oil-rich lauraceous examples (e.g., the avocado is a mega-example) to the "fruit" of the cashew, which is hardly more than a bag of sweet water. In the former example, the large seeds are swallowed intact by frugivorous birds (e.g., the quetzal of Guatemalan fame); their gizzards strip off the thin oily pulp, and a short time later the bird regurgitates the undamaged soft seed. In the latter example, the "fruit" is a fleshy receptacle eaten by monkeys whose molars would be lethal to the oil-rich cashew seed, but who avoid it, owing to the actual fruit wall being of two layers, between which lies copious amounts of an extremely vessicant oil (the very same oil that causes the rash of poison ivy). There are Indian *Terminalia* fruits with as high as 40 percent dry weight tannin concentration (gathered for the tanning trade), which are, however, highly sought after by some Indian rodents and deer. It is of interest in this context that the tannins

are hydrolyzable and, therefore, far more digestible than the nonhy-drolyzable tannins so commonly found in foliage. There are Costa Rican palm fruits that taste marvelously sweet when you eat the first one, but the sore throat sets in about the time that you finish the second one. This would appear to be adaptation to keep any one animal from eating all the ripe fruits, and thereby avoiding having all the seeds defecated in one place. In short, tropical fruits exemplify well that one bird's dinner is another's poison.

2. Dispersal is adaptive, not only in the classical sense of getting to a place where there is a new space into which to grow, but also in the context of getting away from the seed predators that may congregate on the seed crop and at the parent tree. In the more species-rich tropical forests, it is evident that a seed dispersed far from the parent has a great chance of not landing in the middle of a conspecific's seed crop. Being dispersed far from one's sibs is of value not only when the plant is a juvenile, but when the plant is an adult as well. The further the adult plant from other conspecific adults, the less likely it is to share the members of seed predator guilds with the other plants. Of course, there is another side to this coin—the farther apart the adults, the more reliable the pollinators have to be to ensure adequate pollen flow.

Again, the diversity of seed predators in a tropical forest means that no single seed-dispersal pattern is best. To escape from parrots, a seed may have to do nothing more than drop off the tree. To escape from agoutis or pacas searching for seeds beneath the parent tree, the seed may have to be carried not only tens of meters from the parent, but be buried as well. Mango seeds carried hundreds of meters from the parent by bats, and deposited in piles beneath the bat roost, may be found as easily by rodents as mango seeds left below the parent by monkeys when they are feeding in the mango tree. Single *Andira* seeds dropped by bats in Costa Rican deciduous forest may be found by curculionid weevils at distances of 100 meters (330 feet) from the parent. At the other extreme, many legume seeds become unavailable to bruchid beetle seed predators as soon as the green fruit ripens; in short, it may be a race between the development time of the bruchid and the development time of the fruit (seeds).

3. Many tropical seed-dispersal agents are also seed predators (this is, of course, a problem in extratropical habitats as well). There are two forms of double role playing. There are animals such as neotropical agoutis and pacas that eat a number of seeds beneath the parent tree but carry some off and bury them. They may not even eat the seeds at first but, rather, eat the fruit and then go off and bury the nuts. At other times of year, when seeds or fruits are in short supply, they return to the burial sites and attempt to locate the buried nuts by smell (much as do extratropical squirrels). Some will have germinated and some will be missed. On the other hand, there are many animals that are dispersal agents to one species of plant and seed predators to another. Many African forest trees have large seeds in a large woody and fibrous covering which pass unharmed through the elephant's grinding molars, and are not only dispersed but provided with a large dab of fertilizer at

the same time. Other seeds with more fragile hulls are easily ground up by elephants. Costa Rican deer struggle to peel the pulp off the hard seeds of *Hymenaea courbaril* and finally swallow the seed whole along with the pulp. However, the soft seeds of *Sterculia apetala* are ground to a fine pulp (and even if they were not, they would probably be killed by the passage through the deer's gut).

VINES

That vines (Figure 7-6) seem to be most peculiar plants probably stems from the fact that they have received virtually no attention by either ecologists or natural historians of plants. This, of course, applies to extratropical vines as well as tropical ones, but vines make up a much greater portion of the total plant biomass in some tropical habitats than in almost any extratropical habitat, and therefore this peculiarity becomes more obvious. In short, a vine is a plant that grows in response to unused sunlight resources and uses other plants for physical support. This often involves growing up and over other plants, which in turn requires a faster growth rate than that of self-erect plants. This means that vines should be conspicuously absent in certain tropical habitats: areas of high elevation, exceptionally poor soils, and heavily shaded forest understory. On the other hand, they are extremely abundant in early stages of succession on good soil at low elevations. As succession proceeds, many species of vines stay with the enlarging trees, until in a climax tropical forest, there will be some vines with crowns as large as those of the large trees. These large vine crowns often fill in the interstices between large tree crowns and partly overtop several different tree crowns. If we could keep track of one over the years, we might well find it to gradually shift location like a giant ameba. Its connection to the ground is by a trunk as much as 25 centimeters (10 inches) in diameter with water-conducting elements up to 1 millimeter in diameter. As some of the trees fall out from under the vine as they senesce, its trunk may fall to the forest floor in places like a giant rope. As these sections sometimes take root, the vine may effectively shift its entire location, not just that of the crown.

There are even some vines in tropical forests that can make their way to the canopy in a climax forest, simply by growing upward until they reach the light. *Dioclea megacarpa*, a large legume vine in the deciduous forests of Costa Rica, starts out as a seed 2 centimeters in diameter. The new seedling shoot tip grows straight upward for as much as 3 or 4 meters (10 or 13 feet) before it produces its first leaf, and long afterward, it continues to put most of its photosynthate into vertical elongation rather than crown development in the low-intensity light of the forest depths. It is evident that the loss of a single shoot tip to an herbivore could be catastrophic for such a plant, especially if it is relying on its dwindling seed reserves to get it high enough to be large enough to make it through the upcoming dry season. One way to ensure against such a loss is to have a high concentration of secondary compounds in the shoot tip and associated stem. Another is to minimize the period that the

FIGURE 7-6.
Tropical vine (Convolvulaceae or morning glory family) from Villa Colon, Costa Rica. (Photograph by Dan Janzen.)

growing shoot tip is available to herbivores. Several large forest vines have solved this problem by growing first as a forest understory shrub. While apparently sitting inactive and stunted, they are, in fact, growing a large storage organ on the taproot below ground. When a small break appears in the canopy overhead, or perhaps when the reserves attain a certain size, the plant suddenly produces a very rapidly elongating central axis which may grow upward at a rate of 5 centimeters per 24 hours. From this point on, the plant develops as a "proper" vine.

Once a vine has reached the canopy overhead, it is confronted with a number of strategic questions. Is it to grow outward along a number of axes using large open flat areas over which to spread its leaves, to grow off the edge of the tree crown and then hang pendant as a sheet of foliage (with attendant interesting physiological problems of moving fluids in the stem in the opposite direction to those moving up the stem from the ground below), or to grow a condensed crown *in situ* among the branches of an established tree? Tropical vines do all three of these things, and nothing is known of the attendant morphologies and physiologies most appropriate for each choice.

Although we know a great deal about how to grow certain tropical orchids and bromeliads in greenhouses, our understanding of the general ecology and natural history of tropical epiphytic plants is very poor indeed. The almost complete absence of large epiphytes from the branches of extratropical trees is one of the most conspicuous differences between tropical and other forests. And where there are large epiphytes in extratropical forests, such as the Spanish moss (Bromeliaceae) of the southern United States, they generally occur in monospecific stands. Of course, true mosses and lichens occur on tree branches and trunks in both tropical and extratropical habitats, so the real question is why large epiphytes are so abundant and species-rich in at least some tropical habitats. The general answer that is usually attached to this question is that physical "conditions are more favorable" for epiphytes on tropical tree branches, and there may well be nothing more to it than that.

We must immediately emphasize, however, that tropical epiphytes live in a very harsh environment. First, they are exposed to the full blast of a tropical rainstorm and then, minutes later, are exposed to full sun and dry wind. The rain that is not held directly by their roots and the associated pocket of litter runs off immediately. Small wonder that there are cacti growing in the canopies of rain forest trees in habitats receiving 3 to 4 meters (120 to 160 inches) of rain annually. The nutrients they obtain have to come in via rainwater, animal feces and carcasses, ants that nest among the roots, and whatever plant litter they can trap as it falls from above. They have no access to the albeit thin pool of litter nutrients on the forest floor and cannot modify the direction and shape of their root crown on a large scale to make use of local heterogeneity such as a newly fallen log or dead elephant. It is therefore not surprising to again find that they have many small morphological adaptations for harvesting nutrients as they pass by. A bromeliad (a "tank" plant) has a rosette of leaves that traps litter and rainwater, and nutrient uptake may occur through the leaves themselves. Southeast Asian myrmecophytic epiphytes are fed by the ant colony that lives in them. Orchids have mycorrhizal associates that may not only aid in rapid uptake of minerals from the branch surface, but may be parasitic on the living branch.

When all these adaptations are added up, however, the tropical epiphyte still lives in a very nutrient-poor habitat, and the plant morphology itself shows this. The leaves are few and large and may live for many years. When an orchid or bromeliad is starved, it often just slows the rate of its production of new leaves and keeps the old ones rather than be on a fixed leaf-replacement schedule, as appears to be the case with many tropical trees. The leaves are extraordinarily resistant to herbivores if we may reason from the negligible amount of herbivore damage seen on tropical large epiphytes. A leaf damaged by an herbivore is a double loss—there are minerals taken out directly, and there are those that will be lost if the leaf has to be shed and replaced because it is inefficient in its damaged state.

Epiphytic orchids, cacti, bromeliads, and ferns appear to grow at a very slow rate for their highly insolated habitat. However, this may be a strategic consideration as well as a reflection of low productivity of the habitat. An epiphyte that gets large runs the risk of breaking off its branch or ripping itself from its root attachments. The major source of mortality of vascular epiphytes of all ages appears to be falling off the tree rather than starvation, desiccation, or herbivory. Not surprisingly, epiphytes are victims, not only of their own weight and natural branch pruning by the tree, but also of specific adaptations of the tree to shed epiphytes. It is commonplace to see two different species of rain forest tree only a few meters apart, one covered with epiphytes and the other quite clean. While this difference may be in part due to toxic compounds in the bark of the tree, it is quite likely that branch angles, branch shedding patterns, rates of bark shedding, bark micro-morphology, and so on, all influence the rate of establishment of epiphytes. It is quite noticeable that when a tree is transplanted to a rain forest from a tropical deciduous forest site (where epiphytes are relatively scarce and species-poor), it carries far more biomass and species of epiphytes than do the other rain forest trees growing around it.

Compared to trees, epiphytes are very small plants. However, an epiphyte may well be nearly as old as the tree if it has established itself on a major branch or trunk. The epiphyte is simply not putting its photosynthate into the accumulation of much biomass. As it grows at the top, the bottom rots away or becomes part of the mass of roots and litter in which ants nest and which constitutes the epiphyte's own private garden. In fact, once an epiphyte has reached its optimal weight and has a set of nearly indestructable leaves, one wonders why it continues to grow at all, and perhaps it does not.

What, then, does an epiphyte do with its photosynthate? It makes seeds and flowers, and for a perennial plant, epiphytes have enormous ratios of the weight of the inflorescence–infrutescence to the weight of the entire plant. Since an epiphyte seed has the problem of hitting a very small piece of suitable habitat spaced among a very large volume of totally unsuitable habitat (air), we might expect two kinds of dispersal patterns. Where an animal moves accurately from one micro-habitat to another, such as birds do when moving from perch to perch, we might expect bird dispersal coupled with a very sticky seed coat, even after passage through the bird. The best examples of this come from mistletoes, which are, of course parasites and not epiphytes. Further-more, they have slightly different requirements than do epiphytes. A mistletoe seed does best if dispersed to a fairly small diameter healthy branch in the outer reaches of the crown—the sort of place in which a small frugivorous bird is likely to land. An epiphyte would appear to do best if deposited on large-diameter branches with thick bark and deep crevices.

Whatever the reason for not relying on birds, it is clear that the best dispersal pattern for epiphytes, as shown by ferns, orchids, and bromeliads, is to produce millions of propagules the size of sand grains or smaller and wind-disperse them over the habitat in hopes of hitting an occasional good site. What density-dependent population regulation system operates here, however, is a mystery. There are many extremely rare epiphytes in a tropical rain forest. Are we to assume that each has

FIGURE 7-7.
Aerial view from Mt. Ma Ma in Roraima Territory, Brazil, illustrating the vast jungle vegetation. (Courtesy of N.Y. Botanical Garden; photograph by Ghillean Prance.)

extraordinarily specific site requirements and their rarity is because such sites are extremely rare? We would be assuming here that they are competing very strongly for those few best sites. An equally unlikely possibility is that as any one species becomes common, some kind of pest or disease begins to take an exceptional toll. There are no empirical data to suggest this as a possibility.

Summary

Can we come to some kind of summary statement about the tropics, or at least more general generalizations than we have offered up to now? Probably not, and perhaps we should not. We live at northern latitudes. Do we find it useful to make statements that encompass in one generalization the prairies, conifers, and eastern deciduous forest? Any such generalization is only useful if it covers not only these habitats, but tropical ones too, and that is not the intent of this chapter. Rather, our intent has been to give you a feeling for the tropics as other than a green hell of colorful birds and showy flowers.

You should come away from this chapter in a state of mild confusion about what the tropics are and how to characterize them. Unfortunately, there will be no opportunity to provide clearer plant ecology, natural history, physiology, and so on, for many tropical habitats, since most will have fallen to the axe, plow, cow, plantation, and man's insatiable appetite to convert the world to edible usable material goods long before we have the chance to understand them (Figures 7-7 to 7-11). The Costa Rican forest in which the author worked in 1965 is now cattle pasture and rice fields. The dipterocarp forests of Borneo are being clear felled right now by U.S. lumber companies. The Amazon basin at present is being cleared for rice fields and pastureland. Man may not live by bread alone, but he is doing his best to put himself in a position where this may be necessary.

Ashton, P. S. 1969. Speciation among tropical forest trees: Some deductions in the light of recent evidence. *Biol. J. Linn. Soc.* **2**, 155–196.

Bawa, K. S. 1974. Breeding systems of tree species of a lowland tropical community and their evolutionary significance. *Evolution* **28**, 85–92.

Carlquist, S. 1965. *Island Life.* Doubleday & Company, Inc., Garden City, N.Y.

Dodson, C. H., R. L. Dressler, H. G. Hills, R. M. Adams, and N. H. Williams. 1969. Biologically active compounds in orchid fragrances. *Science* **164**, 1243–1249.

Frankie, G. W., H. G. Baker, and P. A. Opler. 1974. Comparative phenological studies of trees in tropical wet and dry forests in the lowlands of Costa Rica. *J. Ecol.* **62**, 881–919.

Gilbert, L. E. 1975. Ecological consequences of a coevolved mutualism between butterflies and plants. In: *Coevolution of Animals and Plants*, L. E. Gilbert and P. H. Raven, eds., pp. 210–240. University of Texas Press, Austin.

Selected References

FIGURE 7-8.
Riverine tropical forest, Rio Javari, Brazil. (Courtesy of N.Y. Botanical Garden; photograph by Ghillean Prance.)

FIGURE 7-9.

Riverine forest, Rio Javari, where vast numbers of trees have been felled, absolutely desecrating this jungle vegetation. (Courtesy of N.Y. Botanical Garden; photograph by Ghillean Prance.)

HOLTTUM, R. E. 1961. *Plant Life in Malaya.* Longmans, Green & Company Ltd., London.

JANZEN, D. H. 1970. Herbivores and the number of tree species in tropical forests. *Am. Nat.* **104**, 501–528.

JANZEN, D. H. 1971. Escape of juvenile *Dioclea megacarpa* (Leguminosae) vines from predators in a deciduous tropical forest. *Am. Nat.* **105**, 97–112.

JANZEN, D. H. 1971. Euglossine bees as long-distance pollinators of tropical plants. *Science* **171**, 203–205.

JANZEN, D. H. 1973. Dissolution of mutualism between *Cecropia* and its *Azteca* ants. *Biotropica* **5**, 15–28.

JANZEN, D. H. 1974. Tropical blackwater rivers, animals and mast fruiting by the Dipterocarpaceae. *Biotropica* **6**, 69–103.

KEAY, R. W. J. 1957. Wind dispersal of some species in a Nigerian rainforest. *J. Ecol.* **45**, 471–478.

KERFOOT, O. 1963. The root systems of tropical forest trees. *Commonwealth For. Rev.* **42**, 19–26.

ODUM, H. T., and R. F. PIGEON, eds. 1970. *A Tropical Rain Forest.* U.S. Atomic Energy Commission, Washington, D.C.

OPPENHEIMER, J. R., and G. E. LANG. 1969. Cebus monkeys: Effect on branching of *Gustavia* trees. *Science* **165**, 187–188.

RAMIREZ, W. 1970. Host specificity of fig wasps (Agaonidae). *Evolution* **24**, 680–691.

Rehr, S. S., P. O. Feeny, and D. H. Janzen. 1973. Chemical defense in Central American non-ant-acacias. *J. Anim. Ecol.* **42**, 405–416.

Richards, P. W. 1952. *The Tropical Rainforest.* Cambridge University Press, New York.

Richards, P. W. 1970. *The Life of the Jungle.* McGraw-Hill Book Company, New York.

Smythe, N. 1970. Relationship between fruiting seasons and seed dispersal methods in a neotropical forest. *Am. Nat.* **104**, 25–35.

Snow, D. W. 1966. A possible selective factor in the evolution of fruiting seasons in a tropical forest. *Oikos* **15**, 274–281.

Stiles, F. G., and L. L. Wolf. 1970. Hummingbird territoriality at a tropical flowering tree. *Auk* **87**, 467–491.

Tomlinson, P. B., and A. M. Gill. 1973. Growth habits of tropical trees: Some guiding principles. In: *Tropical Forest Ecosystems in Africa and South America: A Comparative Review,* B. J. Meggers, E. S. Ayensu, and W. D. Duckworth, eds., pp. 129–143. Smithsonian Institution Press, Washington, D.C.

Vuilleumier, B. S. 1971. Pleistocene changes in the fauna and flora of South America. *Science* **173**, 771–780.

Wilson, D. E., and D. H. Janzen. 1972. Predation on *Scheelea* palm seeds by bruchids: Seed density and distance from the parent. *Ecology* **53**, 954–959.

FIGURE 7-10.
Manaus–Caracarai road, another example of the way in which the Brazilian jungle is being desecrated by man. (Courtesy of N.Y. Botanical Garden; photograph by Ghillean Prance.)

FIGURE 7-11.
*Another view of Manaus–
Caracarai road, here
showing felled forest on
slopes adjacent to the road,
causing quick erosion.
(Courtesy of N.Y. Botanical
Garden; photograph by
Ghillean Prance.)*

Plants and the Aquatic Environment
James A. Weber

8

Questions for Consideration

1. What kinds of plants grow in the water?
2. Is eutrophication only caused by human activity?
3. What factors affect the growth of aquatic plants?
4. What is pH, and how is it related to aquatic plants?
5. Where does most of the photosynthesis occur in lakes and oceans?
6. What areas within a lake or ocean are most productive on a per hectare (acre) basis?

Liquid water covers about 70 per cent of the surface of the globe. The largest portion is in the oceans, less than 1 percent is fresh water, and an even smaller amount is found in saline lakes. Other forms of water are ice, snow, and water vapor. In this chapter, we are going to investigate the role of plants in the various types of aquatic habitats and the factors that influence their survival. The first part of the chapter will include a synopsis of the wide variety of plant forms found in aquatic systems; the second part will consider the various environmental factors that are important for aquatic plants; and the third part will discuss the roles that plants play in aquatic ecosystems. Finally, the interrelationship between people and aquatic plants will be considered.

A Survey of Aquatic Plants

Aquatic plants can be found in nearly all bodies of water except those which are extremely hot (greater than 85°C), saline, acidic, alkaline, or receive no light. Nearly all major groups of plants are represented by at least a few aquatic species; however, algae far outnumber all other groups of plants in terms of species and individual organisms.

There are a number of ways by which aquatic plants may be divided into groups. In Chapter 2, you were introduced to the phylogenetic, or biosystematic, method, which stressed similarities and differences in the anatomy and morphology of various species. Groupings may also be made on the basis of growth form (tree, herb), habitat (desert, forest), period of growth (perennial, annual), and so on. Aquatic plants can also be divided into three categories by noting their relationship to the water's surface: *submergents*—those with leaves only under water (e.g., most aquarium plants), *emergents*—those with leaves and sometimes stems above the water (e.g., cattails; Figures 17-6 and 17-33), and

FIGURE 8-1.
Drainage ditch in Brunswick County, North Carolina, filled with the water lily Nymphaea odorata. (Courtesy of Larry Mellichamp.)

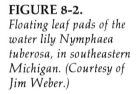

FIGURE 8-2.
Floating leaf pads of the water lily Nymphaea tuberosa, in southeastern Michigan. (Courtesy of Jim Weber.)

floating-leaved forms—those with leaves at the surface of the water (e.g., water lilies; Figures 8-1 to 8-6). Species with floating leaves may be rooted (e.g., water lilies) or free-floating (e.g., duckweeds). Some emergent species, for example, water hyacinth (Figures 8-7 and 8-8), are also free-floating. Finally, some plants have both emergent and submergent type leaves, such as parrot's feather (*Myriophyllum aquaticum*, previously *M. brasiliense*).

Aquatic plants have a number of characteristics which distinguish them from terrestrial plants. First, the leaves of the submerged types are finely dissected and/or very thin. This condition represents an adaptation to the much slower rate of diffusion of compounds in water than in air (e.g., oxygen diffuses about 100,000 times slower in water than in air). *Egeria densa* (the common aquarium plant also known as *Anacharis densa* and *Elodea densa*) has leaves only two cells thick. The alga *Chara*

FIGURE 8-3.
Flower from the water lily Nymphaea odorata, in southeastern Michigan. (Courtesy of Jim Weber.)

FIGURE 8-4.
Cultivated population of plants of Indian water lotus, Nelumbo nucifera, in Union County, North Carolina. (Courtesy of Larry Mellichamp.)

(commonly called char), which imparts a garliclike smell to waters in which it is abundant, has whorls of thin, cylindrical projections. Most algae are so small that a microscope is required to observe them. Second, many species, especially the larger ones, have enlarged air spaces which cause the plant as a whole to be less dense than water. For free-floating species, the reduced density maintains the plants at the surface. In rooted or attached species it causes the plant to extend toward the surface, thus

FIGURE 8-6.
Single plant of one of the world's largest water lilies, Victoria amazonica, or royal water lily, which is native to the Amazon. This plant is in the greenhouse of the Frankfort Botanical Gardens, Frankfort, Germany. (Courtesy of Peter Kaufman.)

substituting in some degree for the support tissue found in terrestrial plants. The air spaces also provide a pathway for the exchange of oxygen between shoot and roots of aquatic flowering plants. This aspect of the air space is especially important because the oxygen concentration found in the sediments where the plants are rooted is generally very low. Third, larger submerged species have little supporting tissue, thus allowing them to sway with the movement of the water.

FIGURE 8-7.
Vegetative plants of the water hyacinth, Eichhornia crassipes. The leaves have inflated bases that act like pontoons, allowing the plant to float. (Courtesy of Larry Mellichamp.)

The algae form the largest subdivision of aquatic plants, both in terms of number and total biomass. They are, by-and-large, submergents, ranging in size from spherical cells only 15 micrometers (1/2000 inch) in diameter to the giant kelps, which may grow to over 60 meters (200 feet) in length. Members of all divisions of the algae may be found either free-floating (phytoplankton) or attached to rocks (Irish moss, *Chondrus crispus*), sticks, and other organisms.

Phytoplankton (i.e., microscopic algae that are free-floating) are the main producers in the oceans and most lakes. The variety of shapes seems nearly infinite, ranging from simple spherical cells (*Chlorella*) and colonies (*Volvox*) through the filamentous (*Oedogonium, Spirogyra*) and platelike forms (*Gonium*) to those with spines and complex geometries (*Ceratium* and *Micrasterias*, Figure 8-9). The diatoms are generally the most numerous in terms of both species and individuals. These tiny organisms live in what could be called glass houses because they enclose themselves in boxes of silica (Figures 8-10, 8-11, and 8-12). Also quite numerous in many waters are the dinoflagellates (e.g., *Gonyaulax*). These organisms are both beneficial and harmful, in that, while they do photosynthesize, some marine members of the group produce a toxin that kills fish. The name "red tide" comes from the red color of the water when there are large numbers (blooms) of these organisms present. At night during red tides, there is an eerie greenish-white glow produced when the plants are disturbed by the breaking of waves. The glow is produced by the breakdown of a compound, luciferin, which is stimulated by agitation. Green and blue-green algae are also important members of the phytoplankton community. Some species of blue-greens (e.g., *Anabaena*) are especially important because they can transform molecular nitrogen (N_2) into forms usable by other organisms.

There are many species, especially of the brown and red algae, which are large enough to be seen with the unaided eye. Some of these algae

reach lengths of over 60 meters (200 feet). These plants, commonly called seaweeds, are typically found in areas of relatively shallow water and are predominantly oceanic. Some red algae have been collected at depths of over 100 meters (300 feet), but the greatest abundance of seaweeds is generally at much shallower depths. A broad division may be made between those plants which inhabit the intertidal zone, the area between the highest and lowest tides, and those which grow in deeper water. Intertidal algae must be able to survive the rigors of crashing waves and drying in the air. A distinct zonation of species develops based on the ability to withstand drying, progressing from those which are most resistant to drying to those which are most sensitive. Below the low-tide line, those plants that cannot tolerate drying and/or the mechanical

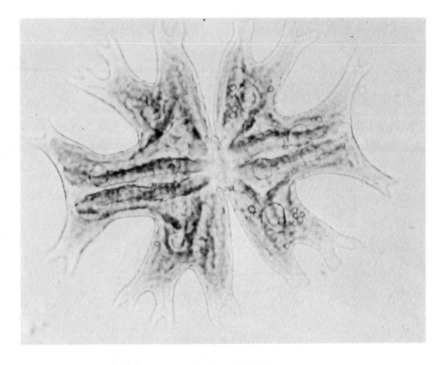

FIGURE 8-9.
Photomicrograph illustrating the beautiful bilateral symmetry of an algal cell, Micrasterias, a desmid in the green algae. (Courtesy of Linda Graham.)

forces of the waves are found. The giant kelps, such as *Macrocystis*, are generally found below the low-tide line, so that they are at most floating at the surface of the water at low tide.

Aquatic flowering plants are essentially restricted to shallow areas, less than 10 meters (35 feet) deep, and to protected areas, such as bays. They vary in size from the smallest duckweeds (*Woffiella*, about 1 millimeter in diameter) to trees such as mangroves and the bald cypress. All these plants require free-standing water in which to grow: growing submerged (many aquarium plants), floating at the surface (water hyacinth, *Eichhornia crassipes*; Figures 8-7 and 8-8), or with their leaves and/or stems above water (rice, *Oryza sativa*). Although most of the species of aquatic flowering plants are found only in fresh water, there are some species, such as mangroves (e.g., *Avicennia* and *Rhizophora*) and eel grass (*Zostera*), which are restricted to salt water.

There are also some species of mosses and ferns which are aquatic and which may be abundant in some areas. For example, *Sphagnum* mosses are important in bogs. Floating ferns of the genus, *Salvinia*, have caused a great deal of trouble in some tropical areas. Other species of ferns (e.g., *Ceratopteris*) are used in garden ponds.

TYPES OF ENVIRONMENTS

The aquatic environment is exceedingly diverse; in fact, about the only thing common to all bodies of water is that they are wet. They range in size from the Pacific Ocean down to the smallest mud puddle. They may be stationary (lakes) or flowing (rivers); essentially permanent (oceans)

The Aquatic Environment

FIGURE 8-10.
Scanning electron micrograph of a cluster of diatoms growing on a red alga. Each diatom cell is enclosed in a siliceous case called a frustule. ×3500. (Courtesy of P. Dayanandan.)

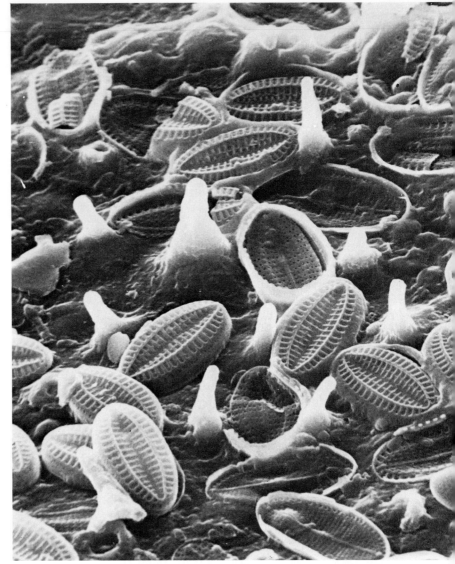

or transient (puddles). They are fresh to very salty, acidic to basic, very clear to very muddy, cold to hot. Thus, it is difficult to talk about the plants that live in these varied environments without knowing something of the habitats in which they occur.

There are four major types of aquatic environments: oceans, lakes, rivers and streams, and swamps and marshes. By far the greatest amount of water is found in the oceans. These large bodies of water are constantly in motion both near the surface through wave action and more deeply through currents. The wave action is almost entirely caused by wind, while currents may be caused by a number of factors, including wind, temperature, and differences in salinity. The movement of currents is especially important in the oceans because currents bring mineral-rich water from the depths to the surface in areas called

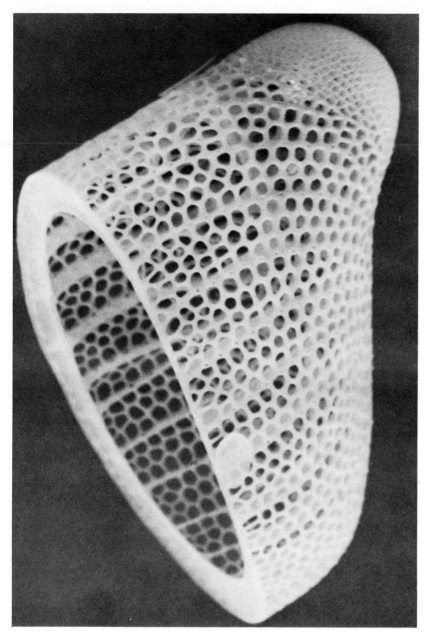

FIGURE 8-11.
Scanning electron micrograph of a single diatom frustule, illustrating the elaborate structure of this glass case. The organism shown here is in the genus Bidulphia. × 1200. (Courtesy of P. Dayanandan.)

upwellings. Some of the most productive areas of the oceans, such as the Pacific Ocean off the coast of Peru, are found in the upwellings.

Lakes are much smaller than oceans. They are generally, but not exclusively, freshwater, and are characteristically temporary in geological terms. Most lakes are too small to develop steady currents such as those found in the oceans. Mixing of water is dependent on the force of the wind, on density changes caused mainly by temperature, and on the movement generated by incoming water from streams.

Rivers, streams, and creeks are moving bodies of water that allow the water to flow to and from stationary bodies of water. Plants

FIGURE 8-12.
Scanning electron micrograph illustrating two spiked cells of the alga Ceratium and two comblike frustules of a diatom. ×3500. (Courtesy of P. Dayanandan.)

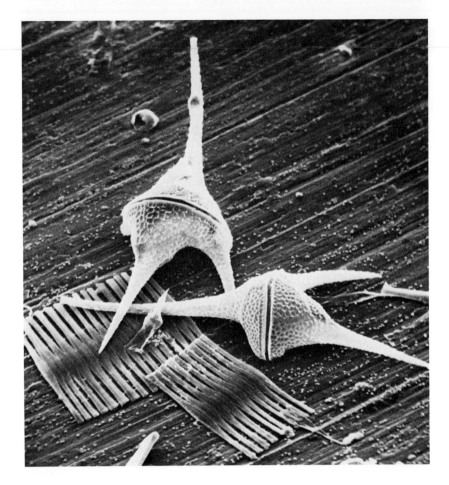

characteristic of rivers attach themselves to stationary objects; otherwise, they would be washed downstream to the next lake. Nearly all free-floating plants (e.g., phytoplankton) found in moderately to swiftly flowing rivers are characteristic of the lake(s), reservoir(s), and so on, upstream from where the sample is taken.

Marshes and swamps are areas of shallow water that can be considered a transition between the aquatic and terrestrial environments. It is in these areas that maximum productivity (i.e., photosynthesis per unit area) can occur. Emergent aquatic flowering plants, such as cattails (*Typha*), are the predominant forms of vegetation.

These are the major types of aquatic environments. Other types of aquatic environments are generally less important to people than those already discussed. However, three peculiar types should be mentioned: the bog, the fen, and the vernal pool. The first two are terminal stages in the filling in of some lakes or ponds and are transitions between aquatic and terrestrial habitats. A *bog* is a small, relatively deep lake or pond with no surface outlet, with a floating mat of mosses (generally *Sphagnum*) growing out from the shore toward the center of the bog. The decomposition of the debris from the mat and material added by mosses causes the water to become acidic and take on a brown tint. The phytoplankton of these waters is much reduced compared to most other lakes. A somewhat similar situation, but without the mat of mosses, is found in a *fen*; however, here the water is basic, owing to the presence of calcareous rocks. *Vernal pools* are seasonally wet areas that form during a rainy period and dry up in a dry period. Plants characteristic of vernal pools must be able to withstand the alternating moisture conditions.

IMPORTANT ELEMENTS OF AQUATIC ENVIRONMENTS

The physical and chemical characteristics of aquatic environments are in many ways quite different from those found in terrestrial ecosystems; for example, changes in temperature in aquatic environments are slower. Major factors that affect the growth of aquatic plants are light intensity and quality (color), temperature, availability of inorganic carbon and other mineral nutrients, and pH. It is unnecessary to go into all the details of each of these factors. However, we shall briefly consider each factor from the point of view of the plant and its needs.

The peculiar properties of water are very important for the maintenance of life, especially in aquatic systems. Unlike most substances, solid water (ice) floats to the surface. This phenomenon is directly related to the fact that water is most dense at 3.94°C (39°F) (1.00 gram per cubic centimeter (g/cm^3) (60.0 pounds per cubic foot). *Below* this temperature the molecules move apart a bit, thus causing water cooler than this temperature to become less dense. When water freezes at 0°C (32°F), it becomes even lighter (0.917 g/cm^3) (54.1 pounds per cubic foot). Furthermore, a layer of ice, being a poor conductor of heat, acts as an insulator, or blanket, and retards the cooling of the water beneath it. It is possible, then, for plants to survive the winter in relatively warm (0 to 4°C) liquid water rather than in the harsher environment above the ice. If it were not for this peculiar change in density, lakes would start freezing from the bottom. Another important attribute of water is its ability to hold many different compounds in solution. In fact, because so many different compounds (solutes) can dissolve into it, it has been called "the universal solvent"; even many oils are slightly soluble in water. The final property that is important is the dissociation of water molecules into hydrogen (H$^+$) and hydroxide (OH$^-$) ions:

$$H_2O \rightleftharpoons H^+ + OH^-$$

As the concentration of H$^+$ increases, that of OH$^-$ decreases, and vice versa. The concept of acidity and basicity is expressed in terms of pH.[1] In general, the pH scale goes from 1 to 14, with 1 being most acidic (highest concentration of H$^+$), 14 the most basic (highest concentration of OH$^-$), and 7 neutral (neither acidic nor basic). The pH is important for the growth of aquatic plants through its effects on the various ions in the water. The direct effect of pH on plants is usually only important at very acidic or very basic pHs, where the high concentration of H$^+$ or OH$^-$, respectively, is injurious to the tissues of the plant.

[1] The pH is approximately equal to minus the log$_{10}$ of the hydrogen ion concentration.

Temperature is important for all living things because of its effect on the rate of metabolism—at higher temperatures, the metabolic rate is faster. Large, rapid fluctuations in temperature can be detrimental to the survival of an organism. However, those organisms which live in the water are fortunate because water can absorb and release a great deal of heat without much change in temperature. In fact, few naturally occurring compounds have as great an ability to absorb heat as water. This property is familiar to those living along the shores of oceans and large lakes, where the air temperature of surrounding land is moderated (i.e., cooled or warmed) when it is above or below the temperature of the water.

The temperature profile of the oceans is rather stable, with cold water at great depth and near the poles, and warmer water near the surface toward the equator. Surface water will be cooler in winter and warmer in summer, except near permanent ice. Upwelling of water from the deep parts of the ocean to the surface will also cool the surface water. Variations in the salinity, as well as temperature, affect the density of water. Variations in density, along with the effects of the earth's rotation, cause currents that circulate water throughout the oceans.

Lakes and rivers, being smaller bodies of water, are more strongly affected by air temperature. In temperate areas, where there is wide variation in air temperature during the year, the temperature of the water can vary from 0 to 25°C (32 to 77°F). In the tropics, where there is little variation in air temperature, water temperature will be correspondingly less variable. Because the cooler water is denser, it tends to remain on the bottom. This change in temperature with depth, called *stratification*, has been used to divide lakes into three portions: the upper part, or *epilimnion*; the lower part, or *hypolimnion*; and the portion where the temperature changes most rapidly, or *thermoclyne* (Figure 8-13). The density difference between the warmer (less dense) epilimnion and the cooler (more dense) hypolimnion produces an effective barrier to the mixing of the two portions of a lake. Only when the whole lake is at one temperature and there are no other factors causing major density differences between layers can the whole lake be mixed. The term used for this phenomenon is *turnover*. Lakes that freeze in the winter will be at a uniform temperature just before and just after ice is on the lake. In lakes that do not reach freezing temperatures, it is much less likely that a uniform water temperature will occur; thus, mixing of the entire lake will only

FIGURE 8-13.

Changes in temperature with depth in a north temperate lake. The data are hypothetical but representative. The thermoclyne (B) is defined as the region in which the change in temperature is equal to or greater than 1°C per meter of depth. A is the epilimnion and C is the hypolimnion.

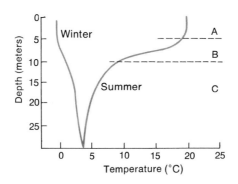

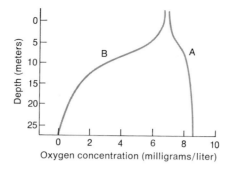

FIGURE 8-14.
Level of oxygen in the summer in an oligotrophic (A) and a eutrophic (B) lake. The increase in oxygen concentration below 5 meters in A is due to the decreased temperature and the increased solubility of gas at lower temperatures.

occur infrequently. The energy required for mixing the water is generally supplied by the wind.

The division of lakes into hypolimnion and epilimnion has many consequences, all of which stem from the isolation of the two parts. First, and most obvious, is the temperature difference, which affects the rate of metabolism of organisms in the lake. Second, there is essentially no exchange of dissolved compounds, just a more-or-less constant rain of debris from the epilimnion to the hypolimnion. If there is little *biological oxygen demand* (BOD), that is, few organisms living in the hypolimnion, there will be only a small reduction in the oxygen level in the hypolimnion (Figure 8-14). However, if there is a large BOD, the oxygen will be rapidly consumed and the hypolimnion may become anaerobic (devoid of oxygen). This usually happens when there is much organic matter to serve as food for organisms living in the hypolimnion. Lack of oxygen also affects the nutrient balance by dissolving minerals from the lake bottom which would, in the presence of oxygen, be precipitated out of solution. [One example is the precipitation of phosphate as $FePO_4$ in the presence of oxygen and the dissolution of it as $Fe_3(PO_4)_2$ in the absence of oxygen.] Nutrients depleted from the epilimnion by the growing algae during the period when a lake is stratified cannot be replaced except by runoff from the surrounding areas; thus nutrient deficiencies may arise that can inhibit the growth of some or all of the species of algae present. The only way for the oxygen in the epilimnion and the nutrients in the hypolimnion to be exchanged is through mixing of the two as a result of turnover.

Light is the source of energy for nearly all life. Only the chemosynthetic bacteria, for example, those which can use hydrogen sulfide (H_2S) for energy, are not dependent directly or indirectly on light energy. As light passes to the earth's surface, various wavelengths are absorbed or reflected by the various components of the atmosphere; for example, ozone (O_3) absorbs much of the harmful ultraviolet radiation (see Chapter 9). When the sunlight reaches the surface of the water at solar noon on a cloudless day, the intensity is approximately 1.2 calories per square centimeter per minute (4.3 BTU per square foot per minute), about half of which is in the visible wavelengths (400 to 700 nanometers). (A calorie is the amount of heat that is needed to raise the temperature of 1 gram of water from 15 to 16°C.) When the sun is directly overhead, very little light energy is lost through reflection at the surface. However, as the angle of incidence of the light decreases, more light is reflected. This phenomenon leads to a prolonged "twilight" when the sun is low in the sky (morning and evening). Once in the water, the reduction in intensity with depth depends on the dissolved and suspended matter. The loss of energy is not equal for all wavelengths; red and infrared wavelengths are more rapidly absorbed than blue because the former are less energetic and because water absorbs much infrared radiation. For this reason, in pure water, blue light reaches the greatest depths. In very clean, colorless portions of the oceans and lakes, blue light can reach to over 200 meters (600 feet). Dissolved and suspended matter will markedly alter this distribution of light. The presence of phytoplankton will cause the green portion of the spectrum to reach the greatest depths, because the red and blue parts of the spectrum are used in photosynthesis, whereas green is not absorbed nearly as much. The presence of suspended clay particles will very effectively block penetration of light more than 1 meter or so and make the water appear brown. Very fine, light-colored particles suspended in the water will cause the water to appear blue-green or turquoise. This situation may be seen in the ocean, where the sand is disturbed by wave action. Dissolved compounds, such as tannins, will also absorb light differentially and change both the quality (color) and quantity of light that can penetrate through the water.

Both lakes and oceans can be divided into regions, depending on the penetration of light and the characteristics of the bottom (Figure 8-15). The *photic zone* is the region from the surface down to the limit (usually defined as 5 percent of full sunlight) of light penetration. It is in this area that plants are most active. Another definition of the photic zone is the region above the depth at which the rates of respiration (energy use) and photosynthesis (energy accumulation) of plants balance each other. The *littoral zone* is the area where the bottom is illuminated; it may be divided into the *eulittoral,* the zone of fluctuating water level, and the *sublittoral,* the zone that is continually submerged. The *pelagic zone* is the part of the photic zone where light does not reach the bottom. Below the photic zone is the *aphotic zone;* here the light is insufficient for significant photosynthesis.

The other major factors, inorganic carbon and mineral nutrients, are chemical in nature and more-or-less interrelated with pH. The relation between pH and inorganic carbon is particularly close. When carbon

FIGURE 8-15.

Cross section of a typical lake, noting the photic zone (Ph), the aphotic zone (A), the littoral zone (L), and the pelagic zone (Pe).

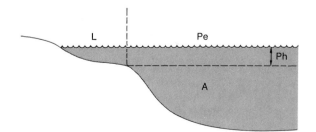

dioxide (CO_2) dissolves in water, it can go through a series of reactions, producing both bicarbonate and, in some cases, carbonate.[2] Carbon dioxide produces an acidic solution in pure water. However, if there are ions in the water that can combine with the hydrogen ions (e.g., hydroxide ions), the pH will not be as acidic, and may in fact be basic. This complex set of reactions is important to plants because it controls the acidity (pH) of water as well as the amount of carbon dioxide that the plants can use in photosynthesis.

Mineral nutrients, such as nitrate and phosphate, are needed for the proper functioning of all plants. There are some groups of plants that have special requirements as well. One example is the silicate requirement of diatoms which build their outer wall covering from this material. Most aquatic plants absorb these nutrients from the water; however, aquatic vascular plants which root in the substrate may take up 50 percent or more of their nutrients from the soil.

Although many of the concepts of terrestrial plant ecology can be applied to aquatic plants, there are some major concepts that are used primarily in reference to aquatic environments. In this section we explore the relationships between aquatic plants and natural waters. For this discussion we shall concentrate on lakes, but it should be noted that other aquatic systems may show similar characteristics.

There is a direct relationship among density of plants, primary productivity, and level of oxygen in the water. Plants, as primary producers, trap light energy and transform it into chemical energy. This process, called *photosynthesis*, involves the uptake of carbon dioxide (CO_2) and the production of carbohydrates, sugars, and oxygen (O_2) by the following overall reaction:

$$6CO_2 + 6H_2O \xrightarrow[\text{chlorophyll}]{\text{light}} C_6H_{12}O_6 + 6O_2$$

Thus, in the light, aquatic plants remove CO_2 from and add O_2 to water. When there is no light present, all organisms use the energy that was originally stored in the sugar molecules through photosynthesis for their metabolism. The process of releasing the chemical energy to carry on metabolism is called *respiration*. Most organisms require oxygen for this process, which is called *aerobic respiration*. There are some organisms, especially bacteria, that can respire in the absence of oxygen (i.e., *anaerobic respiration*). In those regions of a lake where there is no light, the aerobic organisms depend on outside sources of oxygen for their metabolism. If the requirement for oxygen (*biological oxygen demand*) is greater than the supply, the oxygen level of aphotic regions will drop

Plants in the Aquatic Environment

---

[2] The equations for the reactions are

$$CO_2 + H_2O \rightleftharpoons H_2CO_3 \quad \text{(carbonic acid)}$$
$$H_2CO_3 \rightleftharpoons H^+ + HCO_3^- \quad \text{(bicarbonate)}$$
$$HCO_3^- \rightleftharpoons H^+ + CO_3^- \quad \text{(carbonate)}$$

considerably. Food (stored energy ultimately from photosynthesis) for the organisms in the aphotic zone comes from living and nonliving organic material that sinks or is brought down from the photic zone. Some of this organic matter may also come from terrestrial sources. Finally, in areas where massive blooms of algae or large stands of other submerged aquatic plants occur, it is possible for the oxygen produced during the day to be nearly consumed at night by respiration of the plants as well as the animals. With these facts in mind, let us look at the relationship of plants to various types of lakes.

Lakes that have a relatively small phytoplankton population and have oxygen in good supply even at the bottom throughout the year are termed *oligotrophic.* The "picture postcard" lakes with deep blue water are of this type. Lakes that have a large phytoplankton population and have little or no oxygen near the bottom during all or part of the year are called *eutrophic.* The water in these lakes will often take on a greenish tint as a result of the abundant algae and may have a heavy scum of algae and other plants on the surface. Aquatic flowering plants will be most abundant in eutrophic lakes unless the algae are so thick as to shade them out. There are also *mesotrophic* lakes, which are intermediate in character between these extremes. Lakes show a gradual transition through time from oligotrophic to eutrophic and will gradually fill in and cease to exist. This process, called *eutrophication,* is quite natural and usually quite slow. However, we have recently been accelerating the process by adding large quantities of sewage, and therefore nutrients, to some waterways.

There is a continual change in the species of plants and their abundance in lakes. The changes occur in yearly cycles, as well as over many years. The yearly variation in the plant communities is related to the variation in the seasonal environment. Where there is little change, there will be little change in the plant community; where there is large variation, there will be a correspondingly large variation in the plant community. Temperature plays an important role in this process. Different species grow better at different temperatures; thus, there are plants that are characteristically found in abundance in spring and others in summer or fall. After a bloom of algae, the concentration of mineral nutrients is generally reduced from that before the bloom, with particular nutrients often becoming limiting to the most abundant species. For example, when the level of silicates in the water is reduced, the growth of diatoms will be limited, because they require silicates to form their wall cases, or frustules. Algae not requiring silicates (e.g., green algae) can then grow more rapidly, because they will not be competing with diatoms for other nutrients, such as nitrate and phosphate. Thus, at different times, different species may be favored or restricted, depending on their nutrient requirements, response to temperature, and other factors. The changes that occur over the years are generally unidirectional and noncyclic. One plant community will succeed another as the lake goes through eutrophication. In oligotrophic lakes, diatoms are the dominant type of phytoplankton. As eutrophication proceeds, green and blue-green algae become more important, becoming dominant in eutrophic lakes. The changes that occur in plant communities have been and are being studied as measures of the effects of human activity on the aquatic environment.

As would be expected, *productivity* (i.e., rate of photosynthesis per unit area per unit time) is higher in eutrophic than oligotrophic lakes. Refer to Table 8-1, which gives some idea of the productivity of various aquatic communities. From these data, we can see that the most productive areas are the shallows; however, these areas account for only a small portion of the total area of most aquatic ecosystems. Recent data from a very productive lake in Africa show that productivity of phytoplankton may approach that of emergent vegetation under circumstances of high fertility and favorable light and temperature. Phytoplankton are the major producers because they are found over larger areas than the stationary, attached, or rooted plants.

Most animals found in the water depend directly or indirectly on the productivity of algae. One may imagine a typical food chain starting with the algae, which are consumed by herbivorous zooplankton (microscopic animals), which in turn are eaten by zooplankton carnivores. Any or all of the planktonic organisms may be consumed by filter-feeding fish, crustaceans (shrimp), mollusks (clams), or even whales. Again, there may be another level or two of carnivores. Another food chain, not entirely separate from the one just described, starts with *macrophytes* (e.g., seaweeds and water weeds) and goes through the various levels of herbivores and carnivores. Decomposers (fungi and bacteria) complete the cycle by breaking down the organic detritus and remains of organisms. The inorganic compounds, such as CO_2 and phosphate, which are released, are then available to plants for producing more organic matter and fixing energy.

TABLE 8-1 / Comparison of the Yearly Productivity of Different Aquatic Plant Communities on Fertile Sites

	Metric Tons of Dry Organic Matter per Hectare[a]	(Short Tons per Acre)
Marine phytoplankton	1–4.5	.4–1.9
Lake phytoplankton	1–9	.4–3.8
Freshwater submerged macrophytes (water weeds)	4–20	1.7–8.5
Marine submerged macrophytes (seaweeds)	25–40	10.6–16.9
Marine emergent macrophytes (salt marsh)	25–85?	10.6–36.0
Freshwater emergent macrophytes (reedswamp)	30–85	12.7–36.0

[a] Estimates of the upper levels of productivity. *Source:* D. F. Westlake, Some basic data for investigation of the productivity of aquatic macrophytes, *Mem. 1st. Ital. Idrobiol. Suppl.* **18,** 229–248 (1965).

People and Aquatic Plants

The human race benefits in many ways from aquatic plants. Over 50 percent of the energy fixed by photosynthesis is fixed by aquatic plants. They provide food either directly or indirectly for fish and shellfish that we use as food. Also, a number of species are used directly as food or for other products.

Various species of seaweed have been used by people around the world as food, fodder for animals, compost, and sources of soda, potash, and iodine. The Chinese and Japanese have used seaweeds for centuries. The algae may be dried for use as noodles (*Laminaria, Eisenia,* and *Ecklonia*—all brown algae), as sweetening and seasoning (*Laminaria japonica* and *L. angustata*), and as an additive to flour. In the British Isles, *Porphyra laciniata* (laverbread) is used to make English black butter in southern Wales. The Scots use dulse (*Rhodymenia palmata*) as an appetizer. The bladders of *Nereocystis leutheana* have been used to make imitation citron, orange, and lemon peel for candies.

Seaweeds, especially *Fucus* (Figure 8-16) and *Ascophyllum* species, have been harvested for the production of barilla (i.e., ash from burning the plants), which was used in Europe to make soap and glass. When alternative sources replaced seaweed ash in the mid-nineteenth century, the extraction of iodine by a new process became the principal use of seaweed in France. Species of *Laminaria* were found to be the best. However, this use of seaweeds also was discarded when other sources of iodine were found. Today, seaweed is the source of algin, an important emulsifying and thickening agent in foods, first extracted in quantity around 1880. A number of large seaweeds, including *Laminaria* and *Macrocystis,* are used around the world. Another emulsifying agent, a carrageenin (sometimes spelled carrageenan), is extracted from Irish moss (*Chondrus crispus*), a red alga. Many products contain one or both of these emulsifiers, for instance, ice creams, salad dressings, paints, and many more. Finally, agar, an important gelatinlike substance used for

FIGURE 8-16.
View of beds of the brown alga Fucus growing on boulders that are exposed at low tide. (Courtesy of William Randolph Taylor.)

growing microorganisms, is extracted from some species of red algae. The Japanese first extracted agar in the seventeenth century and are today a major source for the world.

The cultivation of seaweed has been practiced in Hawaii, Ireland, and most extensively in the Far East. In the last 100 to 150 years, cultivation of a number of species, primarily red and brown algae, has been tried successfully. *Porphyra* species have been in cultivation in Japan for some time. These plants require high salinity for best germination of spores, but produce the large, thick, tender fronds in waters of low salinity. In order to get both the large fronds and good germination, bundles of twigs or bamboo are placed in a muddy bay so that they are just covered at high tide. Spores of *Porphyra* settle on the bundles and germinate in response to the high salinity. Once sufficient young plants have developed, the bundles are moved to water with lower salinity, where the large fronds are produced. Other species have also been cultivated (*Enteromorpha, Gloeopeltis*); however, cultivation of algae has been generally of local, not worldwide importance. This situation may change in the future. Studies are being conducted on the possibility of growing giant kelp (*Macrocystis pyrifera*) on artificial reefs anchored in deep water [greater than 100 meters (300 feet)] with grids for supporting the plants 12 to 15 meters (40 to 50 feet) below the surface. Mineral-rich water would be pumped up from the depths to serve as fertilizer and to control the temperature. It is estimated that a 2.6-square-kilometer (1-square-mile) seaweed plantation could provide food and energy for 300 people at the present rates of consumption in the United States (Branning, 1976).

Aquatic vascular plants have also been important in the human economy. Various reeds have been used for making dolls, baskets, and boats. Papyrus, an emergent plant, was used by the Egyptians to make large boats as well as paper. Even in this century, reeds have been used to make canoes (Heyerdahl, 1972). In the Gulf of California, Indians have made extensive use of the sea grass *Zostera marina*. Rice (*Oryza sativa*) has been cultivated for centuries as the major grain in many areas of the world. Wild rice (*Zizania aquatica*) was used extensively by North American Indians and is being cultivated not only in the United States, but also in the Orient. Many other aquatic vascular plants are used by people in various ways; for a more complete account, see the chapter on economic uses of aquatic vascular plants in Sculthorpe (1967).

People have also abused the aquatic environment and have had a major impact on certain areas. Recalling the characteristics of the aquatic environment, a number of ways in which human activity can affect aquatic plants will be evident, including addition of heat (thermal pollution) and siltation. It is in the shallow areas, such as estuaries and marshes, where the impact of human activity is usually most severe, and it is these areas that are most productive.

There are three things that can happen to plant communities when the waste products of people and their activities are dumped into the water. First, there may be an increase in the number of plants present (e.g., a "bloom" of algae). Second, some or all of the plants may be killed, such as by a toxic compound. Finally, there may be no visible change. This last consequence is most unlikely unless the waste products that are being dumped have no effect at all on any part of the ecosystem. If there is some effect, there will likely be a change in the species found and/or their relative abundance. The addition of nutrients through sewage and other sources (e.g., runoff from agricultural areas) will usually stimulate a visible growth of plants, which may include species ranging from esthetically undesirable to toxic. However, the effects of exotic chemicals are much more difficult to define. There are some indications that some chemicals, such as DDT, could affect the rate of photosynthesis in oceanic phytoplankton, but more study is needed to assess their impact adequately. The presence of suspended particles in water will effectively block light penetration and will often cover rooted vegetation, thus reducing productivity of these plants. However, where suspended particles settle to the bottom of a lake, growth of emergent vegetation may be encouraged. Addition of heat, *thermal pollution,* may accelerate plant growth if water temperature is raised only a few degrees, but may kill many species if it is raised too high. The effect will depend strongly on the reactions of the individual species.

The effect of human activity on aquatic plants can be dramatic, with repercussions on the human economy. One example is the disappearance of the giant kelp *Macrocystis pyrifera* from coastal areas along the southern California coast. Prior to World War II, large beds, or forests, of this kelp could be found in the coastal waters at a depth of about 15 meters (50 feet). Because the plant is a good source of algin, a large industry developed around harvesting and processing it for this important compound. In the late 1940s, biologists began noticing that the large beds were declining. One bed off Point Loma, California, decreased from about 15 square kilometers (6 square miles) to about 2.5 hectares (6 acres). From extensive study of the problem, it became apparent that a number of things combined to cause the near-destruction of the kelp beds. The key was the sea urchin, which eats through the stem (or stipe) of the kelp plant, thus allowing it to be washed away. Sea urchins also eat the young plants. Under natural conditions, sea otters keep the population of sea urchins in check; however, sea otters had been exterminated from the southern California coast. Without additional factors, the kelp and sea urchins would go through alternating cycles of increase in the sea urchin population until the kelp was nearly gone, then regrowth of the kelp when the sea urchins had died or left the area. However, pollution from the cities along the coast provided the sea urchins with an alternative source of food, thus allowing them to remain, to the detriment of the kelp. The reduction of the kelp beds not only affected the algin industry but also the myriad species of animals, including crab, abalone, and fish, which depend on kelp for food and shelter. Unlike many stories of this kind, there appears to be a happy ending. Studies have shown that careful application of quicklime will eradicate the sea urchins; then the areas can develop new stands of kelp. For a more complete account, see the article by Branning (1976) on reestablishment of the giant kelp.

Luxuriant plant growth in the water can directly affect people in a number of ways. The presence of large amounts of plant matter can clog waterways, interfering with navigation, hydroelectric systems, irrigation projects, and certain types of recreation. Blooms of algae can affect the taste of water and can make swimming unattractive.

The water hyacinth, *Eichhornia crassipes* (Figures 8-7 and 8-8), is one of the classical examples of an introduced plant that became a problem (i.e., a weed). This plant is native to northern Brazil and Venezuela, where, along with four closely related species, it apparently causes no particular problem. Of the six species of *Eichhornia* (one is native to Africa), only *Eichhornia crassipes* has become a weed. The plant gets its common name from the spike of lavender flowers which it produces. Because of the beauty of the flowers, at the turn of the century this was a prize plant for garden pools throughout the world. However, prodigious vegetative reproduction produced so many new plants that a large amount of material was discarded. Some of the excess ended up in waterways, where, under favorable conditions, it prospered. Throughout the tropical and subtropical areas of the world, water hyacinth has become a problem, most notably in the southeastern United States, the upper Nile basin, and southeastern Asia. To give some idea of the growth rate of this plant, in Louisiana during the growing season from March 15 to November 15, 10 plants can produce over 650,000 plants, which will cover 0.4 hectare (1 acre). The edge of the mat may grow at a rate of 0.5 to 0.75 meter (1.5 to 2.5 feet) per month under optimal conditions. Not only does water hyacinth clog waterways, it also increases the loss of water, through transpiration, by as much as three to four times. So you can see that this plant can be, and is, a major pest.

On the brighter side, recently it has been found that water hyacinth may help solve a sewage problem in areas of the world where water is abundant. After secondary sewage treatment, which breaks down organic matter, there are still minerals left which can stimulate unwanted algal growth. Sewage water is piped from the secondary treatment facility to a pond or series of ponds where the water hyacinths remove the minerals. The extensive root systems, which extend into the water below the leaves, and the rapid growth allow the plant to remove both nutrients and some organic compounds from the water. Treated water is discharged and plants harvested. Uses for the harvested material range from animal fodder to soil amendments to methane production; the particular use depends, of course, on whether toxic materials are present.

The free-floating fern *Salvinia auriculata* lacks roots entirely. What may appear to be roots are actually finely dissected fronds. The plants consist of a slender stem with pairs of oblong floating fronds (about 2 by 3 centimeters) and one dissected submerged frond. The floating fronds are covered on their upper surface by water-repellent hairs. This seemingly innocuous plant, native to the New World, has caused serious problems in other parts of the world. In Sri Lanka (Ceylon), it became a problem by clogging waterways and canals, in rice paddies, and in one case by blocking the cooling-water intake for a power plant. The plant was introduced to Sri Lanka in 1939 for use in botanical studies at Colombo University. From the botanical garden of the university, plants somehow found their way into surrounding waters. Since all that is required for growth of the plant is a single node with fronds, small fragments were effective in spreading it. In the 12 years between 1942, when it was first noted outside the botanical garden, and 1954, 810 hectares (2000 acres) of waterways and 89,100 hectares (220,000) acres of rice paddies became infested.

Somehow the plant also found its way into the Zambesi River in Africa. It was first noted in 1949 about 55 kilometers (35 miles) above Victoria Falls. Subsequently, it was found in protected areas along the river for 100 kilometers (65 miles) above the Falls. Plants that went over Victoria Falls (over 100 meters high) (300 feet) were generally killed. The swift current of the river prevented the plant from becoming a problem, and the yearly floods in the river below the Falls probably washed out any surviving plants. However, when the Zambesi was dammed to create Lake Kariba, a large area of still water was created (about 4500 square kilometers (1700 square miles) when the reservoir was full in 1963). In a little more than 1 year, the fern covered over 200 square kilometers (77 square miles) of surface. In 1962, *Salvinia* could be found along most of the 3200-kilometer (2000-mile) shoreline and reached a maximum coverage of 1000 square kilometers (385 square miles). In the open water, wind and wave action prevent the buildup of thick mats; however, in protected areas along the shore, mats may grow to over 15 centimeters in depth. These mats provide a habitat for other plants, such as rushes, which further bind the mats together. Thus, this apparently insignificant fern has caused a great deal of trouble for these two areas. Fortunately, no further major infestations have occurred.

A number of submerged aquatic vascular plants cause problems in lakes and waterways. One of the most troublesome in the United States is European watermilfoil, *Myriophyllum spicatum,* which first appeared in Chesapeake Bay around the turn of the century. However, it did not become a problem until the late 1950s. Between 1960 and 1963, the area overgrown with the plant increased from 20,000 to 80,000 hectares (about 50,000 to 200,000 acres). Many areas in the southeastern United States were invaded by this species of watermilfoil, including the reservoirs of the Tennessee Valley Authority. The rapid growth of the shoots of this plant caused native species to be shaded out. As with many water weeds, all that is needed for a fragment of the plant to survive is the presence of a node and favorable conditions. Furthermore, unlike most other submerged vascular plants, *Myriophyllum spicatum* does not completely die back in the winter, thus providing the plant with a head start on the next year's growth. Also, it can withstand the brackish water of some bays and channels along seacoasts. Herbicides have been most effective in controlling the spread of this species, especially formulations of 2,4-D. Between 1965 and 1967, a 95 percent decrease in the coverage by European watermilfoil was noted in Chesapeake Bay. The cause appears to be a viruslike disease. The ability of this plant to live throughout the year and its general hardiness are characteristics that are being exploited in an attempt to remove mineral nutrients from treated sewage. As with water hyacinth, the plants will be harvested and used in various ways.

Algae are not generally singled out as individual pest species except for some macroscopic forms. The microscopic forms that are generally considered pests are found among the green and blue-green types.

When present in large numbers throughout the water, they give a green color to the water, producing a condition often referred to as "pea soup." A similar effect is produced when ponds are fertilized to encourage the growth of algae, which shade out rooted vegetation. In some very fertile water, a scum several centimeters thick may appear. The mat, which is often too thick to row a boat through, is made up of various filamentous algae (e.g., *Spirogyra* and *Oedogonium*). Of the macroscopic algae, *Chara* (or char) and related species that grow attached to the bottom can grow in such abundance that they clog shallow areas and impart a garliclike smell to a lake. One benefit may be that mosquito larvae are said to do poorly in ponds with much *Chara* in them. Seaweeds have generally been less pesty than the freshwater plants. However, a number of species in their juvenile stages can become attached to the hulls of ships. As the plants grow, they dramatically increase the drag, slowing the ship and increasing fuel consumption.

The preceding are but a few examples of the interactions between people and aquatic plants. People derive many benefits from aquatic plants, ranging from food to industrial products, and the number of these benefits will likely increase. People also affect water plants, often with undesirable results. These ties, although not always obvious, are important both to the human race and to the plants, and bind us to the many aquatic habitats found throughout the world.

Summary

Aquatic plants come in many shapes and sizes, from microscopic spheres to the giant kelps. While most are algae, mosses and flowering plants are also represented. Nearly all bodies of water are inhabited by aquatic plants, including many hot springs. The peculiarities of water itself markedly affect the environment of aquatic plants (e.g., its large heat capacity). Bodies of water with few plants, and therefore low productivity, are termed oligotrophic; those with dense plant growth, and hence high productivity, are eutrophic. People benefit from food produced directly (rice) and indirectly (through fish) by aquatic plants; however, human activity may adversely affect the growth of aquatic plants mainly by dumping wastes into the water.

Selected References

BRANNING, T. G. 1976. Giant kelp: Its comeback against urchins, sewage. *Smithsonian* 7, 102–109.

GOLDMAN, C. R., ed. 1969. Primary productivity in aquatic environments. *Mem. Ist. Ital. Idrobiol. Suppl.* **18.** University of California Press, Berkeley.

HEYERDAHL, T. 1972. *The Ra Expeditions.* Doubleday & Company, Inc., New York.

HUTCHINSON, G. E. 1957. *A Treatise on Limnology:* Vol. 1, *Geography, Physics, and Chemistry.* John Wiley & Sons, Inc., New York.

HUTCHINSON, G. E. 1967. *A Treatise on Limnology:* Vol. II, *Introduction to Lake Biology and the Limnoplankton.* John Wiley & Sons, Inc., New York.

HUTCHINSON, G. E. 1975. *A Treatise on Limnology:* Vol. III, *Limnological Botany.* John Wiley & Sons, Inc., New York.

KING, C. A. M. 1963. *An Introduction to Oceanography.* McGraw-Hill Book Company, New York.

KINNE, O., ed. 1970. *Marine Ecology: A Comprehensive Integrated Treatise on Life in Oceans and Coastal Waters.* 4 vols. John Wiley & Sons, Inc. (Interscience Division), New York.

National Academy of Sciences. 1976. *Making Aquatic Weeds Useful.* Washington, D.C.

NEWTON, L. 1951. *Seaweed Utilization.* Sampson Low, London.

OTTO, N. E., and T. R. BARTLEY. 1972. *Aquatic Pests on Irrigation Systems: Identification Guide.* Government Printing Office, Washington, D.C.

ROUND, F. E. 1965. *The Biology of the Algae.* St. Martin's Press, New York.

RUTTNER, F. Translated by D. G. Frey and F. E. J. Frey. 1963. *Fundamentals of Limnology.* University of Toronto Press, Toronto.

SCULTHORPE, C. D. 1967. *The Biology of Aquatic Vascular Plants.* Edward Arnold (Publishers) Ltd., London.

SMITH, G. M. 1950. *The Fresh-Water Algae of the United States.* McGraw-Hill Book Company, New York.

SMITH, G. M. 1955. *Cryptogamic Botany:* Vol. I, *Algae and Fungi.* McGraw-Hill Book Company, New York.

TAFT, C. E. 1965. *Water and Algae—World Problems.* Educational Publishers, Inc., Chicago.

TIFFANY, L. H. 1958. *Algae—Grass of Many Waters.* Charles C Thomas Publisher, Springfield, Ill.

Atmospheric Ecology and Our Polluted Atmosphere 9
Hazel S. Kaufman

Questions for Consideration

The Atmosphere and Our Weather

An adequate study of the atmosphere and the weather associated within would fill volumes. Much has been written about the weather and its effect on animal and plant life. In this chapter, the emphasis will be on how humans affect the weather.

Even in our highly technological modern world, we are still dependent on climate and weather. Most life on earth is dependent on green plants, which in turn depend on the weather for their growth. The food we eat, the type and location of the house we build, the fuel we burn, the clothes we wear, as well as our health, recreation, and transportation are all influenced by weather and climate. Although the terms "weather" and "climate" are often used interchangeably, there is a significant difference in their meaning. *Weather* is defined as the state of the atmosphere with respect to such things as temperature, humidity, wind, cloudiness, and precipitation. *Climate* is defined as the characteristic weather conditions of a given place averaged over an extended period of time.

When the astronauts walked on the moon, they did not experience weather. The moon has no atmosphere, and an atmosphere is necessary as a medium for weather. The atmosphere is an envelope of air surrounding the earth proportional in thickness to the skin on an apple. Besides supplying air, it protects living things from the harmful radiation of the sun and outer space. The atmosphere is composed of a mixture of gases whose percentage remains relatively constant. Its principal components are nitrogen, 78.1 percent, oxygen, 20.9 percent, and argon, 0.9 percent. It also contains traces of neon, helium, krypton, sulfur dioxide, nitrogen oxides, and varying amounts of water vapor, carbon dioxide, and pollutants.

Figure 9-1 shows the average temperature and heights of the lower layers of the atmosphere. Although the atmosphere extends upward hundreds of kilometers, until it meets the rarified interplanetary medium of the solar system, 99 percent of its mass is found within 30 kilometers (19 miles) of the earth's surface. Most of our weather occurs in the troposphere, which extends an average of 12 kilometers (7.5 miles) above the earth's surface and contains 80 percent of the atmospheric total mass. The outer edge of the troposphere is called the *tropopause.* Its height varies from an average of 16 kilometers (10 miles) at the equator to 8 kilometers (5 miles) at the poles. The next layer, the stratosphere, has been prominent in the news recently because it contains the ozone layer whose destruction has become a topic of great concern.

In order to have weather, there must also be heat. This is provided by the sun. Some of the sun's heat bounces back off the earth's atmosphere, some is reradiated by the earth, some warms the earth and its atmosphere, and some provides motive force for the global weather and energy to change water into water vapor. Heat provides the convective force that lifts masses of air vertically. Since the sun shines more directly

FIGURE 9-1.
Temperature distribution in the atmosphere.

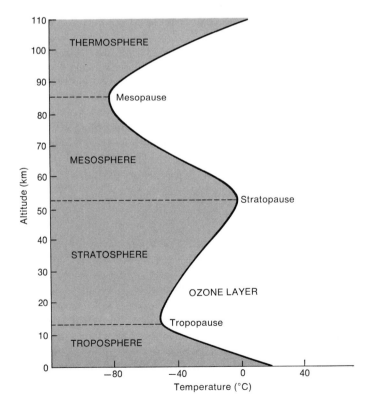

on the equator than on the poles, it heats the earth's surface in the equatorial regions more than in the polar areas. Solar heating and radiational cooling set up temperature differentials between the equator and poles. Temperature differences occur between oceans and land masses, because bodies of water gain and lose heat more slowly than land. Temperature differentials lead to pressure differentials, since cold air is heavier than warm air. Pressure differences generate air movement or winds. Air always moves from high to low pressure, causing both the soft, gentle sea breeze and the violent winds of a tornado. With reference to a global scale, heat increases the buoyancy of the air near the equator. Since there is a horizontal temperature and pressure gradient due to the unequal heating between the equator and the poles, the rising air tends to move poleward and the cold polar air moves toward the equator to fill the vacated space. These wind movements are not directly north and south, because they are greatly affected by friction and the earth's rotation. This deflection of the winds to the right in the northern hemisphere and to the left in the southern hemisphere is known as the *Coriolis force*. In these ways the general global wind and pressure patterns are formed (Figure 9-2). This global wind pattern exists mostly in theory. Many forces are superimposed on it, to change the wind's directions, especially in the temperate zones. These forces include land and sea breezes, day and night unevenness of heating, ocean currents, the uneven surface of the land, regional wind systems such as the monsoons, and the huge migrating air masses with their leading edges known as *fronts*. If a cold mass of air advances toward warm air, it will push itself under the warm air, causing the latter to rise abruptly, forming a line of cumulus-type clouds with frequent showers and thundershowers. This is known as a *cold front*. Squall lines and tornadoes are frequently associated with the cold front when the temperature contrast between the two air masses is extreme and the warm air contains much moisture. When a warm mass of air advances, it moves slowly over the cold air and forms a wide band of stratus-type clouds. This is a *warm front*. It is accompanied by a steady type of precipitation and, frequently, fog. When a front shows little movement, it is called a *stationary front* and usually causes cloudiness and precipitation, often of several days' duration.

Another essential ingredient of weather is moisture. The weather you see as you look out of the window—clouds, fog, rain, snow, or sleet—is some form of water. Moisture exists in the air as invisible water vapor until it is cooled below its saturation point and condenses out on tiny particles to form water droplets or ice crystals. Thus clouds are formed. If a cloud rests on the ground, it is known as *fog*. The water droplets in a cloud may combine to form larger drops, which finally fall as *rain*. If the rain drops fall through a layer of freezing air, they may reach the ground as *sleet*. Ice crystals falling from the clouds may reach the earth as *snow* if they pass through cold air. If they pass through warm air, they may reach the ground as rain. The vertical currents in a thunderstorm cause ice particles being formed in the cloud to rise and fall many times alternately through freezing levels and warm moist ones until masses form of sufficient size to fall out of the cloud as *hailstones*.

Climate Changes

In recent years, there has been an increasing interest in what effect human activity might have on the changes in world climate. Twenty thousand years ago, ice covered the places where many of the world's largest cities now stand. Some of that ice still remains in the polar regions and in the Antarctic. It is estimated that if the Greenland ice cap alone were to melt at a speed comparable with that reached following the last ice age, the sea level would be raised about 7.6 meters (25 feet), coastal regions would be inundated, and much of our present land would be covered with water.

In the past, climate had been generally considered to be relatively static, except on the geological time scale. Recently, it was noted that a definite warming trend existed from the mid-nineteenth century until 1940, when quite an abrupt cooling trend started. A closer study revealed alternate cooling and warming trends since the time of recorded history. A brief outline of the warmer and cooler periods follows:

8000–7000 B.C.	The last major ice sheet disappeared from Scandinavia
5000–3000 B.C.	Postglacial climatic optimum (the world temperature was 2 to 3°C higher than now)
900–450 B.C.	Decline from the climatic optimum
1000–1200 A.D.	Secondary optimum of climate
1430–1850	Little Ice Age
1850–1940	Warming trend
1940–	Cooling trend

The reasons for these worldwide temperature fluctuations remains a matter of conjecture. Some think the cooling trends are due to less radiation from the sun, caused possibly by sun spots. Other theories suggest that an increase of volcanic dust caused more reflection of the sun's radiation back to space, thus causing the earth's cooling. Air pollution has been blamed for recent climate change. The warming trend since 1880 parallels the great increase in the use of fossil fuels, which was accompanied by an increase in CO_2. This suggests that the warmer global temperatures were caused by the *greenhouse effect* of CO_2. Similar to the glass in a greenhouse, CO_2 is largely transparent to short-wave solar radiation, but absorbs much long-wave radiation from the earth and reemits it down to earth, thus causing an increase in the earth's temperature. During the nineteenth century the atmospheric CO_2 concentration was 290 parts per million (ppm), whereas the present value is 320 ppm. Precise measurements made at Mauna Loa Observatory in Hawaii indicate that the increase of CO_2 is now 0.7 ppm annually.

However, although the amount of CO_2 is still increasing, the global temperature has been lowering since 1940. Reports from the Mauna Loa Observatory indicate a continual increase in the earth's turbidity. Therefore, it is theorized by some that this increase in particulate matter, with an accompanying increase in the reflection of the sun's radiation back to space, causes a cooling effect of the earth's surface that exceeds the warming effect of CO_2. The increasing number of jet trails also cause more sunlight to be reflected back into space.

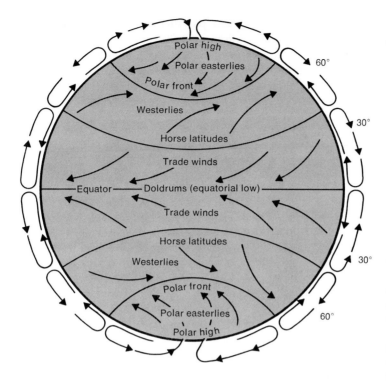

FIGURE 9-2.
Schematic diagram of the wind systems of the world. If the earth did not rotate, surface air would move south from the polar areas directly to the warm equatorial area. Thereafter, heating would cause it to rise and return to the polar regions. As a result of rotation, points on the earth's surface move faster at the equator. At the higher latitudes, it moves slower because it covers less distance during a rotation. Winds blowing toward the equator have lower starting speeds and drag behind, thus becoming easterly in direction. Those blowing from the equator get ahead of the turning earth and become westerlies. The air heats and rises in the doldrums and cools and sinks in the horse latitudes. The stormy polar fronts are caused by the meeting of the warm westerlies and the cold polar easterlies.

Long before people had made any impact on the environment, the world went through alternate warming and cooling periods. What effect our increased pollution has had on recent global temperature is yet to be determined. Many scientists are worried, for only a slight increase in temperature could mean the melting of the polar ice caps, with accompanying extensive flooding, whereas a slight decrease could cause mass starvation as the result of a climate change unfavorable to crops.

The thrust of present weather research is in obtaining a better understanding of climate. The National Science Foundation, doing major research under GARP (Global Atmospheric Research Program), is studying the meteorology and oceanography of the equatorial belt so as to better understand the general circulation patterns. The first global experiment takes place in 1978–1979 when more than 120 nations will provide data for mathematical models. Over 70 scientists met in Stockholm in the summer of 1974 to review the present knowledge of climate and, with that as a base, to plan research. One planned project includes plotting the record of weather backward. Deep cores will be drilled in Greenland to obtain pollen records and other data from past climates, tree rings will be studied, and so on. All these data will be used in computer models to help determine what caused the changes in past climates. This knowledge will help us in predicting what effect our pollution may have on future climates.

Even though we are not certain that pollution affects global climate, we do know that it affects local climate. We know that a city tends to be warmer, cloudier, rainier, and has lower visibility than the surrounding countryside. The first day of frost comes later in the city. Concrete walls and streets absorb heat during the day and then radiate it at night. Home heating and industry also help to increase a city's temperature. La Porte, Indiana, located 48.2 kilometers (30 miles) downwind from the steel mills of Gary, Indiana, and south Chicago, has 31 percent more rain, 38 percent more thunderstorms, and 24 percent more days of hail and considerably less sunshine than nearby communities. Rainfall in La Porte correlates with steel production in Gary, Indiana. The steel mills, with their smoke, produce heat, moisture, and condensation nuclei. More moisture is added from Lake Michigan.

Although heat produced by combustion does have a local effect, especially in the heat islands forming over cities, it appears to have no global significance. Fears have been expressed about the depletion of oxygen because its consumption in the combustion of fuels is thought to be greater than its supply from photosynthesis. Recent studies have found no evidence that the concentration of O_2 is changing. However, if the phytoplankton that supplies most of our O_2 is destroyed by either water pollution or an increase in ultraviolet radiation, we may have real reason for concern.

The Relation of Weather to Air Pollution

Weather determines what happens to pollutants after they reach the atmosphere. Surface winds help scatter pollutants; the greater the speed, the better the dispersal. Convective currents also help. During daylight hours, the earth's surface is warmed by the sun, causing currents of warm air to rise upward, carrying pollutants with them. As the warm air rises, it cools, because the pressure on it is decreased, allowing it to expand and cool. It continues upward until it becomes the same temperature as the surrounding air. The distance between the earth and this altitude is known as the *mixing depth*. It determines the depth to which pollutants will be mixed. Its height is continually changing, and it is higher in the summer than in the winter.

In the troposphere, air normally decreases in temperature with an increase in altitude. However, during a temperature inversion, a layer of warm air lies over the cooler air and acts as a lid, keeping the smoke and other pollutants down near the earth's surface. Temperature inversions occur under several different meteorological conditions. When a warm front moves over static cooler air, a temperature inversion is formed. On clear cool nights, the ground radiates its heat skyward and cools the air at low levels while warmer air prevails aloft. The disastrous smog episode in London in December 1952 was due to a third condition, which is caused by subsidence. This inversion results when upper layers of air descend or subside during a developing anticyclone or high-pressure area. The air warms as it descends, and the warming is greater at higher levels than near the ground. This results in a temperature inversion. If the anticyclone remains stationary, the inversion may persist and the pollutants accumulate until severe conditions develop. Perhaps the most widely recognized temperature inversions are those that occur over Los Angeles. They usually occur on at least 300 days during a year. These inversions are caused by local meteorological conditions augmented by the mountainous topography of the area. The cold Humboldt ocean current flows along the California coastline west of Los Angeles. As the prevailing westerlies with their warm moist air from the South Pacific pass over this current, the lower air layers are cooled, often forming fog, whereas the upper layers remain warm. Thus, when the air moves eastward over Los Angeles, which is surrounded by mountains, the cooler lower air layers with fog and the city's air pollutants become trapped not only by the lid of warm air above it but also by the surrounding mountains.

Since prehistoric times, people have been attempting to modify the weather. The early rainmakers tried rain dances, sacrificial fires, and shooting arrows into clouds. Early men of religion prayed for rain. Modern rainmaking made its first big stride, when in November 1946, Vincent Schaefer of General Electric Company dumped dry ice on a supercooled cloud and snow fell from the cloud. Bernard Vonnegut and Irving Langmuir were two other important figures in early scientific weather-modification work. Shortly after World War II, Irving Krick, for 15 years head of the weather department at the California Institute of Technology, began a commercial cloud-seeding venture. Not only did he do a good business in the United States, but he was also hired in many other countries to set up rainmaking systems. Krick estimated that his operations increased rainfall from 13 to 15 percent. In 1959, we had only 36 weather-modification projects reported by state, local, and commercial operators. Shortly thereafter, weather-making projects increased greatly, and by 1968, there were dozens of commercial firms seeding clouds under contract to farmers for increased rainfall or hail abatement, to power companies for increased snowpack above dams, to cities for additional water in their reservoirs, and to airports for fog dispersal. Hurricane seeding and lightning suppression were also tried. Many law suits ensued, for rain from the seeded clouds did not always fall only on the customer's property, and sometimes no rain fell. The question of who owns the rain in the clouds arose. By 1968, 22 states had laws dealing with at least some aspect of weather modification.

The federal government was slow getting into the field of weather modification, owing largely to the opposition of the U.S. Weather Bureau. The latter demanded statistical proof based on countless tests over a period of years. A committee on weather control appointed by President Eisenhower reported in 1957 that the rainmakers did cause an increase in precipitation from 9 to 17 percent. Nevertheless, even by 1965, only $4 million were spent by the federal government on actual weather modification. Since then, federal funding was increased, and now many federal agencies are involved. Among these are the National Oceanic and Atmospheric Administration, Department of Transportation, U.S. Forest Service, Federal Aviation Agency, and the military. One large project is the Bureau of Reclamation's attempt to increase the water available in the Colorado Basin (Lynn, 1974). This project, known as WOSA (Winter Orographic Snowpack Augmentation), has gained a considerable amount of opposition. Increased rainfall in the Colorado

Modifying Weather

Basin would mean increased sediment in Lake Powell. The water quality of the Colorado River would deteriorate when more lands were brought under irrigation and more industries were developed as a result of the increased rainfall. Increased precipitation might also cause an increase of avalanches and flooding.

A prerequisite to useful weather modification is accurate prediction of the weather. Even though elaborate equipment, including satellites and computers are now used, weather forecasts are still not accurate. The cost of the necessary research and the engineering to successfully modify the weather is prohibitive. Any major modification of the weather would be likely to destroy nature's balance in some way and lead to unpredictable results that might be catastrophic. Instead of trying to modify the weather, perhaps we should try to better cope with what we have by doing such things as improving blind-landing equipment for aircraft, developing better warning systems for approaching storms, refraining from building in places likely to flood, and growing only those crops suited to the prevailing climate.

Atmospheric Pollution in Our Environment

Now that we have presented a brief discussion of the nature of our atmosphere and its weather, it is relevant to find out how seriously our atmosphere is being polluted and how this pollution affects human and plant life in our environment.

When we think of air pollution, we usually associate its source with the activity of humans. Nature actually causes more pollution. Salt from the oceans, gases and debris from volcanoes, terpenes and pollen from plants, and cosmic dust are some of nature's pollutants. Dust storms and forest fires add more debris to the atmosphere. Nature, however, has the power to recover from her own ravages. Following the industrial revolution, humans have caused an upsetting of nature's air ecology. If our rush for increased comforts continues as in the past, we shall not only suffocate ourselves, but also destroy all life around us. Air pollution, basically the presence of foreign substances in the air, has become a serious problem because the concentration and qualities of these substances are injurious not only to property, but also to vegetation and all animal life.

Motor vehicles are by far our worst polluters, followed by industry, power plants, and home heating (Figure 9-3). Carbon monoxide constitutes the greatest mass of any air pollutant, followed by sulfur dioxide, hydrocarbons, nitrogen oxides, and particulates (Figure 9-3).

Major Air Pollutants

When looking at our transportation systems, one would have to agree that their most deleterious effect upon the environment would have to be in their contribution to air pollution. As the automobile is responsible for roughly 60 percent of the pollution found in the United States today (Esposito, 1970; Horowitz and Kuhrtz, 1974), we shall focus our attention most closely upon it. When looking at three of the major pollutants

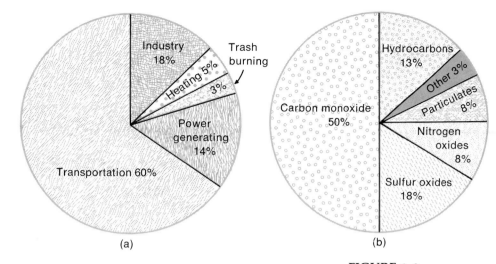

FIGURE 9-3.
Comparison (by percent) of the major sources of man-made air pollution in our environment (a) and the constituent polluting gases and particles (b).

found in auto emission—carbon monoxide (CO), hydrocarbons (HCs), and nitrogen oxides (NO_xs)—the figures for 1971 show 72.34 billion kilograms (160.75 billion pounds) of CO, 8.77 billion kilograms (19.48 billion pounds) of NO_xs, and 12.50 billion kilograms (27.78 billion pounds) of HCs released into the atmosphere annually by cars and trucks in rural and urban areas (1974 National Transportation Report Summary). In addition, cars emit sulfur dioxide (SO_2) and particulate matter, among which is found lead. About 180 million kilograms (400 million pounds) of lead is poured out by automobiles per year (Edel, 1973; Esposito, 1970). Certain other gases are produced indirectly by automobiles. These gases occur when NO_2 reacts with hydrocarbons and some other gases to form what is known as "photochemical smog." In the paragraphs that follow, we shall take up each of the major automobile pollutants, and in addition, air pollutants from other sources.

CARBON MONOXIDE (CO)

Carbon monoxide, a colorless, odorless, poisonous gas, is the product of incomplete combustion of carbon and its compounds. Created largely by the automobile, CO constitutes more than half the pollution made by humans. Carbon monoxide is very dangerous to humans because it combines with hemoglobin in the blood to form carboxyhemoglobin. If taken in sufficient quantities, CO prevents the hemoglobin from carrying the needed oxygen to the tissues. Since hemoglobin binds CO about 200 times as strongly as it binds oxygen (Esposito, 1970), it can readily be seen that even in moderate concentrations, CO is a very effective inhibitor of O_2 binding to hemoglobin. People with heart disease are especially in danger when exposed to CO gas (Esposito, 1970; Fri, 1973). Even for healthy persons, CO in a concentration below the toxic level can cause dizziness, headaches, and slow our driving reactions, thus increasing traffic accidents. Even though CO levels of 10 parts per million (ppm) are thought to cause an increase in mortality in heart-disease patients, this level is often surpassed in large cities, and concentrations exceeding 87 ppm have been measured in underground garages and in tunnels.

SULFUR DIOXIDE (SO$_2$)

The sulfur dioxide in our atmosphere comes largely from the combustion of fuels containing sulfur. Sulfur dioxide is a colorless, nonflammable gas, highly soluble in water. It has an irritating odor at concentrations above 3 ppm. In the atmosphere, it oxidizes readily to sulfur trioxide (SO$_3$), which in the presence of water becomes sulfuric acid (H$_2$SO$_4$). The latter is then often absorbed by fine particles of fly ash, which also result from combustion. These particles cause irritation to the delicate membranes that line the eyes and respiratory tract. Sulfur dioxide was the devastating pollutant associated with the smog disasters in Donora, Pennsylvania, and London, England. It causes the acid rainfall (with pH values as low as 3) that has been falling over northeastern United States and Scandinavia.

HYDROCARBONS

Automobiles produce more than half the hydrocarbons in our atmosphere. Another heavy contributor is the petroleum industry. The most potent hydrocarbon, benzopyrene, is also present in cigarette smoke and is associated with lung cancer. Hydrocarbons are thought to be responsible for the increased number of deaths of birds and animals in the Philadelphia Zoo, which is located between two heavily traveled streets. The importance of hydrocarbons to air pollution lies almost entirely in their formation of photochemical smog.

NITROGEN OXIDES

Automobiles are responsible for about half of the nitrogen oxides dumped into our atmosphere. Chemical plants producing or using nitric acid and power plants are other principal sources. Although many oxides of nitrogen exist in the atmosphere, only nitric oxide (NO) and nitrogen dioxide (NO$_2$) are considered to be pollutants. They cause a brown haze which irritates the eyes and nose and lowers visibility. Their greatest harm is done, however, when in the presence of sunshine and other exhaust gases, they are transformed into photochemical smog.

PHOTOCHEMICAL SMOG

Photochemical smog is formed from the exhaust gases of automobiles by a complex process of photochemical reactions in the presence of sunshine. In the initial reaction, ultraviolet energy ($h\nu$) is absorbed by nitrogen dioxide (NO_2). The highly energized molecule (NO_2^*) then decomposes or photolyzes into nitric oxide (NO) and atomic oxygen (O). The latter reacts with molecular oxygen (O_2) to form ozone (O_3). If no energy-absorbing molecule (M) is present, the ozone will rapidly decompose. With a third body (M) present, a stable ozone molecule is formed. If nitric oxide is present, it reacts with the ozone to form more NO_2 and a molecule of oxygen. These reactions are shown by the following chemical equations:

$$NO_2 + h\nu \rightarrow NO_2^*$$
$$NO_2^* \rightarrow NO + O$$
$$O + O_2 + M \rightarrow O_3 + M$$
$$O_3 + NO \rightarrow NO_2 + O_2$$

In the presence of certain hydrocarbons, other reactions take place. Some of the atomic oxygen, ozone, and nitric oxide react with these hydrocarbons to form a variety of products and intermediates. One of the products formed is peroxyacetyl nitrate (PAN), which is considered one of photochemical smog's most harmful components. Photochemical smog, characterized by an unusual odor, irritates the eyes and throat, reduces visibility, and does much damage to plants. California's agricultural loss of $130,000,000 per year is due largely to it. Nearly all urban areas show some effect of photochemical smog, but the most intense effect is found in sunny, subtropical areas with heavy traffic at times when stagnant air prevails.

OZONE

Ozone is a colorless, pungent gas formed by the photochemical breakdown of NO_2 (as we have just shown). It plays a dual role. In the stratosphere, it is our friend, shielding us from the ultraviolet rays of the sun; at the earth's surface, it is an enemy. At levels as low as 10 ppm, it has been found to cause headaches, eye irritation, coughing, and chest discomfort. It causes rubber to crack. An early analytical method for the determination of ozone was the depth of the cracks it produced on stretched rubber bands. The deterioration of the sidewalls of automobile tires in Los Angeles in the 1940s was a serious problem. Addition of an antioxidant alleviated the situation. Ozone also deteriorates fabric, causes colors to fade, and does great damage to vegetation.

PARTICULATES

The air pollution that we see as dust, smoke, fumes, and haze is composed of tiny solid and liquid particles. Most of these are formed by industry and the home burning of fuels. Coarse dust particles, larger than 10 micrometers in diameter, settle out quickly and cause trouble only near their source. Smaller dust particles travel farther, and particles less than 1 micrometer in diameter, generally referred to as *aerosols,* remain suspended in the air for long periods of time. They scatter light and may act as nuclei on which water vapor condenses. Aerosols may also reflect the sun's heat and prevent it from reaching the earth. Particles less than 2 or 3 micrometers in size can penetrate deep into the lung, where it is unprotected by mucus. These particles may carry harmful chemicals such as SO_2 with them, whereas SO_2 alone would most likely dissolve on the mucus of the respiratory tract before it reaches the vulnerable non-mucus-lined areas of the lungs. Thus we find that particulates do more serious damage than that which is readily seen, such as soiling our clothes, buildings, and other property.

FLUORIDES

Estimates by the National Research Council and the EPA suggest that 108,000 to 136,000 metric tons (120,000 to 150,000 tons) of fluoride are emitted into the atmosphere annually. The main sources are aluminum smelting, ceramic manufacturing, and phosphorus-processing plants. According to the U.S. Department of Agriculture, fluoride has done more damage to livestock worldwide than any other pollutant (Lillie, 1970). Some forage plants can accumulate many times the level of fluorides present in the air. Livestock eating these contaminated plants may develop skeletal deformations. Fluoride emissions from manufacturing plants, such as those cited above, also cause serious damage to vegetation.

LEAD

Starting in 1923, tetraethyllead began to be used as an antiknock additive to gasoline. Now, on a typical day, the automobiles of Los Angeles pour out over 13,000 kilograms (28,000 pounds) of lead every day! Nearly 50 percent of the lead we breathe is absorbed by the body. In experiments with mice and rats, lead at human tissue levels is toxic to the animals. Some think that lead may be one of the causes of the high infant-mortality rate in our country. Bamboo grown near a busy street in Washington, D.C., was found to have too high a lead content to risk being fed to the National Zoo's pandas given to us by the People's Republic of China.

PESTICIDES

Pesticides are a major source of air pollution in some agricultural regions (see Chapter 10). The use of highly toxic agricultural chemicals in California has made field work one of the most hazardous occupations in the state. Crop dusting is not only dangerous but also wasteful, as only about 10 percent reaches the specific target. Each year 270 million kilograms (600 million pounds) of pesticides are sprayed, dusted, and fogged in the United States, the equivalent of 1.35 kilograms (3 pounds) for every man, woman, and child in the country. Residues drift through the air to remote areas. Pesticides are found in the tissues of reindeer in Alaska and in penguins in the Antarctic.

AEROALLERGENS

Aeroallergens are airborne pollutants, primarily pollen and spores. They usually range from 2 to 50 micrometers in diameter. Their major characteristic is that they cause allergic reactions in humans (e.g., allergic rhinitis, often misnamed "hay fever"). To act as an aeroallergen, pollen of a given species must occur in the atmosphere in large quantities close to centers of human population and possess strong antigens that are capable of sensitizing and causing allergy symptoms (Sheldon, Lovell, and Mathews, 1967). Some of the most prevalent pollen aeroallergens come from members of the grass family (*Poaceae*) and from ragweed (*Ambrosia* spp., Figures 3-5 and 3-6) and walnut (*Juglans* spp.). Pollen from many other plant species may also act as aeroallergens. Among the fungi, spores of only a few dozen genera have been implicated as aeroallergens. However, these are quite common genera that produce immense quantities of spores, often outnumbering airborne pollen. In the Midwest and northeastern United States, the most important fungal genera appear to include *Cladosporium* and *Alternaria* (both are sac fungi, or Ascomycota; see Chapter 2); spores of *Helminthosporium, Aspergillus, Penicillium, Fusarium,* and *Rhizopus* are of secondary importance (Sheldon, Lovell, and Mathews, 1967).

Other air pollutants, which are more dangerous but do not appear in as great a volume, include asbestos, beryllium, and mercury. The EPA has designated these three substances as hazardous air pollutants, and on December 7, 1971, published proposed emission standards for them. Radioactivity is becoming an increasing hazard with the increasing number of nuclear power plants.

Property Damage

Cleopatra's Needle in Central Park is a sad monument to air pollution. This tall granite obelisk dating from 1570 B.C. was in perfect condition when the Egyptians gave it to us in the early 1880s. Now, the hieroglyphics covering it are dimmed and all but obliterated on the west or windward side. Buildings are damaged by air pollutants that erode marble, limestone, and slate. Steel deteriorates two to four times faster in the cities than in rural areas. Paint discolors and peels. Dirt in the air increases our laundry and dry cleaning bills and causes fabric to deteriorate. Harmful chemicals rot rubber and leather. The Council on Environmental Quality has estimated that air pollution costs the people of the United States $16 billion annually for property damage; this corresponds to $80 for every man, woman, and child in the country.

Effects of
Air Pollution
on Plants

Since plants are sensitive to air pollutants, they are often used as warning devices to indicate the presence of a pollutant. For the past 20 years, scientists have been growing plants under controlled conditions and exposing them to various concentrations of common air pollutants. Characteristic injuries to certain plants can be interpreted to show not only the presence, but also the relative concentration, of the air pollutant. Tobacco is an effective ozone indicator. Pinto bean plants are used to detect peroxyacetylnitrate (PAN) and gladiolus to determine fluoride accumulations (Figure 9-4). Excellent detectors of SO_2 are dahlias,

FIGURE 9-4.
Hydrofluoric acid damage to gladiolus leaves. (Courtesy of the U.S. Department of Agriculture.)

FIGURE 9-5.
*Sulfur dioxide damage to
an ash (Fraxinus) leaf.
(Courtesy of the U.S.
Department of Agriculture.)*

petunias, alfalfa, and cotton. The leaf is the primary indicator of injurious effects. Low concentrations of gases may cause injury that is not readily apparent, such as growth suppression, dwarfing, early maturation, and a decreased crop yield.

Sulfur dioxide was perhaps the first air pollutant found to do extensive damage to vegetation (Figures 9-5 to 9-8). Early in this century, SO_2 fumes from a copper smelter in Ducktown, Tennessee, killed 2833 hectares (7000 acres) of vegetation. Soil erosion washed away most of the top soil, leaving a desert in the midst of a lush forest. After antipollution measures were taken, and the SO_2 was converted to useful H_2SO_4, the return on the once-wasted SO_2 proved greater than that from copper! SO_2 injury appears as clearly marked dead tissue between the veins and on the margins of leaves on broadleaved plants, and as brownish discoloration on the tips of pine needles. Some of the SO_2 in the air

FIGURE 9-6.
Sulfur dioxide damage to
a sumac (Rhus) leaf.
(Courtesy of the U.S.
Department of Agriculture.)

changes to sulfur trioxide, which, in the presence of moisture, forms H_2SO_4. Damage caused by the latter appears as definite spots on the leaves.

Photochemical smog is a relatively recent hazard to vegetation. Extensive injury to leafy vegetable crops in the Los Angeles Basin was first noted in 1944. The injury was quite different from that of SO_2. A shiny, oily effect that appeared on the leaf's lower surface developed into a glaze and turned silver. Extensive investigations determined that the culprit was photochemical smog. The cost of smog damage to vegetable production is approximately $10 million per year in the United States.

Two different types of injury have been identified as smog damage: one caused by ozone (Figures 9-9 and 9-10) and the other caused by PAN (Figure 9-11). Ozone damages the palisade layers of the leaf and appears as flecking or stippling on broad leaves, as streaks on the leaves of cereal crops, and as browning on the tips of pine needles. Tobacco, beans, tomatoes, and white pine are especially susceptible. Flecking of tobacco leaves reduces their value, particularly as cigar wrappers. Ozone has been responsible for the destruction of many acres of ponderosa pine in the San Bernardino Mountains when smog drifted out of the Los Angeles area. PAN injury appears as bronzing, silvering, and glazing on lower leaf surfaces.

Chlorine damage (Figure 9-12) is often found near swimming pools and sewage disposal plants where chlorine is used as a disinfectant. Its damage to leaves is similar to that caused by SO_2. It is very difficult for the untrained eye to determine the cause of plant injury, since not only

FIGURE 9-7.
Sulfur dioxide damage to peach leaves (two leaves at left). Undamaged leaf (not exposed to SO_2) appears at the far right. (Courtesy of the U.S. Department of Agriculture.)

FIGURE 9-8.
Sulfur dioxide damage to alfalfa leaves (middle and to the right). Undamaged leaf (not exposed to SO₂) appears at the far left. (Courtesy of the U.S. Department of Agriculture.)

do the various air pollutants cause similar-appearing damage, but natural agents such as drought, frost, mineral deficiencies, and disease produce symptoms very similar to those caused by air pollutants.

Since different plant species differ greatly in their sensitivity to air pollutants, programs for the breeding and dissemination of pollution-resistant cultivars of crop plants and species of street trees are now under way. Unfortunately, this method of control would result in loss of the use of several of our important and favorite plants.

Human Costs

In the United States alone, it is estimated that air pollution is at least partially responsible for 110,000 deaths annually. Its major effects on health appear to be the result of irritating material acting on the respiratory tract. Air pollutants greatly increase the symptoms of emphysema. They cause the constriction of the bronchioles, which lead to the tiny air sacs of the lungs (alveoli). During exhalation, more air remains in the alveoli than should. When new air is inhaled, they stretch and burst. With a decrease in alveoli, there is a decrease in the number of capillaries, and the oxygen exchange is lowered. This leads to the shortness of breath of the emphysema patient. Pollutants also inhibit the movement of the cilia lining the respiratory tract. Particles that would ordinarily be removed by the cilia remain in the respiratory tract and cause damage. Although lung cancer cannot be attributed to a single cause, the death rate from lung cancer is twice as great in large cities as in rural areas. The heart is also affected by air pollution. All chronic respiratory diseases involve the heart. It must work harder to pump enough blood to compensate for the loss of oxygen due to respiratory disease. The death rate for heart disease is also greater in metropolitan areas than in rural areas.

According to Dr. Rene Dubos of Rockefeller University, "The increase in chronic and degenerative diseases is due in part at least, and probably in a very large part, to the environmental and behavioral changes that have resulted from industrialization and urbanization" (EPA, 1973).

Noise Pollution

Sound, which is a vital part of our existence, has grown to such disagreeable proportions that today it is becoming a very real threat to our health. An estimated 16 million people in the United States suffer from some degree of hearing loss directly caused by noise. Short exposure to intense sound waves may lead to temporary hearing loss; long exposure, to permanent hearing loss. Continual exposure to loud sounds could lead to deterioration of the inner ear and subsequent deafness. Noise also causes pain, stress, headaches, nervousness, and irritability. Recent studies suggest that existing noise levels may be a cause in the rising rate of heart disease, ulcers, and mental illness, and may even adversely affect the unborn child.

FIGURE 9-9.
sycamore (Platanus) leaves, showing varying degrees of damage due to the air pollutant ozone. (Courtesy of the U.S. Department of Agriculture.)

FIGURE 9-10.

Spinach plant, showing ozone damage. (Courtesy of the U.S. Department of Agriculture.)

Sound is measured by pressure levels caused by sound waves and is expressed in decibels. The 0 on the scale in Table 9-1 is based on the lowest sound the healthy ear can detect. Scientists tend to agree that the noise level for potential hearing loss begins at about 70 decibels. This is cause for concern, because our normal daily life regularly exposes us to higher levels, even in our homes. The noise of dishwashers, garbage disposals, mixers, and blenders may drive the kitchen din up to 80 or 90 decibels. In the living room, the television set may put out 70 to 80 decibels, and the hi-fi set may run beyond 100 decibels. Outside in traffic, 70 decibels is a typical level, with trucks and motorcycles going up to 100 decibels. At work, an office averages 50 decibels; a factory, 85; a print shop, 95; a construction site, 100; a lumbering site, 125; and a jet runway, 130.

Clearly, something must be done soon or we will permanently damage the hearing of much of our population. Fortunately, the knowledge and technology to control noise already exists. There are two practical means of control: (1) reduction of noise at its source, such as making the sound producer quieter; and (2) alteration of the sound path by distance or shielding, such as constructing better walls between apartments. More communities are adopting zoning ordinances that attempt to separate noisy airports and factories from residential districts. Aircraft is increasingly being required to use reduced power, and noise-abatement procedures are being implemented around airports. Much more could be

FIGURE 9-11.
PAN (peroxyacetyl nitrate) damage to romaine lettuce seedling. (Courtesy of the U.S. Department of Agriculture.)

TABLE 9-1 / Examples of Noise Levels and Human Response

Source of Noise	Decibels	Human Response
Jet takeoff (nearby)	150	
Carrier deck jet operation	140	Threshold of pain
Hydraulic press (1 meter)	130	Limit of amplified speech
Jet takeoff (60 meters)	120	Maximum vocal effort possible
Construction noise (3 meters)	110	
Subway station	100	Very annoying
Heavy truck (15 meters)	90	Hearing damage (8 hours)
Alarm clock	80	Annoying
Freeway traffic (15 meters)	70	Telephone use difficult
Air conditioning unit (6 meters)	60	Intrusive
Quiet residential street	50	Quiet
Living room	40	
Soft music	30	Very quiet
Broadcasting studio	20	
Rustling leaves in breeze	10	Just audible
	0	Threshold of hearing

FIGURE 9-12.
Chlorine damage to leaves of corn (Zea mays). (Courtesy of the U.S. Department of Agriculture.)

done to reduce noise at its source. For example, mufflers or other noise-abatement devices could be added to air compressors, pneumatic riveters, and other noisy equipment. Truck tires could be made with quieter treads. Many other noises could be reduced but are not, because of the apathy and desires of the American public. An example of the latter is the problem a power-lawn-mower manufacturer encountered when he designed and marketed a quieter mower. Sales were poor and purchasers returned the mowers complaining that they were "under-powered." The mowers were just as powerful as competing mowers, but too many Americans equate noise with power. Since both government and industry respond to public demands, the public must become aware of the dangers of noise and then demand a quieter environment. When the public demands quieter machines, industry will begin to compete on the basis of how quiet their machines can be.

Recently, several noise-control laws were passed. The 1970 Occupational Safety and Health Act (OSHA) requires administrative or engineering controls to be applied when noise levels exceed 90 decibels for 8 hours in any day. The Federal Noise Act of 1972 gives EPA the power to set limits on acceptable noise levels and to regulate products that exceed them. The EPA has defined 70 decibels as the maximum 24-hour level. This act also required EPA to identify major sources of noise and to provide information on techniques to control them. EPA encourages cities and communities to set up their own local noise ordinances, and at least 300 communities have done so. A nationwide citizens' group called NOISE (National Organization to Insure a Sound Environment) is working to combat the increasing menace of noise pollution caused by jet aircraft.

In the stratosphere, there is a layer of ozone which acts as an umbrella shielding the earth from much of the sun's ultraviolet rays. This layer is concentrated at 20 to 25 kilometers (12 to 16 miles). Without the shielding effect of this layer, it is doubtful if life would ever have evolved on the earth.

Ozone is simultaneously formed and destroyed in the stratosphere by naturally occurring chemical reactions. When oxygen molecules (O_2) absorb sunlight (light energy, hv), they dissociate into free oxygen atoms (O), which then combine with O_2 to form ozone (O_3):

$$O_2 + hv \rightarrow 2O$$
$$O + O_2 \rightarrow O_3$$

Sunlight is also responsible for the destruction of O_3. It dissociates the ozone molecule into oxygen atoms and molecules. The free oxygen combines with ozone to form more oxygen molecules:

$$O_3 + hv \rightarrow O + O_2$$
$$O + O_3 \rightarrow 2O_2$$

That our technology might be affecting the ozone layer was not considered until the early 1970s. Then, during the investigations into possible environmental consequences of a fleet of supersonic transports (SSTs), it was discovered that nitrogen oxides (NO_x) emitted in the exhaust fumes of the aircraft could accelerate the ozone layer's break-down:

$$NO + O_3 \rightarrow NO_2 + O_2$$

H. Johnston at the University of California, Berkeley, estimated that a fleet of SSTs would emit enough NO_x into the stratosphere to pose a threat to the ozone layer. Under the Climatic Impact Assessment Program of the Department of Transportation, several studies produced a general consensus that a 50 percent increase in nitric oxides from the proposed SST fleet would reduce the ozone level by 7 to 12 percent.

Another source of ozone-destroying nitrogen oxides was brought to our attention as a result of the SST investigations. During nuclear explosions, oxygen and nitrogen are broken down and large amounts of nitrogen oxides are formed.

Early in 1974, F. S. Rowland and M. J. Molino at the University of California, Irvine, announced that our common aerosols were a far more dangerous threat to the ozone layer than the nitrogen oxides. These fluorocarbon compounds, commonly known as Freons, break down in the stratosphere and produce chlorine atoms, which are six times more effective than the nitrogen oxides in destroying the ozone layer. In their experiments, they used a vertical diffusion model, which examines the upward drift of molecules in the atmosphere.

These compounds are, more specifically, dichlorodifluoromethane (CF_2CL_2), commonly known as Freon 12, and trichloromonofluorometh-ane ($CFCL_3$), known as Freon 11. About 1.5 billion aerosol cans containing Freon were used annually in the United States for personal-care items, such as perfumes, hair sprays, and deodorants. Another 190 million cans were used for household products, such as window cleaners, air fresheners, and oven cleaners. Spray paints, insecticides, and food products helped to fill the remaining 700 million cans. According to the U.S. Tariff Commission, more than 376,000 metric tons (415,100 tons) of the principal fluorocarbons were sold in the United States in 1973.

The Freons are popular propellants because they are easily liquefied at relatively mild pressures, and they vaporize readily at room temperatures. Because they are relatively inert in the lower atmosphere, they have been considered valuable as tracers of atmospheric motions. It is their quality of stability that makes them a threat to the ozone layer. Once they are released, they gradually ascend unchanged through the troposphere, perhaps over a period of years, until they reach the stratosphere. There, the strong ultraviolet light from the sun causes them to dissociate:

$$CF_2Cl_2 + h\nu \rightarrow CF_2Cl + Cl$$
$$CFCl_3 + h\nu \rightarrow CFCl_2 + Cl$$

The Cl atom then causes the following chain of reactions, which leads to the breakdown of the ozone layer (Bartell and Ritz, 1974):

$$Cl + O_3 \rightarrow ClO + O_2$$
$$\underline{ClO + O \rightarrow Cl + O_2}$$
$$O_3 + O \rightarrow 2O_2 \text{ (net)}$$

Studies by R. J. Cicerone and colleagues at the University of Michigan, M. B. McElroy at Harvard University, and P. J. Crutzen at the National Center for Atmospheric Research in Boulder, Colorado, all reported that the Freons can destroy the ozone layer. Their estimates differ as to the rapidity of its destruction. These vary from 5 percent destruction by the end of the century if we do not increase our use of fluorocarbon, to 40 percent destruction if we increase our use of fluorocarbons at the rate we have been increasing them in the past few years.

Measurements made from 1971 to 1974 show that there has been a definite increase in fluorocarbon concentration in the troposphere (Wilkniss et al., 1975). More recently, measurements of Freon 11 were made in the stratosphere. In the lower stratosphere, there was a concentration of 60 parts per trillion (ppt), whereas at higher altitudes, the concentration dropped to 20 ppt. This decrease was consistent with its estimated decomposition by ultraviolet light. Hydrochloric acid has also been measured in the stratosphere. Its concentration increased with altitude, as was predicted, because of its formation following the breakdown of fluorocarbons.

The task of finding alternative substances for the Freons will not be an easy one, as they have been found to be superior propellants. Several nonchlorinated fluorocarbons tested release fluorine very readily. This, when combined with hydrogen, is damaging to lung tissue. Hydrocarbons and carbon dioxide now used with other products do not produce a fine spray like the fluorocarbons, and therefore tend to feel wet and cold on the skin, making them undesirable for personal-care items. Some companies are using roll-ons and atomizers as replacements for aerosols.

Further studies were made by several scientific groups including the National Academy of Science. The results of their investigations indicated that the Freons deplete the ozone layer. In 1977, the EPA, together with the Food and Drug Administration and the Consumers Product Safety Commission, imposed a ban on all fluorocarbons by April 15, 1979.

With the increased study of the stratosphere, more culprits are being found. A new suspect is bromine (Hammond, 1975). It is thought to act to break down the ozone layer in a manner similar to chlorine. Methyl bromide (CH_3Br), widely used to fumigate agricultural land, is thought to be the source of stratospheric bromine.

It is well documented that excessive exposure to sunlight can cause skin cancer in humans. The effect of ultraviolet radiation on the human skin is a cumulative process; that is, malignancies may be induced with repeated exposure to ultraviolet radiation. Studies made with mice indicate that ultraviolet wavelengths in the range 280 to 320 nanometers are most effective in tumor development. Since loss of stratospheric ozone would lead to increased intensities in precisely this range, it is clear that increased occurrence of human skin cancer could result from destruction of the ozone layer. It is generally accepted that a 10 percent loss in the ozone layer would cause a 20 percent increase in the incidence of skin cancer. A recent survey by the National Center for Health Statistics indicates that nearly 600,000 cases of skin cancer are diagnosed annually in the United States.

Summary

Moisture and heat are the primary factors that account for the weather in the earth's atmosphere. Even though people have inadvertently modified the weather, their attempts to control it have met with little success. A better understanding of the weather will help us to cope more successfully with our environment, particularly in connection with managing our crops, controlling soil and wind erosion, and combating air pollution.

Air pollution has greatly increased since the advent of the industrial revolution, especially since the invention of the automobile, which has become our worst polluter. Our major air pollutants are those we cannot see, such as carbon monoxide, sulfur dioxide, hydrocarbons, and nitrogen oxides. Besides being a serious health hazard, air pollution does considerable damage to property and vegetation. Noise pollution not only leads to deafness but is detrimental to our health in other ways. Our entire air pollution problem has been made more serious by the finding that fluorocarbon compounds known as Freons can destroy the protective ozone layer in the stratosphere. The resultant increase in amount of ultraviolet radiation reaching the earth's surface could lead to a greatly increased incidence of skin cancer and deleterious effects on both phytoplankton and zooplankton and the growth of plants in terrestrial habitats.

EDINGER, J. G., W. D. BONNER, and M. NEIBURGER. 1973. *Understanding Our Atmospheric Environment.* W. H. Freeman and Company, San Francisco.

GATES, D. M. 1972. *Man and His Environment: Climate.* Harper & Row, Publishers, New York.

HALACY, D. S., JR. 1968. *The Weather Changers.* Harper & Row, Publishers, New York.

LAMB, H. H. 1966. *The Changing Climate.* Methuen & Company Ltd., London.

LAMB, H. H. 1972. *Climate: Present, Past, and Future:* Vol. 1, *Fundamentals and Climate Now.* Methuen & Company Ltd., London.

LYNN, I. 1974. Whither weather? *Environmental J.,* February.

McGraw-Hill Staff. 1974. *Encyclopedia of Environmental Science.* McGraw-Hill Book Company, New York.

AIR POLLUTION

AHMED, A. K. 1975. Unshielding the sun—Human effects. *Environment* **17**(3), 6–14.

BARTELL, L. S., and C. L. RITZ. 1974. Stratospheric ozone destruction by man-made chlorofluoromethanes. *Science* **185,** 1163–1164.

CICERONE, R. J., S. WALTERS, and R. S. STOLARSKI. 1975. Chlorine compounds and stratospheric ozone. *Science* **188,** 378–379.

Edel, M. 1973. Autos, energy, and pollution. *Environment* **15**(8), 10–17.

EIGNER, J. 1975. Unshielding the sun—Environmental effects. *Environment* **17**(3), 15–18.

ESPOSITO, J. C. 1970. *Vanishing Air.* Grossman Publishers, Inc., New York.

FAITH, W. L., and A. A. ATKISSON, JR. 1972. *Air Pollution.* John Wiley & Sons, Inc., New York.

FRI, R. 1973. Statement to the Public on Clean Air and the Automobile. Environmental Protection Agency, Washington, D.C.

GROTH, E. III. 1975. Fluoride pollution. *Environment* **17**(3), 24–38.

HABER, G. 1974. The crumbling shield. *The Sciences* **14**(10), 21–24.

HAMMOND, A. L. 1975. Ozone destruction: Problem's scope grows, its urgency recedes. *Science* **187** (4182), 1181–1183.

HODGES, L. 1973. *Environmental Pollution.* Holt, Rinehart and Winston, New York.

HOROWITZ, J., and S. KUHRTZ. 1974. *Transportation Controls to Reduce Automobile Use and Improve Air Quality in Cities.* Environmental Protection Agency, Washington, D.C.

LEDBETTER, J. O. 1972. *Air Pollution—Part A: Analysis.* Marcel Dekker, Inc., New York.

LIKENS, G. E., and F. H. BORMANN. 1974. Acid rain: A serious regional environmental problem. *Science* **184**(4142), 1176–1179.

LILLIE, R. J. 1970. *Air Pollutants Affecting the Performances of Domestic Animals* (Agriculture Handbook 380). Agriculture Research Service, U.S. Department of Agriculture, Washington, D.C.

Mansfield, T. A., ed. 1976. *Effects of Air Pollutants on Plants.* Cambridge University Press, New York.

McClellan, G. S. 1970. *Protecting Our Environment* (The Reference Shelf, Vol. 42, No. 1.) H. W. Wilson Company, New York.

McGraw-Hill Encyclopedia of Environmental Sciences. 1974. McGraw-Hill Book Company, New York.

National Tuberculosis and Respiratory Disease Association. 1971. *Air Pollution Primer.* The Association, New York.

Reitze, A. W., Jr., and G. L. Reitze. 1975. Law. Living with Lead. *Environment* **17**(3), 2–3.

Sheldon, J. M., R. G. Lovell, and K. P. Mathews. 1967. *A Manual of Clinical Allergy,* 2nd ed. W. B. Saunders Company, Philadelphia.

Stapp, W. B., and J. A. Swan. 1973. *The Environment and the Citizen: Unit 12. Air Pollution,* University of Michigan, Ann Arbor, Mich.

Stern, A. C., H. C. Wohlers, R. W. Boubel, and W. P. Lowery. 1973. *Fundamentals of Air Pollution.* Academic Press, Inc., New York.

U.S. Department of Health, Education, and Welfare. 1970. *Air Pollution Injury to Vegetation.* Government Printing Office, Washington, D.C.

U.S. Department of Transportation. 1974. *1974 National Transportation Report Summary.* Washington, D.C.

U.S. Environmental Protection Agency. 1973. *Action for Environmental Quality.* Washington, D.C.

U.S. Environmental Protection Agency. 1972. *Noise Pollution.* Office of Public Affairs, Washington, D.C.

U.S. Environmental Protection Agency. 1973. *Environmental Health Effects of Pollution.* Washington, D.C.

Van Brachle, R. D. 1967. The farmer's stake in air pollution. *Crops and Soils Mag.,* October.

Wagner, R. H. 1978. *Environment and Man,* 3rd ed. W. W. Norton & Company, Inc., New York.

Wilkniss, P. E., J. W. Swinnerton, R. A. Lamontagne, and D. J. Bresson. 1975. Trichlorofluoromethane in the troposphere—Distribution and increase, 1971–1974. *Science* **187,** 832–833.

Agricultural Ecology and Pest Control
J. Donald LaCroix

Questions for Consideration

1. Why is corn the most productive grain?
2. What is the relative importance of the three major U.S. crops: corn, wheat, and soybeans?
3. How important is environment in the cultivation of a plant?
4. How does the nutritional value of corn, wheat, and soybeans compare?
5. What is the explanation for the meteoric rise in soybean production in the United States?
6. How would one proceed to increase a crop's yield per hectare (acre)? What are the prospects of doing so for the world's major foods?
7. What are the advantages and disadvantages of pesticides?
8. What is the ecological importance of persistent pesticides in the environment?

The world's present human population of approximately 4 billion people depends largely on a dozen major foods. Although the majority of these foods are grasses, of which corn and wheat are the primary crops, an equally important source of nutrition is a leguminous plant, soybean. Because of our great dependence on these cereal grains and legumes as sources of carbohydrates, fats, proteins, oils, vitamins, and minerals, this chapter will deal with various aspects of the culture, products, and pests of these three food-producing plants.

Corn

CLASSIFICATION

Indian corn or maize (*Zea mays*) is an annual, herbaceous, flowering monocotyledonous plant and a member of the grass family (Gramineae-Poaceae). Important relatives are wheat, oats, rice, barley, rye, sorghum, millet, sugar cane, and bamboo. Varieties of corn include the common sweet corn, popcorn, flint, flour, and dent corn, which differ in the consistency, shape, and content of their kernels (Figure 10-1).

ORIGIN

The ancestry of corn, as we know it today, is shrouded in mystery. A number of theories as to its beginnings have been set forth. Among them is the theory proposed by Paul Mangelsdorf and R. G. Reeves in 1939 that maize originated from the hybridization of wild gamagrass, *Tripsacum*, with a primitive pod-type corn called teosinte. Later unions between teosinte and maize produced better varieties of corn. As a result of new evidence, this hypothesis has fallen into disfavor. Now many people believe that teosinte itself, a wild grass found in Mexico, is a natural species and the progenitor of modern corn. Teosinte is a tall, robust annual, closely related to maize, in which the culms or shoots branch from the base of the plant and can attain a height of 1.8 to 4.8 meters (6 to 16 feet).

STRUCTURE

The mature corn plant is also a tall grass which can attain a height of 0.6 to 6.1 meters (2 to 20 feet), depending on the variety. The adult plant produces an extensive fibrous root system, as well as aerial prop roots which arise at the base of the stem or soil line. These adventitious roots may spread laterally or penetrate vertically; they serve to support the plant.

FIGURE 10-1.
Varieties of corn. Left to right: popcorn, sweet corn, flour corn, flint corn, dent corn, and pod corn. (Courtesy of the U.S. Department of Agriculture.)

The stalk of the plant, which supports the 10 to 18 leaves and the inflorescences, consists of a solid, jointed stem. The elongate, narrow leaf blades exhibit wavy margins, are alternately arranged, and are two-ranked. The leaf sheath arises from the node, surrounds the internode above, and gives rise to the ligule at its tip. The leaf, which is parallel-veined, bears a prominent midvein.

Although the corn plant itself is monoecious, the staminate or male flowers are contained in separate structures called tassels, or the terminal panicle; the pistillate, or female inflorescences, are located on the cobs, or ears, which are borne on short branches that emanate laterally from the corn stalk and are enclosed by a husk consisting of leafy bracts. Projecting from the open end of the cob is a cluster of fine threads or long styles, the corn silk, each of which is attached to an ovary, or immature corn grain. The grains are arranged in rows, perhaps as many as 30 per cob. This mature corn grain, or caryopsis, containing approximately 12 percent protein, is a simple, dry, indehiscent fruit and consists structurally of the hull, aleurone layer, endosperm, and embryo (see Chapter 3).

CULTIVATION

The availability of the world's food supply has always been regulated by climate, and adverse climatic conditions have had devastating effects on the kinds and amounts of food harvested. Our current world food shortage can be blamed, in part, on drought, frost, excessive rains, and typhoons that struck much of the world's cropland in 1972. Also, during the last 35 years, there has been a very gradual cooling trend, resulting in a lowering of the mean global surface temperature by approximately 0.55°C, which has the effect of shortening the growing season by as much as 10 days (see Chapter 9).

Air and soil temperature and precipitation are the two most important weather variables affecting the growth of corn. Germination of sweet corn will take place in 4 to 7 days at a temperature of 20 to 30°C. A mean day temperature of 24°C and a night temperature of not less than 14°C result in optimum growth. An average annual precipitation of 50.8 centimeters (20 inches) is ideal, with the greatest demand occurring during the period of accelerated growth.

The sowing of seed 2.5 to 10 centimeters (1 to 4 inches) deep in a well-drained, loamy soil that contains sufficient nitrogen, and the judicious use of pesticides, will encourage maximum yield. Depending upon various factors, such as variety and environmental conditions, the growing season for corn varies from slightly less than 100 days up to 180 days.

The corn belt of the United States, a midwestern region extending from Ohio westward through Indiana, Illinois, Iowa, Missouri, Kansas, Nebraska, and the Dakotas, is especially well suited to the production of corn, since climatic conditions of temperature and moisture and soil characteristics are ideal for high yields.

PRODUCTION AND USES

Areas of corn production are widely scattered both in the United States and around the world. However, because of more favorable environmental conditions in specific geographic areas, production is more concentrated in those areas.

With the development of hybrids, corn has become the most productive of grains, and the United States ranks first in its production. The average annual United States production for the 3-year period 1972 to 1974 was 5,295,000,000 bushels, with an average yield per acre of 86.6 bushels grown on 61,557,000 acres. The principal producing states in order of production in 1974 were Iowa, Illinois, Indiana, and Nebraska. Those states produced 54.8 percent of the crop. Forty years ago, for the 3-year period 1932 to 1934 the average annual yield was 2,615,713,333 bushels,[1] with an average yield per acre of 24.7 bushels.

The average annual world production for the period 1967 to 1974 was 290,925,000 metric tons,[2] of which the United States produced 130,968,000 metric tons, or approximately 45 percent of the total crop. For the year 1974, world production was 279,200,000 metric tons, while U.S. production was 118,100,000 metric tons or 42.3 percent of the crop. In 1974, the principal producing countries in order of production after the United States, were the People's Republic of China (FAO estimate), Brazil, USSR, South Africa, Argentina, France, Yugoslavia, and Mexico. Although the United States still produces vast quantities of corn, world production trends have been altered over the years (Figure 10-2).

Increased agricultural productivity is due to several factors, among which are improvements in crop breeding (e.g., the development of hybrid corn), more extensive use of fertilizers, pesticide applications, and better mechanization.

For the period 1972 to 1974, approximately 76 percent of the corn produced in the United States was used domestically, while the other 24 percent was exported. In many parts of the world, corn is consumed directly by the people, while in the United States a large percentage of the crop is only indirectly eaten by the population.

The greatest amount of corn grown in this country is utilized as a food source for livestock: hogs, cattle, sheep, horses, poultry, and other animals consume the grain, as well as the remainder of the plant, which is used as fodder and silage. The direct human consumption of corn and the utilization of corn and its by-products by industry account for the remainder of U.S. production of corn that is not exported.

The per capita consumption of cornmeal, cornstarch, corn cereal, and hominy is approximately 7.3 kilograms (16 pounds) per year, while corn syrup and sugar per capita consumption has increased over the years to approximately 12.7 kilograms (28 pounds) per year.

Continued support for research has resulted in the development of corn containing higher protein, and the discovery of mutants with

---

[1] 1 bushel = 2150.42 cubic inches = 35.24 liters.
[2] 1 metric ton = 1000 kilograms = 2204.6 lbs = 1.111 short tons.

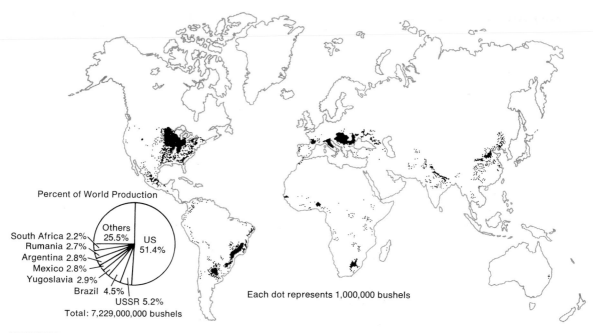

Percent of World Production

South Africa 2.2%
Rumania 2.7%
Argentina 2.8%
Mexico 2.8%
Yugoslavia 2.9%
Brazil 4.5%
USSR 5.2%

Others 25.5%
US 51.4%

Total: 7,229,000,000 bushels

Each dot represents 1,000,000 bushels

FIGURE 10-2.
World corn production, average, 1957 to 1961. (From A Graphic Summary of World Agriculture, *U.S. Department of Agriculture Misc. Publ. 705.)*

increased starch content and higher amounts of the amino acids lysine and tryptophan. These achievements have strengthened the already eminent status of the corn plant as a major food crop.

PESTS OF CORN

Corn has long been susceptible to both animal (insect) and plant (fungal) pests. Fortunately, many of these organisms, such as the corn borer, can now be controlled by the cautious use of pesticides and by breeding varieties of corn that are resistant to many diseases.

Some common fungal diseases, affecting all parts of the corn plant, are listed in Table 10-1.

Wheat

CLASSIFICATION

Common, cultivated wheat, *Triticum aestivum,* is another annual grass belonging to the family Gramineae (Poaceae). It is a hexaploid, having 21 pairs of chromosomes. On the basis of the number of chromosomes, the cultivated species are classified into three groups: diploid (*T. monococcum,* einkorn wheat), tetraploid (*T. dicoccum,* emmer wheat; *T. durum,* durum wheat; *T. persicum,* Persian wheat; *T. polonicum,* Polish wheat; *T. turgidum,* poulard wheat), and hexaploid (*T. compactum,* club wheat; *T. spelta,* spelt wheat plus *T. aestivum*). Closely related wild species such as goat grass (*Aegilops speltoides* and *A. squarrosa*) will hybridize with

TABLE 10-1 / Common Fungal Diseases of Corn

Name of Disease	Geographic Distribution	Causal Organism	Site of Infection	Control
Brown spot	Southeastern United States	*Physoderma zea—maydis*	Blade, sheath, culm	Resistant varieties
Diplodia disease	Southern half of corn belt and worldwide	*Diplodia maydis, D. macrospora*	Seedling, root, stalk, ear	Seed treatment with captan; resistant varieties
Downy mildew	Tropic and subtropic	*Sclerospora graminicola*	Leaves, inflorescence	Resistant varieties
Seedling blight; root, stalk, ear rot	East, central, southern United States and worldwide	*Gibberella roseum, G. moniliforme*	Grain, seedling, root, stalk, ear, husk	Sanitation, seed protectant, resistant varieties
Corn smut	Worldwide	*Ustilago maydis*	Seedlings, stem, leaf, axillary bud, ear, male flower	Resistant varieties
Root, stem, stalk rot	Worldwide	*Pythium* spp., *Phytophthora* spp.	Root, stem, stalk, petiole	Seed protectant, resistant varieties
Corn blight	Southern United States	*Helminthosporium maydis*	Leaf	Unknown

species of *Triticum* and result in the production of polyploids. Wheat also hybridizes with rye, *Secale cereale,* to produce triticale, a new grain, richer in protein than either parent and used for livestock feed.

ORIGIN

Wheat, from which is made our "staff of life," bread, has long been of great importance and value in the nutrition of mankind. It is one of the earliest known cultivated crops. Evidence of wild wheat has been found in excavations in the Near East, in the vicinity of Turkey, that date back 9000 years. Those primitive wheats were vastly different from the polyploid and disease-resistant varieties cultivated today, and were probably utilized as food, hull and all, in the form of a thin liquid, some of which was allowed to ferment before consumption. The evolution of the wild wheat into the later cultivated types probably was the consequence of many natural crossings, with resultant strains having chromosome numbers 14, 28, and 42. From its point of origin in the Near East, wheat was transported to India, China, Afric, Europe, and the Americas by the migration of those peoples, and it continues as an important and basic source of food.

The wheat seedling produces a seminal (seedling) root system consisting of six to eight slender branching roots. The adult plant, which may attain a height of from 0.6 to 1.2 meters (2 to 4 feet), is supported by an extensive, secondary fibrous root system. The secondary root system spreads laterally and penetrates so deeply into the soil that it may measure twice the combined length of the aerial portions of the plant.

The stem or culm, which produces lateral shoots called tillers at the base, is erect and is jointed, solid at the nodes, and usually hollow in the internode. Approximately six two-ranked foliage leaves arise alternately. Each foliage leaf consists of the sheath, which encloses the culm, the elongate, parallel-veined blade, which is 1 to 2 centimeters ($\frac{1}{2}$ to 1 inch) wide, and the ligule at the juncture of the blade and sheath.

The inflorescence of wheat is terminally produced and is called a spike. It is usually 5 to 12 centimeters (2 to 5 inches) long and consists of 15 to 20 solitary, sessile, awned or awnless, alternately arranged spikelets, each of which bears one to nine flowers, called florets. The usual number of mature grains or kernels produced per spikelet is two.

As in corn, the mature ovary or fruit is called a grain or caryopsis. It consists of the husk or bran, which is rich in carbohydrates, vitamins, and minerals; the aleurone layer of protein and phosphorus; the massive endosperm containing carbohydrates, primarily starch; and the embryo, composed of carbohydrates, fats, protein, vitamins, and minerals, which constitutes only about 6 percent of the entire grain (Figure 10-3). Small-grained wheats are higher in protein content, containing approximately 13 to 16 percent protein, and are called "hard wheats." The large-grained wheats are lower in protein, 8 to 11 percent protein, but are higher in starch content and are termed "soft wheats."

CULTIVATION

Important factors affecting the growth of wheat are temperature, precipitation, humidity, and soil composition. Although requirements vary somewhat depending on varietal differences, wheat is adaptable to a broad range of environmental conditions.

Optimum growth occurs in a cool, dry climate with moderate rainfall. Those conditions having the most deleterious effect on growth are high temperature, excessive rainfall, and humidity; indeed, annual precipitation must be in the range 25 to 76 centimeters (10 to 30 inches).

Sowing time depends upon the type of seed being planted. Spring wheat, planted in the spring and harvested in late summer, requires a growing season of not less than 90 to 100 days. Winter wheat, on the other hand, is sown in the fall, harvested in early summer, and is less susceptible to disease, thereby giving a higher yield.

The seed is planted 2.5 to 8.0 centimeters (1 to 3 inches) deep in a clay, loam, or slightly sandy soil that contains adequate amounts of nitrogen, phosphorus, potassium, sulfur, and calcium. The formation of tillers and

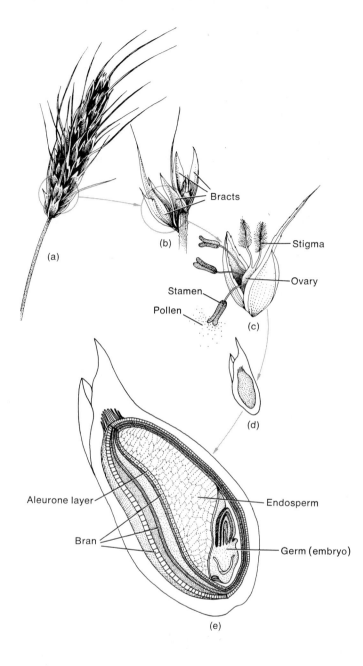

FIGURE 10-3.
*Flowers and grain of wheat:
(a) spike with spikelets;
(b) spikelet consisting of
three flowers; (c) parts of
single flower; (d) developing
ovary or fruit; (e) mature
ovary, fruit, or grain. [Parts
(a) to (d) from "Hybrid
Wheat," by Byrd C.
Curtis and David R.
Johnston. Copyright © 1969
by Scientific American,
Inc. All rights reserved.
Part (e) from the Wheat
Flour Institute.]*

secondary roots depends on the depth of planting, with fewer tillers developing when the seed is planted deeply.

Because wheat can adapt itself to such a wide range of growing conditions, it is one of the most widely cultivated crops in existence. Major areas of cultivation include the central, southern, northern, and northwestern United States; western Canada; Argentina; the Mediterranean countries; northwestern Europe; India; Pakistan; the southern USSR; the People's Republic of China; and southern Australia.

The average annual U.S. production of wheat, for the 3-year period 1972 to 1974 was 1,682,000,000 bushels, with an average yield per acre of 30.6 bushels grown on 55,589,000 acres. The principal wheat-producing states, in order of production, are Kansas, North Dakota, Oklahoma, and Washington. These states produced 43.7 percent of the U.S. crop in 1974.

The average annual world wheat production for the period 1972 to 1974 was 356,633,000 metric tons, and the United States produced 45,008,000 metric tons, or approximately 12.6 percent of the total world yield. For the year 1974, world production was 346,800,000 metric tons, while U.S. production was 46,400,000 metric tons, or 13.4 percent of the crop. In 1974, the principal producing countries in order of production were the USSR, the United States, the People's Republic of China (FAO estimate), India, France, Canada, Australia, Turkey, Italy, West Germany, and Pakistan. Indeed, during the last decade, the major wheat-producing nations of the world have remained relatively unchanged, except that India and Pakistan have assumed a more important role in the production of this essential grain (Figure 10-4).

Because of the research efforts of Norman E. Borlaug, which resulted in new, high-yielding strains of wheat, the application of higher amounts of fertilizer and pesticides, and the development of disease-resistant varieties, worldwide wheat yields have increased dramatically. This increase in yield, however, is accompanied by other problems, brought about by the excessive use of fertilizers and pesticides (see Chapter 11). When these difficulties associated with the production and planting of hybrid wheat can be resolved, the "King of Grains" is certain to retain its stature.

FIGURE 10-4.
World wheat production, average 1957 to 1961. (From A Graphic Summary of World Agriculture, *U.S. Department of Agriculture Misc. Publ. 705.)*

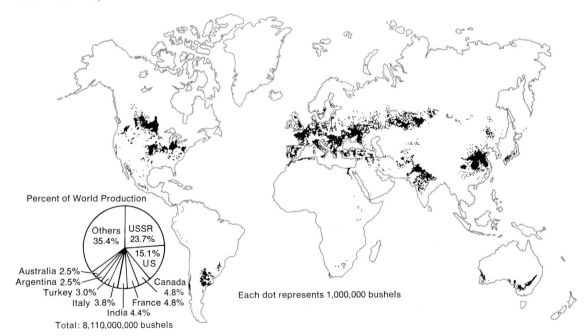

Percent of World Production

Others 35.4%
USSR 23.7%
15.1% US
Australia 2.5%
Argentina 2.5%
Turkey 3.0%
Italy 3.8%
India 4.4%
France 4.8%
Canada 4.8%
Total: 8,110,000,000 bushels

Each dot represents 1,000,000 bushels

Much of the wheat produced is consumed by humans, although some is used as animal feed. Most of the product (endosperm) of the protein-rich hard wheats is converted into flour for making bread. Although the flour from soft wheats, which contain more starch, is also used for human foodstuffs, soft wheats are, in large part, fed to livestock.

Although it has declined in the last 30 years, the per capita consumption of wheat flour is approximately 48.08 kilograms (106 pounds) per year, while the per capita consumption of wheat breakfast cereals has remained steady at approximately 3 pounds per year.

PESTS OF WHEAT

Susceptibility to the wheat rust fungus *Puccinia graminis* has long been a problem for the farmer. Plant breeders are constantly attempting to develop varieties of wheat that will prove continually and completely resistant to this pest. Although wheat rust is, by far, the most serious enemy of wheat, it is not the only threat to bountiful harvests (Table 10-2).

CLASSIFICATION

Soybean

Soybean (*Glycine max*) is a member of the very large and economically important legume family (Fabaceae), a group that includes annuals, perennial herbs, vines, shrubs, and large trees. Familiar legume relatives of soybean are beans, peas, sweetpeas, alfalfa, clover, peanut, honey and black locust, and redbud trees.

ORIGIN

Soybean originated in southeastern Asia and has been cultivated in the Orient for many centuries, where it was, and still is, an extremely versatile and valuable source of food. It is believed that *G. max*, the cultivated soybean, is descended from the wild form, *G. ussuriensis*, which exhibits a long and twining stem. The soybean was introduced to Europe in the seventeenth century and to the United States in 1804. Despite its relatively late entry into world food markets, it has become so economically important that it now ranks as one of the most economically rewarding crops for the farmer.

STRUCTURE

The soybean plant is a small, erect, bushy-branched annual which attains a height of 0.9 meters (3 feet). Occasionally, it tends to twine and be prostrate. The short or elongated branches may be spreading or ascending from a well-defined main stem.

TABLE 10-2 / Common Fungal Diseases of Wheat

Name of Disease	Geographic Distribution	Causal Organism	Site of Infection	Control
Wheat rust	Worldwide	Puccinia graminis	Leaf, culm, spike, grain	Resistant varieties, eliminate alternate host
Browning root rot	Worldwide	Pythium spp.	Root	Seed protectant, resistant varieties
Bunt (stinking smut)	Worldwide	Tilletia caries, T. foetida	Spike, grain	Seed treatment, resistant varieties, cultural practices
Flag smut	Worldwide	Urocystis tritici	Seedlings, culm, leaf, seed	Resistant varieties, shallow planting and wet soil
Head blight, seedling blight, foot rot	Worldwide	Gibberella roseum	Spikelet, grain, embryo, endosperm, seedling internodes, roots	Sanitation, seed protectant
Leaf rust	Southern central U. S. plains	Puccinia recondita	Leaf	Unknown
Loose smut	Worldwide	Ustilago tritici	Spike	Resistant varieties, hot-water treatment of seed, solar treatment of seed
Powdery mildew	Worldwide	Erysiphe graminis	Aerial structures	Resistant varieties

The pinnately compound leaves are alternately arranged on the stem, and there is much variation in their shape, size, and color. The leaves usually abscise prior to fruit maturity.

The small, inconspicuous white or purple flowers, usually 8 to 16 in a cluster, are borne on short axillary or terminal inflorescences called racemes. The flower, which is characteristic of the legume family, is self-fertile.

Each flower produces a fruit called a pod or legume. This fruit is pendulous, long-stalked, 4 to 8 centimeters ($1\frac{1}{2}$ to 3 inches) long, and varies in color from light to dark green. Each legume contains two to three seeds, and in most varieties they are unicolored. The pods, leaves, and stems of soybean are covered with hairs (Figure 10-5).

CULTIVATION

Since the climatic requirements of soybeans and corn are practically identical, it is not surprising that the corn belt is one of the major areas of soybean cultivation in the United States, but acreage also exists in the lower Mississippi valley and the Southeast (Figure 10-6).

Although warm, summer temperatures and rain are desirable for optimum growth of soybeans, the soil requirements are less rigid, and can range from the preferred warm sandy loams, which are well drained, to clay loams, which are not well drained.

Seeds may be sown in early spring, about the same time as corn, for those varieties with a long maturation period. Depth of planting in sandy soils can vary between 5 and 10 centimeters (2 to 4 inches), but in heavier clay-type soils, it should not exceed 5 centimeters. Germination time is 5 to 8 days at a temperature of 20 to 30°C. Soybeans are not tolerant of drought and frost, and some may not react as expected to fertilizers, although high-nitrogen fertilizers and preemergence herbicides are applied to attain top yields.

PRODUCTION AND USES

Soybeans can be cultivated under quite variable conditions, but optimum production occurs in temperate, rather than tropical or subtropical, climates.

Over the last 25 to 30 years, soybeans have become a crop of major importance in the United States. Since 1956, the increase in acreage harvested, production, and farm value has been phenomenal. During the period 1956 to 1960, the average acreage harvested, production, and farm value were 22,351,000 acres,[3] 520,000,000 bushels, and $1,075,000,000, respectively. In 1974, these figures rose to 52,368,000 acres, 1,215,000,000 bushels, and $8,070,000,000, respectively. The principal soybean-producing states in 1974, in order of production, were Illinois, Iowa, Indiana, and Missouri. In the United States, the three crops, in order of value that year, were corn, soybeans, and wheat. For that same year, 1974, soybeans were a major crop in 20 states and the principal crop in Missouri, Tennessee, Alabama, and Arkansas.

World production of soybeans in 1974 was 51.5 million metric tons, of which the United States produced 33.6 million metric tons, or 65.2 percent. Estimated U. S. exports were 14.0 million metric tons, or 41.7 percent of the total produced. From 1960 through 1974, world production increased from 30 to 63 million short tons.

The tremendous growth in world soybean production was paralleled by a significant increase in U. S. production, and this was due principally to the harvest in the United States; increased yield in other major soybean-producing countries, such as China, was minimal.

[3] 1 acre = 0.405 hectares or 1 hectare = 2.471 acres.

Soybeans have been utilized as food for many centuries in China and Japan, and they are now valued as an important source of nutrition in other parts of the world. Soybeans are extremely rich in protein, containing in excess of 40 percent. For this reason, they have broad application, being used as a source of food for humans and livestock, to supplement the protein content of foods naturally low in this important substance, and as an extender for meat and meat products. Soybeans, which contain 18 to 20 percent fats and oils, provide an extremely valuable source of these substances. They are utilized by both the food industry and in the manufacture of such diverse products as linoleum, plywood, soaps, inks, caulking compounds, paints, and varnishes. The vegetable oil from soybeans is of an extremely high quality and is finding ready acceptance in such products as margarine, cooking oil, and shortening. Indeed, since 1950, the per capita consumption of margarine and shortening has risen from 6.1 to 11.3 and 11.0 to 17.0 pounds,[4] respectively.

Unlike corn and wheat, however, yield per acre of soybeans has not seen a dramatic increase. In 1974, it was only 23.2 bushels as compared to 20.0 bushels per acre in the early 1950s. Successful growth and production of protein-rich soybeans in other parts of the world, and increasing yield per acre through hybridization or pollination by honeybees, might help relieve the world food crisis for 4 billion people.

PESTS

Many species of insect pests injure soybeans, small plants being especially susceptible; however, they can be controlled with insecticides and good farming practices. Some common rot diseases caused by fungi are listed in Table 10-3.

Pest Control

Because millions of people in the world are so dependent on such crops as corn, wheat, and soybeans, the careful use of pesticides is a necessity. Although the use of these chemicals is not solely responsible, constant increases in production, acreage harvested, yield per acre, farm value, and ability to export huge quantities of food to hungry people have been made possible because of their utilization.

People have long used chemicals to control undesirable plants—algae, fungi, weeds; and animals—nematodes, insects, ticks, mites, rodents. As is so often the case, however, the development and use of a beneficial product also produces harmful effects. In many ways, our environment, our homes as well as the outdoors, has been improved through the use of pesticides. At a time when millions of people are starving to death, pesticides can be effectively utilized to control plant diseases and insects that annually destroy billions of dollars worth of crops, grain, and livestock. Pesticides can also be beneficial to mankind in that they are

---

[4] 1 pound = 453.6 grams = 0.454 kilograms.

TABLE 10-3 / Common Fungal Diseases of Soybeans

Name of Disease	Geographic Distribution	Causal Organism	Site of Infection	Control
Root rot	Midwest United States and Canada	*Phytophthora megasperma*	Root, stem, leaves	Cultural practices, resistant varieties
Black root rot	Midwest United States and Canada	*Thielaviopsis basicola*	Root, underground stem	Resistant varieties
Seed and seedling rot	Midwest United States	*Fusarium* spp.	Seed, seedling	Unknown

helpful in the control of many injury-causing insects and in the prevention of the growth of undesirable organisms in certain commercial products. However, evidence continues to mount that the flagrant and irresponsible use of pesticides and their indiscriminate distribution into the air, soil, and water can be very injurious to plants, people, other animals, and the environment; therefore, other methods of pest control must be utilized (see Chapter 18).

CLASSIFICATION OF PESTICIDES

A pesticide is any substance that controls or kills a pest. The pests may be lower plants, weeds, nematodes, insects, mites, rodents, or some other destructive organism.

The principal types of pesticides are: *acaricides*—control mites and spiders that feed on plants and animals; *fungicides*—control plant diseases caused by fungi; *herbicides*—control weeds or undesirable seed plants; *insecticides*—control insects; *nematocides*—control plant-destroying soil nematodes or eel worms; and, *rodenticides*—control mice, rats, and other rodents.

TYPES AND USES

FUNGICIDES

A great deal of recent research has dealt with herbicides and insecticides rather than with fungicides, although the use of fungicides as protectants has been steadily increasing since the latter half of the nineteenth century. The discovery of two major fungal diseases of plants, the late blight disease of the potato in Ireland in 1845 and 1846, caused by *Phytophthora infestans,* and a downy mildew disease of grape, caused by *Plasmopara viticola* in the vineyards of France in 1878, accelerated interest in fungicides. Furthermore, efforts to control mycotic infections of the hair, skin, and mucous membranes, and systemic mycotic infections of animals and human beings caused by species of imperfect fungi such as *Microsporum, Trichophyton,* and *Epidermophyton,* also demonstrated that the development of effective fungicides was essential.

Some important fungicides include the following: Bordeaux mixture, which is a mixture of copper sulfate, lime, and water; captan; pentachlorophenol; cycloheximide; maneb; and methyl mercury dicyandiamide.

Although the U.S. production of fungicides in 1970 was only approximately 90.72 million kilograms (200 million pounds), and represents less than 20 percent of total pesticide production, fungicides are used extensively as foliar protectants, for seed treatment, as turf sprays, and as soil fumigants to control "damping-off" of seedlings caused by such soilborne fungi as species of *Pythium* and *Rhizoctonia*. Fungicides also control the growth of fungi when propagating plants from corms, cuttings, and lily bulb scales.

HERBICIDES

Although inorganic compounds (ammonium sulfamate, potassium cyanate, sodium arsenite, sulfuric acid) and organic compounds have been used to kill plants for years, the pesticide industry was given great impetus by the discovery of 2,4-D (2,4-dichlorophenoxyacetic acid) in 1944. When used at low concentrations of 500 to 1000 parts per million, it acts as a selective herbicide, killing many undesirable broad-leaved dicotyledonous plants while leaving intact desirable narrow-leaved monocotyledonous plants. However, complete plant kill can be achieved with 2,4-D at higher concentrations. It also exhibits a characteristic of plant hormones since, when applied to a plant at a very low concentration, it behaves as an auxin, resulting in the promotion of root growth and cell enlargement. A compound closely related to 2,4-D, yet more effective, is 2,4,5-T (2,4,5-trichlorophenoxyacetic acid).

Other well-known herbicides are dalapon, MCPA (2-methyl-4-chlorophenoxyacetic acid), MH (maleic hydrazide), and TCA (trichloroacetic acid).

The control of weeds by the use of herbicides can be effected by preplanting treatments (application before the crop is planted), preemergence treatments (application before the crop emerges) and postemergence treatments (application after the crop emerges). Herbicides have been of inestimable value in agriculture, home gardening, and right-of-way clearing. The ability of 2,4-D to selectively kill broad-leaved species with no side effects on narrow-leaved species, such as grasses, and of dalapon and TCA to kill narrow-leaved plants without damaging broad-leaved plants, has been essential to the farmer. Weed control in home gardens and lawns has helped individual homeowners in their efforts to combat high food costs by "growing their own," and in their effort to improve and maintain the appearance of the immediate neighborhood. Using herbicides to clear the right-of-way along highways to improve visibility has undoubtedly saved many lives. Herbicides, such as 2,4-D, 2,4,5-T, and cacodylic acid have been misused for highly destructive purposes in Vietnam.

INSECTICIDES

Protecting people and their crops from insect infestations has long been one of our major concerns, and the use of both organic and inorganic chemicals has played a vital role in the control of such pests.

Sucking insects, such as aphids, can be controlled with contact insecticides, which are absorbed by the insect directly through the body covering or through the respiratory or sensory system. The famous, or infamous, DDT (dichlorodiphenyltrichloroethane), developed during World War II, is a prime example of a contact insecticide. It was used with much success to control typhus and malaria. Stomach insecticides, which are ingested by the animal, are employed to control biting or chewing insects; included in this group are the arsenic and fluorine products.

Among the contact insecticides are the chlorinated hydrocarbons (also called organochlorine compounds or organochlorides) and the organophosphates (or organophosphorus compounds). Chlorinated hydrocarbons include aldrin, chlordane, DDT, dieldrin, heptachlor, lindane, and methoxychlor. Common organophosphates are malathion and parathion. It should be stated, however, that the Environmental Protection Agency has banned aldrin, DDT, and dieldrin, and suspended chlordane and heptachlor.

The utilization of DDT and other pesticides has increased agricultural productivity and resulted in larger crop harvests, thus providing food for millions of people who would otherwise have starved to death. The number of deaths attributable to such serious diseases of humans as malaria and yellow fever has also been drastically reduced by the use of DDT to control the causative insects.

Summary

Corn, wheat, and soybeans are vital to the world as a source of food both for humans and for livestock. The development of hybrids and better-yielding strains, an increase in the nutrient content of these food crops, greater disease resistance, better farming practices, and the utilization of pesticides have resulted in gigantic increases in their production. Although the misuse of pesticides has caused untold damage in the ecosystem, these accomplishments, together with a realization of the importance of the environment to these and other crop plants, offer some hope to a hungry world.

Selected References

BROWN, A. W. A., T. C. BYERLY, M. GIBBS, and A. SAN PIETRO, eds. 1975. *Crop Productivity—Research Imperatives.* Michigan State University Agricultural Experiment Station, East Lansing, Mich.

DOVRING, F. 1974. Soybeans. *Sci. Amer.* **230**, 14–21.

EDWARDS, C. A. 1974. *Persistent Pesticides in the Environment.* CRC Press, Inc., Cleveland, Ohio.

Food and Agriculture Organization. *Production Yearbook,* Rome.

GOULD, R. F. 1966. *Organic Pesticides in the Environment* (Advances in Chemistry Series No. 60). American Chemical Society, Washington, D. C.

HEISER, C. B. 1973. *Seed to Civilization.* W. H. Freeman and Company, San Francisco.

Hitchcock, A. S. 1950. *Manual of the Grasses of the United States* (USDA Misc. Publ. No. 200). 2nd ed., revised by Agnes Chase.

Inglett, G. E., ed. 1970. *Corn: Culture, Processing, Products.* AVI Publishing Company, Westport, Conn.

Janick, J., R. W. Schery, F. W. Woods, and V. W. Ruttan. 1974. *Plant Science: An Introduction to World Crops.* W. H. Freeman and Company, San Francisco.

Lewert, H. V. 1976. *A Closer Look at the Pesticide Question.* The Dow Chemical Company, Midland, Mich.

Lukens, R. J. 1971. *Chemistry of Fungicidal Action.* Springer-Verlag, New York.

Mangelsdorf, P. C. 1974. *Corn: Its Origin, Evolution and Improvement.* Harvard University Press, Cambridge, Mass.

Matsumura, F. 1975. *Toxicology of Insecticides.* Plenum Publishing Corporation, New York.

O'Brien, R. D. 1967. *Insecticides: Action and Metabolism.* Academic Press, Inc., New York.

Peterson, R. F. 1965. *Wheat.* John Wiley & Sons, Inc. (Interscience Division), New York.

Pierre, W. H., S. R. Aldrich, and W. P. Martin, eds. 1964. *Advances in Corn Production.* Iowa State University Press, Ames, Iowa.

Quisenberry, K. S., and L. P. Reitz, eds. 1967. *Wheat and Wheat Improvement* (Monograph 13). American Society of Agronomy, Madison, Wisc.

Reitz, L. P. 1970. New wheats and social progress. *Science* **169,** 952–955.

Scott, W. O., and S. R. Aldrich. 1970. *Modern Soybean Production.* The Farm Quarterly, Cincinnati, Ohio.

Siegler, D. S., ed. 1977. *Crop Resources.* Academic Press, Inc., New York.

Sprague, G. F., ed. 1955. *Corn and Corn Improvement.* Academic Press, Inc., New York.

U.S. Department of Agriculture. *Agricultural Statistics.*

U.S. Department of Agriculture. 1948. *Grass.* Yearbook of Agriculture.

Whyte, R. O., G. Nilsson-Leissner, and H. C. Trumble. 1953. *Legumes in Agriculture.* Food and Agriculture Organization, Rome.

Wolf, W. J., and J. C. Cowan. 1975. *Soybeans as a Food Source.* CRC Press, Inc., Cleveland, Ohio.

11 The Pros and Cons of the Green Revolution
Peter B. Kaufman

Questions for Consideration

1. What have been the primary social and economic consequences of the green revolution for peoples in Mexico, India, the Philippines, and southeastern Asia?
2. What are some of the alternatives to the green revolution in underdeveloped countries?
3. What has wiped out nearly all gains in crop yields achieved with green revolution crops, such as wheat in Mexico and rice in the Philippines?
4. How has a country like China achieved self-sufficiency in production of most of its crops since 1971, in spite of its huge population of over 800 million people?

One of the most important "happenings" in agriculture during the past 25 to 30 years has been the occurrence of the "green revolution." This "revolution" refers to the very substantial yield increases obtained by plant breeders resulting from the development of new crop varieties. Such "breakthroughs" have been especially notable with corn (*Zea mays*), wheat (*Triticum durum*), and rice (*Oryza sativa*). In this chapter, we are going to trace how these developments occurred, examine their impact on agricultural practices in countries other than the United States, and look at the pros and cons of the green revolution itself. Needless to say, there has been much publicity given to the "marvels" of the green revolution! There are negative aspects as well. Any account of the green revolution would be incomplete without presenting both sides of this "happening" and its short-term and long-range effects on agricultural ecology, food production, and population levels.

What Is the Green Revolution?

Five major crops have contributed primarily to the green revolution. They are sugarcane, corn, wheat, rice, and soybeans. The first breakthrough occurred in the 1920s with the production by plant breeders of new, high-yielding varieties of sugarcane (*Saccharum officinarum*). Then, using highly inbred parents, plant geneticists in the United States developed new strains of hybrid corn, which gave spectacular increases in yield. The methods by which they produced hybrid corn are illustrated in Figure 10-19 of Janick et al. (1969). The basic phenomenon involves the use of highly inbred parental lines. These are crossed in a double-cross protocol. The resultant hybrid produces very large ears with many more kernels (Figure 11-1). This phenomenon is called *heterosis*, or hybrid vigor. In 1964, a corn variety called Opaque 2 was introduced by plant breeders (Figure 11-2). This particular variety, in contrast with other varieties of corn, has a much higher content of an essential amino acid, lysine. Hence, Opaque 2 is called a lysine-rich corn variety. Its introduction, as with high-lysine sorghum in 1973, has been of major significance in helping to provide more and better-quality protein from carbohydrate-rich crops.

The widespread use of hybrid corn signaled the beginning of a trend toward a monoculture type of agriculture in the United States. Monoculture refers to the practice of farmers planting only a few varieties of a given crop instead of many diverse types of varieties of the crop; these few varieties are typically planted year after year over large acreages. The implications of monoculture are important to understand. One recent case history is especially helpful in this connection. In 1970, the U.S. corn crop was nearly decimated by the southern corn blight disease, caused by the fungus *Helminthosporium maydis*. In this instance, corn with "Texas male-sterile cytoplasm" had been selected and was widely used because of the ease of breeding it (Strobel, 1975). However, such corn proved to

FIGURE 11-1.
Hybrid dent corn (left and right). In the center is an ear from an inbred line of corn. (Courtesy of Erich Steiner, Department of Botany, University of Michigan.)

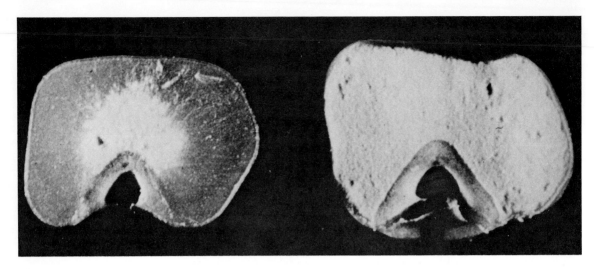

FIGURE 11-2.
Left is a kernel of ordinary corn; it has a yellow color in the endosperm tissue, making the kernel appear yellow. Right is a kernel of Opaque 2; it has a white appearance in the endosperm tissue. This is a high-lysine cultivar of corn. (Courtesy of Erich Steiner, Department of Botany, University of Michigan.)

be especially susceptible to a single race of the southern corn blight fungus, which nearly wiped out the crop in a single year. Such a story may be repeated in the future with some other crop monocultures which are products of the green revolution. This, indeed, is the prediction of a committee of the National Academy of Sciences. They conclude that the high-yielding semidwarf wheats developed in Mexico (see discussion of these below) are among those varieties which represent "extreme potential genetic vulnerability" (Wade, 1974).

After the development of hybrid corn, the next "break" in the green revolution occurred in Mexico, starting in 1943. At that time, the Rockefeller Foundation established the International Maize and Wheat Improvement Center. By 1954, under the leadership of Norman Borlaug, new, high-yielding varieties (HYV) of wheat were developed and were widely disseminated for planting in Mexico. These wheat varieties are semidwarf (and thus more sturdy) and produce immense yields compared with the other taller wheat varieties used earlier. As a result, Mexico, which was once importing one-third of its wheat supply, began to export wheat in the 1960s.

A comparable breakthrough occurred with rice in the Philippines and India. Work by plant breeders at the International Rice Research Institute (IRRI) at Los Baños in the Philippines, sponsored by the Ford and Rockefeller Foundations, resulted in the introduction of IR-8 and other new rice varieties. These new varieties are characterized by maturing earlier, making it possible to grow more than a single crop in a year in some areas. They also have stiffer straw and hence do not lodge as easily. They produce more grains per head (flower shoot or inflorescence; see Figure 11-3), giving far greater yields than the traditional varieties. The other feature about IR-8 rice, often not cited, is that its protein content is *less* than that of the earlier-used traditional rice varieties. As with wheat, the new HYV rice requires more fertilizer.

One consequence of the development of new rice varieties is that a handful of HYVs can replace literally hundreds of traditional rice varieties. In Bangladesh, the latter number up to 1200; in Indonesia, up to 600; in India, several thousand. However, in spite of HYVs lending

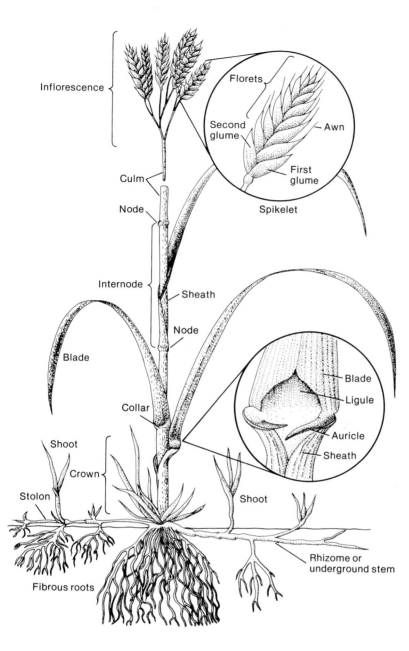

FIGURE 11-3.
Basic morphology of a grass shoot. In the text, we refer to the flowering portion, here labeled the inflorescence. (Courtesy of Erich Steiner, Department of Botany, University of Michigan.)

themselves to mechanized harvesting, packaging, and marketing, they provide much greater risks to the farmer, both financially and culturally. The monocultures that result might be wiped out by epidemic insect or disease attacks, as cited above for the southern corn blight.

Almost at the same time that HYV rice and wheat made their appearance, the production of another crop, soybeans, began to increase dramatically in the United States (Figure 11-4). In 1930, about 0.405 million hectares (1 million acres) were planted to soybeans, with a yield of about 14 million bushels.[1] In 1973, 56 million acres were planted; the

[1] 1 bushel = 2150.42 cubic inches = 35.24 liters.

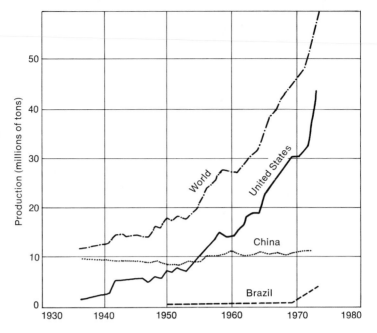

FIGURE 11-4.

Production of soybeans for the past four decades is charted. Curves portray world production and yields of the United States, China, and Brazil. [From "Soybeans," by Folke Dovring, Scientific American, *230 (2) 1974, p. 18. Copyright © by Scientific American, Inc. All rights reserved.]*

harvest that year was estimated to be more than 1.5 billion bushels (Dovring, 1974). In 1950, the exports of soybeans were insignificant, but by 1973 their value exceeded $3 billion. This was equivalent to almost 5 percent of all U.S. export income. Soybeans and soybean products are largely exported to Japan, China, and Europe.

In contrast to corn, wheat, and rice, soybeans produce much more protein (up to 40 to 45 percent of the weight of the beans). Moreover, soybeans produce vast quantities of fat (18 to 20 percent), extracted as soybean oil. Because of its high yields, soybeans can produce more oil and protein per acre than other oil seeds, such as flax and safflower, or high-protein pulse (legume) crops, such as peas and beans. Soybeans, being such a rich source of protein, are now being used as a meat additive and a meat analog.

What have been the environmental and agricultural impacts of the green revolution? Perhaps one of the most important impacts is an alteration in the nutritional quality of the HYV corn, wheat, and rice varieties, compared to the earlier-used, traditional varieties. Let us examine the record: *wheat*—the protein content of Kansas wheat decreased by 44.7 percent from 1940 to 1969 (U.S. Department of Agriculture); *corn*—the U.S. Department of Agriculture indicates that of 4000 samples of corn from midwestern states, the prehybrid corn varieties (open-pollinated) contain an average of 82 percent more crude protein, 37 percent more copper, and 113 percent more manganese than the "superior" hybrid corn varieties; and *rice*—the protein content of IR-8 and other HYV rices tends to be lower than that of traditional rice varieties (7 to 8 percent versus 9 to 10 percent). Such decreases in levels of protein or essential nutrients in the seeds of these green revolution crop varieties raise serious doubts as to whether we have really made a significant gain, in spite of the greater yields.

The new HYV rice, wheat, corn, sugarcane, and soybeans all require more fertilizer, pesticides, and irrigation water (where used). In many countries, such practices are prohibitive in their costs to farmers, such that many cannot afford to buy expensive fertilizers and pesticides. Furthermore, the demand for fertilizers is rapidly exceeding the available supply. Far worse is the fact that increased usage of pesticides (insecticides, fungicides, and herbicides) with HYV crops loads up the ecosystem with more pollutants. The deleterious effects of such pollutants on wildlife are well documented. Lewis Batts (1973) has said: "It has become apparent that the green revolution is not the panacea hoped for nor claimed. Even the most avid proponents have begun to qualify their earlier claims and to maneuver to shift blame for its failures."

Nicholas Wade (1974) has succinctly summarized the case: "The green revolution may lose even the momentum it has unless plant breeders can adapt to other crops, to unirrigated lands, and to the maybe permanent constraints of the energy crisis. These obstacles are probably more likely to be overcome if the green revolution can be continued as a native growth, rather than as a Western implant."

There are also important social implications of the green revolution. The imposition of this Western technology relies on the use of large amounts of water, pesticides, and fuel. With shortages of fuel and high prices for fertilizer and pesticides, how will farmers in the developing nations be able to afford this technology? Indeed, how many people will be displaced from their farms with the increased mechanization that follows from the switch to green revolution crops? The prediction is, according to Wade (1974), that "the mechanization that in practice accompanies the green revolution is in itself double-edged; the use of tractors for rapid land preparation, for example, may create extra jobs by giving time for a second crop. But, overall, mechanization is job destroying." In the final analysis, the green revolution technology may favor the more wealthy farmers, who possess the largest land holdings, as has occurred, and is continuing to occur, in the United States. The social disruption that would result for the small, poor farmers is a tragedy that we cannot dismiss lightly.

The green revolution, because of its severe demands on our scarce energy supplies, its immense costs, its disruption of the ecosystem by increasing pollution, and the social upheavals that it engenders, must be reexamined critically. The environmental impact of this revolution in underdeveloped countries is far from satisfactory. Solutions that cause far less social disruption, provide new varieties of higher- rather than lower-quality (more protein and essential nutrients) food, and allow for traditional agriculture to flourish (not monocultures, but crop diversity) will, in the long run, provide the basis for a true green revolution.

How Long Will the Green Revolution Last?

The green revolution has resulted in only temporary gains in food production of the order of 15 to 20 years. The glowing reports of the 1960s were short-lived. In Mexico, which exported 10 percent of its grain crop between 1965 and 1969, the world's fastest population growth rates soon overwhelmed the gains in food production seen in the 1950s and 1960s. By the mid-1970s, Mexico *imported* one-fifth of its total grain needs (Brown, 1975). The result is that in Mexico, and in other green revolution countries, burgeoning populations are rapidly outpacing any gains achieved by green revolution technology. Populations everywhere will have to be controlled through adequate family planning; otherwise, we shall be faced with more-or-less chronic food shortages. Taken altogether—the negative ecological aspects of the green revolution; the high costs of its technology, fertilizer, and energy; and rapidly increasing populations—all these parameters point to Brown's prediction that "in the future, scarcity may be more or less persistent, relieved only by sporadic surpluses of a local or short-lived nature."

How Has China Achieved Its Own Green Revolution?

Since 1971, China, with a population of over 800 million people, has attained agricultural self-sufficiency, with a record 225 million metric tons (250 million tons) of food produced that year. How has this remarkable record been achieved? The most salient reasons (see Wortman, 1975, for more details) are as follows:

1. The Chinese have employed methods of multiple cropping, where, for example, three crops of rice are harvested each year in South China. This has been possible through a combination of extended irrigation, increased application of fertilizer, and introduction of high-yielding varieties of dwarf rice, wheat, and other crops. The dwarf rice varieties were developed in China independently of those in the Philippines between 1956 and the early 1960s. China is now the leading rice producer in the world [over 34.2 million hectares (84 million acres) planted in 1971].

2. China does not rely on monocultures but, instead, uses many local crop varieties that are adapted to the local growing conditions (soil, climate, topography). Their primary food crops include rice, wheat, sorghum, millet, maize (corn), barley, oats, rye, soybeans, buckwheat, white potato, sweet potato, and yams. Many of these crops are grown in rotation (e.g., rice → rice → wheat; rice → rice → barley; late rice → wheat → early rice—each of these sequences is for 1 year). In such schemes, the land does not lie idle. In fact, often two crops are planted in the same area simultaneously, as, for example, planting rows of maize between rows of spring wheat about 1 month after planting the wheat (in northern China), or soybeans intercropped in orchards. Even "waste" space is utilized; one sees soybeans planted along roadsides, stream banks, and margins of fields. Essentially every bit of arable agricultural land is put to intensive use throughout most of the year in very effective cropping sequences and multiple cropping schemes. The Chinese apply the same methods to the growing of vegetable crops.

3. China has a huge agricultural labor force, 680 to 765 million people, estimated to be 80 to 85 percent of the population. In contrast, in the United States, the agricultural labor force numbers only 4.2 million people, which is equal to 2 percent of the U.S. population of 210 million people. This "human factor" explains why so much hand labor is

available in China for planting, weeding, irrigating, harvesting, and threshing.

4. Another important factor that has influenced China's remarkable gain in agricultural productivity is the marked increase in use of chemical fertilizers. China, by 1973, was one of the world's leading importers of fertilizers; in 1973, the total amount of fertilizer imported amounted to 1.6 million metric tons (1.8 million tons). In China itself, small factories produce over 50 percent of the nitrogenous fertilizer, mostly ammonium bicarbonate, derived from lignite or brown coal. In 1971, it is estimated that of 16.9 million metric tons (18.6 million tons) of fertilizer produced in China, 7.2 million metric tons (7.9 million tons) was ammonium bicarbonate. Ammonium bicarbonate is, incidently, a preferred nitrogen source for rice, in contrast with other inorganic forms based on nitrates.

5. China has also greatly improved and enlarged its irrigation systems through the construction of new diversion canals, catchment basins, and wells with mechanical pumps. It is estimated that over 38 million hectares (93 million acres) of agricultural land is now irrigated in China [out of a total of 107 million hectares (267 million acres) being cultivated in a 1964 estimate]. Because of the extensive use of irrigation, it is rare that one sees agricultural land lying fallow in any part of China in areas where there is adequate rainfall and suitable temperatures for growing crops.

The use of a combination of irrigation, intercropping and mixed cropping, fertilizers, and high-yielding, quick-maturing crops, coupled with a huge labor pool, have all made possible China's remarkable gains in food production, to the point of self-sufficiency after 1971. If this living standard is to be maintained, or improved, China will have to avoid continuous population growth. With present incentives for keeping families small through work-point production systems, restricted allocation of land for private land holdings, and health care (plus residences and recreation programs) for the elderly in each commune, it may be that China can keep its population in bounds and continue to maintain its self-sufficiency in food production.

In many parts of the world, particularly in the humid tropics, it is impossible to import the technology of the green revolution. This technology—fertilizers, pesticides, machinery for tillage and harvesting, irrigation, and access to markets—is simply not available to small farmers in the tropics and subtropics, who have little or no education for such methodology and who have land holdings that are much too small for such operations. Most farmers in these regions have farms less than 2 hectares (4.9 acres), in size, and they typically occur on soils that are nutritionally poor and highly subject to erosion.

What is the alternative for such farmers to improve their crop yields? The answer lies in modifying their traditional methods of *shifting cultivation.* This type of cropping involves farming a given plot of land for 1 or 2 years, then abandoning the area to farm a new area. It has been called "slash-and-burn" agriculture in some tropical areas, and it has

Alternatives to the Green Revolution in the Tropics and Subtropics

wreaked havoc on the landscape in mountainous areas of the tropics, owing to erosion, loss of soil fertility, and slow recovery of vegetation. Much of this destruction can be avoided and higher crop yields attained through modified cropping practices (Greenland, 1975): "(i) zero tillage and plant residue mulches; (ii) mixed crops of high-yielding varieties that are disease- and pest-resistant; (iii) fertilizers to replace the phosphorus and possibly other nutrients removed in produce sold 'off the farm'; (iv) legumes with highly active nitrogen-fixing rhizobia to supply nitrogen to the soil and other crops; and (v) control of acidity by means of ash or mulches of deep-rooted species." Much progress has been made recently in producing high-yielding, pest-resistant maize, cowpeas, cassava, rice, and other crops at the International Institute of Tropical Agriculture (IITA) in Nigeria. The other elements of this alternative program—fertilizers, nitrogen-fixing bacteria, inexpensive tools, and improved cropping sequences—should follow rapidly to attain higher crop yields, less soil erosion, shorter fallow periods, longer cropping sequences, and less detrimental effects on the landscape where shifting cultivation is practiced. These inexpensive, practical recommendations offer a far better alternative to the poor farmer on small land holdings in the tropics than any large-scale technology that characterizes so much of agriculture in North America and in Europe. As you can see now, the tropics has its own special problems, crop varieties, and social conditions. Large-scale importation of Western technology is not the kind of green revolution that will work in these areas. Fortunately, there is a workable blueprint for instituting another kind of green revolution there.

Summary: Prospects for a New Green Revolution

We have already alluded to the diminished quality of certain grains that has arisen as a product of the green revolution, resulting in a decrease in level of protein in the seeds. We have also mentioned the fact that pulse (legume) crops are noted for their high protein content. One of the primary aims of "genetic engineers" is now to attempt to get nitrogen-fixing capacity into cereal grain crops. The prospects for doing this through somatic cell hybridization (see Chapter 4) are rather problematical. However, it is known that several bacteria (*Spirillum lipoferum* and other species of bacteria) already *do* fix nitrogen on the roots of several grasses, including maize, in the tropics. The amount of nitrogen fixed in these grasses can be as high as that fixed by *Rhizobium* (a bacterium) in typical leguminous plants. If bacterial mutants could be obtained that could grow in the roots of grains and fix nitrogen in more northern climates (present nitrogen-fixing bacteria on these grasses will only fix nitrogen at temperatures betwen 31 and 40°C, with little fixation below 25°C), this might be a more promising approach. This approach, coupled with an all-out effort by plant breeders to increase the protein content in cereal grains, and an increase in the use of pulse seeds in our diet, may have promise for instituting a new kind of green revolution that results in a significant improvement in the quality of the food we eat—something that is long overdue!

Batts, H., Jr. 1973. Environmental considerations of the green revolution. *Kalamazoo College Rev.* **35**, 1–3.

Borgström, G. 1974. The food-population dilemma. *Ambio* **3**, 109–113.

Brown, L. R. 1975. The world food prospect. *Science* **190**, 1053–1059.

Brown, L. R., and E. P. Eckholm. 1974. Buying time with the green revolution. *DuPont Circle* **4**, 13–17.

Commoner, B. 1974. Improving resources' productivity: A way to support the growing world population. *Ambio* **3**, 136–138.

Dovring, F. 1974. Soybeans. *Sci. Am.* **230**, 14–21.

Greenland, D. J. 1975. Bringing the green revolution to the shifting cultivator. *Science* **190**, 841–844.

Heiser, C. B. 1973. *Seed to Civilization.* W. H. Freeman and Company, Publishers, San Francisco.

Janick, J. 1974. *Man, Food, and Environment.* W. H. Freeman and Company, Publishers, San Francisco.

Janick, J., R. W. Schery, F. W. Woods, and V. W. Ruttan. 1969. *Plant Science, An Introduction to World Crops.* W. H. Freeman and Company, Publishers, San Francisco.

Strobel, G. 1975. A mechanism of disease resistance in plants. *Sci. Am.* **232**, 80–88.

Wade, N. 1974. Green revolution (I): A just technology, often unjust in use. *Science* **186**, 1093–1096.

Wade, N. 1974. Green revolution (II): Problems of adapting to a Western technology. *Science* **186**, 1186–1190.

Wortman, S. 1975. Agriculture in China. *Sci. Am.* **190**, 1053–1059.

12 Urban Ecology
Spencer W. Havlick

Questions for Consideration

1. What are the major reasons why human communities exceed their carrying capacity causing pollution, decay, and other environmental malfunctions, when so much is known about the dangers and results of population explosions in other biological communities?
2. What factors make up a dynamic, harmonious urban setting for human fulfillment without negating ecological principles?
3. How can dwellers of cities be directly linked to seemingly remote resources such as wilderness areas, mineral deposits, forests, or wild life?
4. Why are the costs of acquiring more national park or national forest probably related to reduced revenues for the inner-city poor?
5. What are the prospects for "new towns" as a solution to the urban problems of the past?
6. How does the ecological impact of urbanization on the local flora and fauna give a clue to the future health of human inhabitants of a proposed city?

An Urban Area as an Ecosystem

With a rate of world urbanization that is unprecedented in human history, it is imperative that city dwellers have a keen sense of what relationships exist between an urban system and the ecosystems that support it. Scattered rural populations of tropical and temperate regions show signs of continual migration into metropolitan centers in the foreseeable future. The impacts of intensified human settlement on vegetation pose grave problems not only in the city but in the sprawl and hinterland areas, where influences of urban people are stronger than ever before.

The entire concept of an urbanized region has begun to evolve after Jean Gottmann's description of megalopolis, especially as it applies to the Great Lakes, California, and the East Coast urban corridor (Figure 12-1). An urban area is seen by many observers as a unique ecosystem dependent on the flora, fauna, energy, and other resources required to nourish and sustain human activities. The transportation system of a city has been compared with circulatory systems of biological organisms (Figure 12-2); the political and other institutional arrangements function somewhat like a nervous system; and, of course, there are continual catabolic and metabolic processes in cities similar to those in biological communities. The survival or extinction of cities (like that of plants and

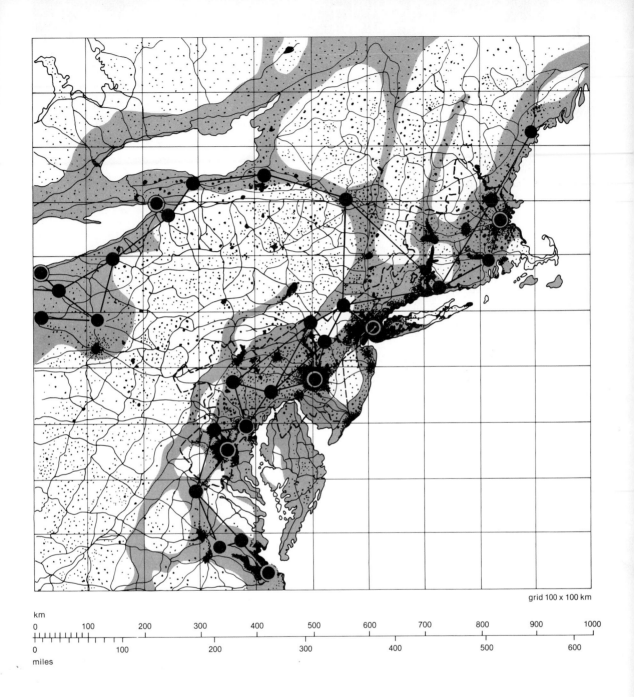

km
0   100   200   300   400   500   600   700   800   900   1000

0   100   200   300   400   500   600
miles

grid 100 x 100 km

North-Eastern American Branch of Ecumenopolis

Ecumenopolis: (urbanized area A.D. 2100)

Deteriorating centers existing in 1960

Network of new centers and new lines of transportation

 Built-up area in 1960

1950 Megalopolis after J. Gottmann

FIGURE 12-1.
The northeastern megalop-olis with major networks.
[*From C. A. Doxiadis,*
Ekistics—The Science
of Human Settlements.
*Hutchinson Publishing
Group Limited,
London, 1968.*

FIGURE 12-2.

Transportation systems in cities are like biological systems of plant communities. Even novel transit systems such as San Francisco's cable cars keep the center city healthy and appealing. Automobiles are the affliction of every urban center. (Courtesy of Spencer W. Havlick.)

animals) is dependent upon certain limiting factors, such as temperature, oxygen, wind, light, humidity, the input and output of energy, and a balance between producers and consumers.

It is with increasing frequency that the various urban pathologies which Ian McHarg, Jane Jacobs, C. A. Doxiadis, Patrick Geddes, Ebenezer Howard, and Lewis Mumford have addressed are the result of citizens not understanding and not following the ecological imperatives as they apply to human settlements. Even less is known about the social and psychological requirements of successful urban living. The disease and decay of the city and the suburb are not so much a fault of the people who live there as they are a failure of those who plan and finance and build cities. Biological communities evolve over tremendous periods of time and tend to assume a degree of stability or climax despite periods of fluctuation in physical constraints and climatic differences.

For the most part, large human settlements are not designed with an equilibrium goal in mind. Quite to the contrary, urban areas grow in uncoordinated surges stimulated by huge developers, annexation and zoning changes, and the ability of single individuals to invest in or buy homes and commercial enterprises. New residential land is developed with the myth that the new property taxes will pay for public investments such as schools, sewers, streets, parks, and police and fire protection needed by the new residents, as well as helping to pay for the bankrupt or underfinanced services of present residents. In short, there has never been a major American urban area that was designed to serve a specific population which represented the carrying capacity of that particular urban container. Consequently, urban areas tend to grow beyond manageable size, and the costs to maintain an urban organism of vastly obese proportions tend to be excessive, especially for the inner-city poor left behind when the more affluent escaped to the suburbs and the rural estates.

The breakdown of urban functions can also be explained by the fact that there is no real synthesis and integration between the professional ecologist and those who build a city. Ecological principles of diversity, predator—prey relationships, territoriality, community stability, niche and habitat requirements, and homeostasis are never thought of by subdivision investors and industrial-site planners. Perhaps the city forester or city geologist (if there is one) thinks about some of these ecological principles, but seldom does the professional ecologist work closely with the professionals in architecture, highway engineering, floodplain management, banking, or land-use zoning. There seem to be disciplinary barriers and petty jealousies that inhibit a good flow of communication among the various professionals and citizens who should be involved in urban design and urban processes. Without the important dialogue and implementation of the concepts that produce a dynamic, fulfilling urban environment, breakdowns soon occur between the integral components of the system. The urban ecosystem, not unlike the traditional biological ecosystem, is essentially made up of the abiotic environment, populations of heterotrophic and/or autotrophic organisms (including man), nutrient cycling, and energy transfers. The continuum from a heavily human-dominated environment to a less heavily human-dominated environment depends upon the sensitivity that is used as postindustrialized society urbanizes.

Some of the Scandinavian new towns (Figure 12-3), such as Farsta, Vallingby, and Tapiola, blend harmoniously with the natural habitat. Other new towns, in France, Great Britain, Australia, the Netherlands,

FIGURE 12-3.
New towns in Sweden have tried to integrate the natural features of the environment with human needs in a postindustrialized urban society. (Courtesy of the Department of Housing and Urban Development.)

and the greenbelt-era towns of the United States, have replanted and reproduced as much of the endemic vegetation as possible. There are towns and parts of older cities that begin to approach interesting and dynamic characteristics which attract people instead of forcing their residents to escape whenever a weekend, a vacation, or a block of leisure time occurs. Some examples that demonstrate an exciting mix and interplay of biotic and abiotic environments, architectural and human diversity, citizen involvement, a sense of pride in the community, and a consciousness about carrying capacity include Atlanta, Georgia; Santa Cruz, Petaluma, and San Francisco, California; San Antonio, Texas; Boulder, Colorado; Minneapolis, Minnesota, and certain areas of Oregon, Vermont, Michigan, Maryland, and Montana that have progressive zoning and land-use regulation. An index to help judge the steady-state condition in a human-dominated urban environment would include a list of services that are within walking distance or easy access by public transportation. Items on the list should include:

1. Retail food stores and restaurants.
2. Medical, dental, and other professional services.
3. Library, theater, and/or museum facilities and schools.
4. Commercial goods and service outlets, such as hardware, shoe repair, sporting goods, and wearing apparel.
5. Governmental offices and community center.
6. Access points to link into the city center, and high-speed transit systems to connect with other urban areas.
7. Parks and open spaces with functional and amenity considerations.

In terms of a dynamic, human-oriented town or neighborhood, these seven basic requirements improve the attractiveness for urban living. Well-known and sought-after neighborhoods that qualify with most of the listed criteria include Georgetown in Washington, D.C.; Beacon Hill in Boston; the French Quarter in New Orleans; Burns Park in Ann Arbor, Michigan; the Hill in Boulder, Colorado; Lake Park in Milwaukee, Wisconsin; and many others, particularly in established areas of older cities. One finds a rich diversity of architectural styles, socioeconomic groups, and abundant gardens and greenery.

It cannot be overemphasized that historical precedents play an important role in the desirability of an urban area. With the agricultural and industrial revolutions, villages and towns grew rapidly into cities that usually could not accommodate the hoards of new residents. In underdeveloped countries, the population increase continued in urban and agricultural areas, thus reducing the ability of urban planners to cope with rapid growth and decreasing standards of living. Within the last two centuries, military forts and trading posts have grown to metropolitan areas with up to several million in population. The northeastern and Great Lakes megalopolitan areas are becoming huge, polynucleated cities. In time, densities will probably increase and frustrate the search for unused land—space for broad freeways, sprawling shopping centers, vast school grounds, and residential lots.

Undoubtedly, with the help of such urban crises as poverty (Figure 12-4) (which is linked to poor health, poor housing, poor education, and

low employment) and pollution, poor utilization of urban resources will force a comprehensive reassessment of urban form and urban functions. The United States is nearly 80 percent urbanized, and if the succession from village to town to city to metropolis to megalopolis to an urbanized continent does not produce an excess of benefits over costs to its residents, some severe changes are in order.

Those who are still able to afford the luxury of analysis will take stock of the process and ask how the urban ecosystem became what it is, why it has begun to collapse, and what can be done to restore its vitality for the benefit of all its residents: the young, the poor, the elderly, the handicapped, even those who are reputed to be the most productive. The thoughtful urbanite will look around the world to see where urban succession has produced a dynamic, satisfying, and homeostatic environment for its residents. From the lessons available in the Greek and Roman cities of antiquity; from the urban arrangements of Africa, South America, and Micronesia; from the urban efforts of China and Japan or from the new towns of western Europe, we have clues about settlements that were sustained and duplicated under various levels of economic development. There is hope for improved urban environmental management. The author believes that part of that hope resides in the ability of mustering the best talents of citizens to press for the best synthesis of human needs blended with the life-support systems of the biotic and abiotic environments that constitute an urban ecosystem.

The Support System of Cities

The biomes or major ecosystems of the world are the result of an interplay of physical, chemical, and biological forces over a time span that approaches hundreds of millions, if not several billion, years. Human evolution has been a very recent event—made possible only by the physical, chemical, and biological events that preceded our paradoxical species. The advent of cities approximately 10,000 years ago has subsequently imposed demands and pressures on the natural ecosystems that heretofore had never been known. Still more recent, and perhaps more risky, is the human invention of metropolises, with over 1 million inhabitants, and megalopolises, with their 25 to 100 million population.

When villages and towns were scattered at points seldom closer than 1 day away from each other, the productivity of the land or the sea helped to determine the size and prime activities of those human settlements. Only the fishery or farmland that was needed to support a particular village was developed. As long as the town was reasonably self-sufficient, the food and fiber supply was obtained locally. The energy sources were confined to forms that were at hand, such as water, wind, animal, or human. A severe change in climate, a depletion of nutrients, or an epidemic that exhausted the animal energy could force a relocation of the town. The more common occurrence, however, is that with industrialization and automation, towns became more specialized and began to exchange goods and services.

If a drought or a flood threatens the long-term survival of a city, the customary approach today is not to move to a more water-abundant site or out of the hazardous floodplain. Instead, technology is imported to try to solve or at best, quick-fix, the problem by water-importation schemes or levees and flood-control dams (Figure 12-5). The same philosophy prevails when a city uses more energy or food or raw materials than it can produce for itself. The additional energy, food, water, or whatever life-support commodity is imported to the metropolitan population, which in time becomes totally dependent on the resources and ecosystems of the periphery and hinterland areas.

Increased standards of living in urban areas tend to be more energy-consumptive, land-consumptive, and mineral-consumptive; and overall, further removed from the resource base that sustains the entire city. The settlement, over time, has essentially moved from a relatively self-sufficient system to a very complex, very specialized entity, which tends to be almost totally dependent on life-support systems that are far beyond the city limits. Lumber and building stone may come from forests and quarries about 1600 kilometers (1000 miles) away. Fresh fruit, fish, and vegetables are flown in to the urban marketplace from around the world. Energy in its various forms is brought from great distances by high-voltage transmission lines, tankers, pipelines, unitized trains, barges, or mixtures of these. If the multitude of lifelines were cut off to any huge metropolitan area, its inhabitants probably could not survive more than 48 hours.

Perhaps more unsettling than the fact of total dependency on external resources (and the networks that transport them to urban consumers) is the fact that very few urbanites recognize how precarious their existence is and that future projections of urban growth promise more of the same. It has been rather thoroughly forgotten by the average urbanite or suburbanite that plants are the original converters of solar energy. Very few individuals think about the linkages between plants and animals and

FIGURE 12-5.
The Corps of Engineers frequently applies quick-fix technology to flood problems. In the long run, floodplain zoning and application of other ecologically based principles are the best solutions to human intrusion into natural phenomena. This Vicksburg, Mississippi, subdivision was under water more than 4 months in 1973. (Courtesy of the U.S. Army Corps of Engineers.)

subsequent transformations that have made possible the contemporary uses of petrochemical products and fossil fuels. Finite fossil fuels are being squandered away in terribly inefficient automobiles, poorly engineered industrial processes, and crudely designed space-heating systems. As nonrenewable resources such as minerals, wilderness areas, and fossil fuels are depleted, costs to use them will become prohibitive and substitutes may be less satisfactory.

One of the greatest challenges for cities in the future will be to use renewable resources more effectively than has been the case in the past. Here is where human ingenuity can play a role never before attempted in the nonhuman biological communities. The reuse of finite resources and the maintenance of reusable or renewable resources in an urban ecosystem is easily understood if we visualize an urban area as a special kind of resource which behaves in many ways like the traditional biological community. The intent here is to demonstrate that the city, as a totality, is a mixed cultural and natural resource, even though a second order of complexity one "trophic level" above the traditional resources described in most textbooks and references written before the 1970s. Instead of viewing an urban area as a static portion of the landscape or as dead-end dumping ground for biological and mineral cycles, a large urban area is considered here as a continually resynthesizing component of the human ecosystem.

The trophic-level concept is borrowed from ecology—the study of interrelationships and consequences of an organism, the species population (autecology), or a community of organisms (synecology) affecting an environment, and vice versa. Different members of a biological community occupy different positions or trophic levels of production, consumption. energy transfer, and energy loss. The most fundamental members, known as primary producers, contain photosynthetic pigment (e.g., chlorophylls a and b). These primary producers provide foodstuffs basic for the next trophic level, which in turn provides it for the next.

In the presence of sunlight, water, and carbon dioxide, these producers photosynthesize a potential food supply for organisms unable to manufacture their own. This basic or primary trophic level is occupied by plant kingdom members ranging in size from the unicellular algae to the towering angiosperms and gymnosperms of the Pacific Northwest forests. The urban analogs of producer representatives are, of course, the fishing, farming, logging, or mining towns, whose major activity is making basic resources available to larger urban areas at the next-highest trophic level.

Just as there are herbivores or organisms that depend directly on the primary producers, so there are larger urban areas that are not involved in extractive activities but act as consumers of raw materials. The raw materials are "produced" by primary-producer-type settlements, such as the ranching, mining (Figure 12-6), logging, or fishing villages.

As the second trophic level of urban organisms evolves, many employment activities shift toward service, reprocessing, wholesale operations, and other middleman functions. With the growth of this kind of urban organism, diversity of production and complexity increases. Illustrations that immediately come to mind include Atlanta, Boston, Chicago, Denver, New Orleans, Minneapolis—St. Paul, Fresno, and Omaha. These are all cities whose primary raison d'être has changed and expanded through the years. Second-, third-, and fourth-level transformation activities become common as food, fiber, metal, and other commodities are reprocessed, refabricated, reused, and then used again. In a considerably more thorough transformation, biological communities recycle and lose energy as one predator is eaten by another predator, who in turn falls victim to perhaps a parasite and eventually to a decomposer in the form of bacteria or fungi of decay.

Thus the fate of natural resources can be traced from original sites of extraction or production to secondary and tertiary consumers. Usually, the first step of refinement takes place near the origin of the resource; fish are cleaned and frozen; petroleum is deemulsified; iron ore is sintered and upgraded; trees are cut and scaled; and gravel is washed and sorted. The next step in the utilization of a resource produces an array of specified processed raw materials in the form of fishmeal, paraffin, ingots, lumber or fiberboard, and concrete mix.

Up to 90 percent of the resources extracted from the hinterland eventually find their way to urban areas as the city becomes a concentrator of resources (Figure 12-7). As the resources are converted from their raw form into commodities such as automobiles, frozen TV dinners, and touch-tone telephones through a series of reprocessing steps, considerable waste of basic resources takes place. This resource waste occurs from rust and corrosion and from the disposal of packaging materials and losses from other noncyclical procedures, and it is similar to the matter lost through nonrecycling in natural systems. The increased knowledge and sophisticated technology required to produce such commodities form information resources that can be used to prevent

FIGURE 12-6.
Coal mining towns of the southern Appalachian mountains are examples of "producer" communities that export their resources (coal, in this case) to large cities up the trophic pyramid of human settlements. (Courtesy of Spencer W. Havlick.)

breakdowns in the biophysical, economic, and social systems that produce and use the highly processed original resources.

People themselves demonstrate the result of energy transformation as they participate in a predator role in the world's natural food chains. It is impossible to avoid the loss of energy as carbohydrates, fats, and proteins move up through the trophic levels from producer to consumer. The second law of thermodynamics shows that no energy transformation can be 100 percent efficient. We always end up with less usable energy after the transfer, with most of the lost energy being transformed into heat.

The example of one predator eating another will serve as a dramatic illustration. With the salmon or any high-level carnivore, a rather complete sequence of energy loss can be shown from primary producer to ultimate consumer. Let us assume that 1 percent of the solar energy available to the algal "food factories" in the ocean is converted or incorporated into the plant. Inasmuch as the algae required some energy for "housekeeping" functions such as respiration, growth, and reproduction, an immediate energy loss begins with the primary producer. It should be noted that terrestrial plants are usually somewhat less efficient in their use of radiant energy than freshwater or marine vegetation. Most aquatic plants, especially the floating and submergent species, are spared the need for extensive supporting structures as well as the work needed to acquire and transport nutrients; that is, they require less energy for maintenance functions than terrestrial relatives, whose leaves, roots, trunks, bark, and attractive floral parts are specialized accessories not required by the phytoplankton of the lakes and oceans.

To continue with the "biological pecking order," the algae are eaten by members of the next trophic level, such as the snails and other mollusks.

FIGURE 12-7.
New York is the ultimate example of a concentrator, transformer, and consumer of energy and other resources. This view is looking south from the observatory roof of the RCA Building. (Courtesy of the New York Convention and Visitors Bureau.)

Vast numbers of invertebrates, from the protozoans to the crustaceans, are also primary consumers who feed upon primary producers. The taxonomic group, however, has nothing to do with who eats whom or what. Vertebrates from almost all families have some members who are primarily vegetarians. Some of the largest mammals, the baleen whales, sift and screen zooplankton and phytoplankton from the sea, illustrating how the intermediate steps of the food chain (and subsequent trophic-level energy losses) are bypassed. The more common energy route is for the algae to be eaten by microcrustaceans. The microcrustaceans, in turn, may be eaten by other crustaceans or small fish, which are eaten by larger fish who eventually feed the predacious salmon species. Energy loss takes place during normal activities. Obviously, when the salmon is consumed by the fisherman or the gourmet, it is a miniscule fraction of the energy originally synthesized by the algae that served as primary producer. A fair estimate is that about 10 percent of the total energy consumed by one trophic level is synthesized into one biomass of the next.

Too often the illustration stops at this point. The important factor to remember is that regardless of where one consumer is on the food chain, there is another link. The chain is a continual loop where the decomposers are always available or waiting to perform the vital task of reconverting and recycling inorganic elements as a part of the decomposition process. Perhaps it is misleading to speak of a biological pyramid because regardless of whatever or whoever is at the top of the pyramid, that individual salmon, man, or hawk eventually succumbs to the relentless consumers or decomposers who prepare the basic ingredients for use by the primary producers. Thus the endless loop continues with entropy taking its gradual harvest from the total system. When a given genus no longer plays an active role as producer and consumer, atrophy sets in.

Similar catabolic processes affect the urban organism. Net loss of population, diminution of the percentage of young adults, a decrease in new business development, and increased residential and commercial vacancies are among factors that over long periods of time determine the viability of a community. In some cases, the city becomes a nonfunctioning vestige; in other examples, it becomes a vital producing organism.

The Impact of Metropolis on Hinterland Environment

Up until recently, the millionaire cities were not viewed as devastators of hinterland environments. Perhaps it has taken projects like the Alaskan pipeline, offshore drilling, vast forest clearcutting, huge water importation projects, mammoth flood-control efforts, and the multibillion dollar agribusiness manipulators of the family-sized farm to finally recognize what the modern postindustrialized city requires under the current economic and political practices. The sphere of influence of a millionaire or multimillionaire settlement is awesome if one attempts to construct an energy or a nutrient input–output model.

The disposal of solid waste is an urban ecological problem of ponderous proportions. The ocean is no longer seen as an appropriate dumping ground; abandoned mines have nearly been filled by garbage shipments, and other landfill sites are scarce. It is generally accepted that our urban solid wastes will be recycled and reused (as happens in the natural ecosystem), but the exact time in the future is yet to be determined. One of the more subtle impacts of metropolis on its hinterland is in the form of leisure time or recreation impacts.

Too frequently, the configuration of a metropolis separates the individual from unspoiled natural environments. As a result, there tends to be a desire to escape the sterile concrete and glass cityscape (Figure 12-8).

As the recreation problems that await the urbanite who hungers and thirsts to "get back to nature" are multiplying, the time approaches when the United States may need to reevaluate her national recreation policy. Could it be that our national goal cannot be to get every urban American into the hinterland? Traditional recreation management strategies cannot guarantee the safekeeping of the natural environment with the burgeoning population of backpackers overflowing even the most remote areas. Pack trips are more frequent, and improved equipment extends the hikers' season and the length of his visit. It now has even become socially desirable and prestigious to get into a wilderness area.

In an attempt to provide for this increased demand for remote areas, Congress passed the Wilderness Preservation Act of 1964, which authorized the creation of a National Wilderness Preservation System. Wilderness was defined as "an area where the earth and its community

FIGURE 12-8.

Backpacking is one of America's fastest-growing recreation activities. Urbanites are trying to escape the sterile suburb and city in order to "get back to nature." Wilderness areas face unprecedented pressures from overuse. (Courtesy of Spencer W. Havlick.)

of life are untrammeled by man, where man himself is a visitor who does not remain." Motorized equipment, roads, and other structures are not allowed but, as in the national forests, hunting and fishing are permitted, in compliance with state regulations. Mining and grazing continue, and mineral exploration is permitted up to 1984. (One wonders what is so magic about 1984 and 1985 when the provisions of the Wilderness Act and those of the Federal Water Pollution Control Act—1972 Amendments go into effect.) To the consternation of preservationists, future mining from patents and claims established before 1984 is also permitted. For the benefit of recreation uses, the provision exists that patents granted in the wilderness areas will not convey rights to any surface mineral resources. Undoubtedly, the courts will struggle with that language as mining and recreation conflicts increase.

In 1964, 9.1 million acres were set aside from 54 areas of national forest land in 13 states. California with over 581,000 hectares (1,435,700 acres) included was one of three largest beneficiaries of the Wilderness Preservation System (Figure 12-9). Congressional approval is required for additions to the system. The first additions since the act came when 323,750 hectares (800,000 acres) of federally owned land were added in 1968. During the Johnson administration, the Redwood, North Cascades,

FIGURE 12-9.
Federal forests and national parks provide some of the most desirable sites for wilderness seekers from urban areas (this photograph was taken in Sequoia National Park). In recent years, federal funds have been totally inadequate to safeguard the national recreation facilities and parks from ecological damage due to overuse. (Courtesy of Spencer W. Havlick.)

and Guadalupe Mountains National Parks were added. Special bonuses to urban dwellers were the National Wild and Scenic Rivers System and a nationwide system of trails.

Urbanites create more than the problem of sheer ecological impact when the remote recreation areas are invaded; they raise the entire socioeconomic—political question of whether federal dollars, that is, U.S. Forest Service, Bureau of Land Management, and U.S. Park Service, should subsidize a tiny fraction of the American public. A close inspection of who uses the wilderness areas reveals that they are certainly not the urban poor.

Tragically, the urban dweller who has been forced to live in the most heavily polluted sections of the city has a proportionately smaller amount of leisure time, a considerably less reliable means of transportation, and a strikingly more modest portfolio of skills and attitudinal

FIGURE 12-10.

In addition to a lack of interest in funding national parks by the Administration in the 1970s, private concessionnaires, such as the Music Corporation of America, which "runs" Yosemite National Park, want to build up the spectrum of commercial enterprises, which would convert this glacial valley into the city of Yosemite. Currently, up to 60,000 persons visit this park each weekend.

motivations for national park or wilderness use than those who are living in the more pollution-free suburbs.

Perhaps there is even a disparity between poor urban residents of the Far Western states and inner-city urbanites in other parts of the United States. The slum dweller living in Boston, New York, Philadelphia, Baltimore, or Washington seldom plans a vacation to Yosemite or Sequoia, because it is just unrealistic, for reasons of distance, time, and other economic constraints. Consider the frustrations, however, for a poor family interested in camping and the treasures of the national parks who lives 3 to 5 hours from Yosemite (Figure 12-10), Kings Canyon, or Sequoia but who is prevented from taking advantage of the park system by the same economic reasons as their eastern city counterparts.

The poor white, black, or Chicano, or other ghetto dwellers simply do not have the required equipment or the flexibility of vacation schedules to enable long treks into remote areas or the kind of transportation options to get there even if their daily concerns about job, medical services, educational equity and personal efficacy, and three nondehydrated meals a day for their families were resolved. As "high-priority" recreation areas are created in remote areas, the proportion of leisure-time facilities and open space for urbanites is decreased.

In other words, the opportunity costs or foregone opportunities of buying, staffing, and maintaining a new national park—let us say North

Cascades, or Redwoods—are that those millions of dollars were not spent in urban vestpocket parks, waterside parkland development, or other outdoor recreation benefits that might have been directly available to the inner-city poor of Seattle, Spokane, Tacoma, Portland, or, for that matter, San Francisco or Los Angeles. The federal recreation budget is not divided equally for all people in all places. The less obvious point is that whenever a particular environment is saved or a specific public is served, it is usually at the cost of someone else's project and probably in another (urban) environment.

The pastoral landscape at the periphery of metropolitan areas becomes less available to the inner-city poor because of subdivision sprawl. Even the swimming hole or pond of the ruralite becomes the focal point of an exclusive tract developer's country estate. The merger of city after city into a megalopolitan expanse swallows up the resistant rancher, farmer, and orchard owner. Rising property taxes force those three and other land-dependent operations to "sell out" to subdividers because of a squeezed profit margin. Los Angeles, the San Francisco Bay area, and (eventually) Puget Sound are examples of low-density urban extension with miniscule fractions of landscape dedicated to public open space or recreation facilities. Huey Johnson's San Francisco-based Trust for Public Lands is an innovative approach to obtaining open space lands in or near urban areas (Figure 12-11) where the critical needs exist now and where they will increase in the future. Yet here is where the daily demand for the use of the out-of-doors exists. How can we continue to lavish our national recreation budgets on remote high-technology-demanding facilities when the basic needs of the vast numbers of metropolitanites have hardly a vest-pocket park?

The Prospect of New Towns as an Alternative Urban Environment

In the late 1960s and early 1970s it dawned on American urban planners that the 100 million Americans whose births are expected in the next 30 years will not be housed, employed, transported, schooled, and contented in the overcrowded and decaying cities of the present. It was decided that a national effort would be launched very similar to the Scandinavian, French, British, Dutch, and German plans, which called for the construction of new communities either within established urban areas, around the periphery of present cities, or as "free-standing" new towns located some distance from the major metropolis.

The small light that Ebenezer Howard lit at the turn of the century took many decades to become a torch in British new towns. It was Howard's dream that a garden city in the countryside could provide the vitality of active town life in the surroundings of the natural environment. Stevenage, Welwyn Garden City, and many others have provided an alternative to those who felt that living in London was more than they could bear.

Several kinds of urban developments have been called "new towns" in the United States and elsewhere. What Howard meant by his garden city new-town concept was a self-contained human settlement predesigned for a specific size and with various urban functions and employment

FIGURE 12-11.
More shoreline and coast-line near metropolitan areas should be acquired, especially for urbanites who are unable to visit the far-away national parks and wilderness areas. Too often the public resources along the waterways and coastal areas are inaccessible to the public. One unique ex-ample of an urban ocean park is Golden Gate Na-tional Recreation Area, shown in this photo from San Francisco's Golden Gate Bridge. (Courtesy of Spencer W. Havlick.)

opportunities that could operate in relative isolation from other nearby towns or the metropolis. Great stress was placed on a rural or pastoral surrounding and easy accessibility by the resident population to work, shops, schools, recreation areas, and civic activities. There is also the ever-present hope that heterogeneity of employment, social classes, and cultural events could be incorporated into the new town. The "new community" concept in the United States has many similarities in theory, except that most of the new community projects and applications show a more conspicuous dependency and relationship to a nearby metropolitan area than do the European examples.

The garden city that Howard envisioned would be separated from the metropolis of London in cultural and political alliances once a string of satellite towns combined efforts to support a specialized hospital, university, museum, or symphony. Ten new towns of 35,000 people each, linked by high-speed transportation, could enjoy cultural ameni-ties now possible through intensive organization instead of high-density population on a single site. Palo Alto and Berkeley serve as this kind of "social city" or "regional city" in an informal way for special cultural functions in the San Francisco Bay metropolis. Lewis Mumford has portrayed the biological implications of the city better than anyone else. For those who would ask if growth and saturation have been considered in Howard's approach, Mumford has this reply:

Howard's greatest contribution was less in recasting the physical form of the city than in developing the organic concepts that underlay this form; for though he was no biologist like Patrick Geddes, he nevertheless brought to the city the essential biological criteria of dynamic equilibrium and organic balance: balance as between city and country in a larger ecological pattern, and balance between the varied functions of the city: above all, balance through the positive control of growth in the limitation in the area, number, and density of occupation, and the practice of reproduction (colonization) when the community was threatened by such an undue increase in size as would lead only to lapse of function. If the city was to maintain its life-maintaining functions for its inhabitants, it must in its own right exhibit the organic self-control and self-containment of any other organism.

Howard sought, in other words, to give to the new kind of city all the advantages that the big city possessed before its inordinate expansion put them beyond the means or beyond the reach of its inhabitants. He saw that, once it has achieved an optimum size, the need for the individual town is not to increase its own area and population, but to be part of a larger system that has the advantage of large numbers and extensive facilities.[1]

Recent events in America have refuted William H. Whyte's charge that it "is an impossible vision" for the new town to have the texture and fabric of the city. It is true that the new town—garden city experiment in America began slowly with Radburn, New Jersey, and the federal green-belt towns of Greenbelt, Maryland; Greenbrook, New Jersey; Greendale, Wisconsin; and Greenhills, Ohio; and then faded until the experiments of Columbia, Reston, and the rush of new communities in the late 1970s. Most of these efforts in the United States and the United Kingdom were intended to relieve urban congestion. However, sprawl of residential developments and scattered business and housing units on low-cost land within 60-minute commuting distance of the city center brought a peculiar backlash effect. The residents of the sprawled suburb began to miss the normal functions of a city that provided for human contacts, a mixture of activities, and some personal involvement in community decision making. There was an expressed need for a better educational climate, various civic functions, and other community-wide activities. Curiously, an effort seemed to be gaining momentum that called for an urbanization of the suburb. The sprawlites, or, as they are more commony called, suburbanites, wanted a branch library, a neighborhood school, a shopping center that had "downtown" stores represented, a community center, and even (despite the increased lot size) a neighborhood park and recreation program.

Scattered communities and housing projects grew and still are growing without thoughtful consideration of the tax base and without adequate regard for air, water, open space, wildlife, or land resources. Furthermore, with the speculative form of growth that has characterized the 1950s through 1970s, a very limited choice has existed for many people as to where they could live and the types of housing and

[1] Lewis Mumford, *The City in History* (New York: Harcourt Brace Jovanovich, Inc., 1961), p. 516.

environment in which they could live. Moreover, employment and business opportunities for inner-city residents are reduced, and the distances between the places people live, where they work, and where they find suitable recreation are greatly increased.

Columbia, Maryland, and Reston, Virginia (Figure 12-12), are examples of midcentury new towns that were begun with private capital, primarily from Connecticut General Life Insurance and Gulf Oil, respectively. The financial difficulties of early-day Reston and Columbia have been well publicized. Most of the financial trouble is created by the need for a mammoth outlay of capital for roads and general community services before the financial returns come back to the private owners at a level that maintains solvency. Ordinarily, the national government will underwrite long-term capital expenditures when social overhead costs are involved that help reduce a national problem such as a housing shortage. In some European countries the national government participates in a very supportive way in new community development. In the Netherlands and several Scandinavian countries, the government has even adjusted interest rates to favor investment in new towns or diverted labor to high-priority new-community construction through preferential building and work permits. Reston and Columbia in this country depended entirely on private capital from oil and insurance companies. For the private sector to have to wait as long as the public treasury, for a return on investment is something most stockholders are reticent to undertake.

The economic life of American new towns was given a new lift under Title IV of the Housing Act of 1968, which authorized the Department of Housing and Urban Development (HUD) to issue loan guarantees to private developers of large-scale new communities. Previously, the argument was made that no private entrepreneur could lay out the necessary capital costs for public services, such as roads, sewers, schools, and recreation centers, as well as quality residential units. The concern was that a minimum of 20 to 30 years would be needed before the developer could break even when the new town finally filled to capacity and then generated adequate marginal revenue (notwithstanding inflationary trends or other unhappy market conditions). As a result of Title IV, developers found their obligations backed by the U.S. Treasury, and the necessary long-term capital could be borrowed through private debt placements or public offerings of a government note of indebtedness and at much lower interest rates than was possible previously. Up to $25 million was available for a single project, plus a total of $250 million for all loan guarantees. Title IV also provided for supplemental grants to local governments for open space and for installation of water and sewer lines in HUD-guaranteed new towns.

Thus the economic building blocks are being put in place, one by one, in the United States in ways rather similar to the public–private financing of new towns and new communities in Europe. Title VII of the Urban Growth and New Community Development Act of 1970 raised the ceiling on total loan guarantees to $500 million and extended the loan possibilities to public agencies such as New York State's Urban Development Corporation. HUD was authorized to issue loan guaran-

FIGURE 12-12.
Reston, Virginia, is often mentioned as an example of new town development in the United States. A major difficulty for American new towns is that of obtaining long-term private capital. This, in turn, raises the cost of housing beyond the reach of low- and middle-income residents. (Courtesy of Spencer W. Havlick.)

tees covering all costs of acquiring and developing land incurred by government authorities for projects they sponsor and 85 percent of the costs carried by private developers. By 1973 HUD had approved 10 new communities for federal guaranteed assistance programs, amounting to $226.5 million. Seventy more applications and preapplications were in the "hopper" that year alone. But by 1976, very little progress had been made. The first 10 new communities being built under this program are Jonathan, Minnesota; Saint Charles Communities, Maryland; Park Forest South, Illinois; Flower Mound, Texas; Maumelle, Arkansas; Cedar-Riverside, Minnesota; Riverton, New York; San Antonio Ranch, Texas; The Woodlands, Texas; and Gananda, New York. These projects are expected to house 687,000 persons in 223,452 units during the next 20 to 30 years.

As those of us who live in a metropolis gaze into the proverbial crystal ball, the future does not appear clearly. Some individuals predict grave disasters and eventual doom. Others see continued improvement in human conditions, along with continual growth. Still others describe sporadic setbacks in the quality of cities and in the kind of lives urbanites have, these being balanced out by incremental leaps forward, with the net result unchanged from the status quo we tolerate at the present.

With the automobile and suburbanization, the city of tomorrow may be a twin city with morning and evening migrations of people moving back and forth from their bedroom communities to their offices, schools, factories, and other activities. The trend seems to be toward placing civic functions and domestic functions in two geographical zones that are thoroughly separated from each other. The contrived student council or typical neighborhood civic meetings are increasingly balms to soothe the restless citizenry instead of meaningful, efficacious steps toward inventing, planning, and implementing goals for the public welfare. Localized blocks or neighborhoods become islands within the megalopolis. For their survival they often feel that they must build walls or ramparts of one kind or another in order to protect their own ideas and quality of life. In the process, they forget some of the ecological imperatives—imperatives that apply equally to the human ecosystem, such as succession, the strengths of diversification, regional climax, evolutionary opportunity, tolerance, limiting factors, habitat adaptability, and competition—cooperation.

An Annotated Outline of Contemporary Case Studies of Urban Ecological Problems

There has never been a major decision to modify the urban environment on purely biological or physical considerations. It is particularly important for enthusiasts of ecological improvement to learn about and reckon with the formidable human forces who believe that economic, social, and political criteria for decision making are much more important than any ecological guidelines. Obviously, trade-offs need to be made among economic, social, political, *and* ecological goals. The difficulty with ecological goals is that they take much longer to achieve than the others ordinarily and, if they are not achieved, the deterioration of the human environment tends to proceed at a slow, subtle pace compared to the other three factors. The truth of the situation, of course, emerges with the realization that all four goals are dependent on one another and no one can be properly achieved without the fulfillment of the others.

Effective environmental management depends, to be sure, on understanding chemical and biological cycles, population dynamics, and many other strictly physical phenomena. But one must also have a keen understanding of the "ecology of the bureaucracy." It is the dynamism of human institutional arrangements that often baffles the dedicated biologist, who cannot understand why a marsh is drained for a housing project or why a forest is bulldozed for a new highway. Critical components of the institutional arrangements of a city are the legal, political, economic, and administrative mechanisms that permit or prevent sound environmental management to take place. Zoning ordinances, water and air pollution control laws, floodplain management, and open-space policies all play an important part in managing any given urban ecosystem and its contiguous ecosystems of forest, lake, prairie, or estuary. One of the most important pieces of federal legislation that forces analysis of ecosystem impact is the National Environmental Policy Act (NEPA) of 1969. The curious feature about NEPA is that it requires proponents of significant federal programs and projects to make an assessment of what impact the proposed activity will (or might) have on the *human* environment. In almost every federal environmental impact statement that is prepared under the NEPA guidelines, the impacts to or by urban dwellers are involved. Over 20,000 impact statements have been filed and are on public record.

For students of urban ecological issues, environmental impact reports or statements usually contain insight into possible impacts on the human and nonhuman environment. These publications are available to the public. The lead governmental agency has copies for public inspection. It is suggested that you may wish to obtain an environmental impact

FIGURE 12-13.
An environmental impact statement would be required on an airport expansion. (Courtesy of Spencer W. Havlick.)

statement (EIS) of a federal project or activity which has been proposed in or near your urban area. Concentrate on a federal EIS written since 1970, when NEPA went into effect. In fact, the more recent documents are more comprehensive, thanks to citizen pressure and court decisions that have forced "full disclosure." Every urban area in the United States has had an environmental impact statement prepared for proposed projects of one kind or another. Wherever federal monies make up a major funding source, an ecological assessment of the project is required. Some examples that may give you a clue about nearby projects that probably required an EIS are as follows:

1. An airport expansion (Figure 12-13).
2. A flood control project (Figure 12-14)

FIGURE 12-14.
Morganza control structure, 20 bays open. (Courtesy of the Corps of Engineers, U.S. Army, New Orleans District.)

3. Acquisition of a federal wilderness area (Figure 12-15).
4. Construction of a pipeline or transmission line (Figure 12-16).
5. A permit to operate a nuclear power plant (Figure 12-17).
6. A new wastewater treatment facility or enlargement (Figure 12-18).
7. Building an extension of the interstate highway system (Figure 12-19).
8. Granting a permit to strip mine coal from land controlled by the Bureau of Indian Affairs (Figure 12–20).

These and many other current project proposals would show how much or how little is known about current skills in ecological assessment. In the specific urban influence area of the proposed project, data may be insufficient to do an adequate analysis. In another situation, perhaps the lead federal agency would be embarrassed by the ecological impact of a project, and so the EIS does not portray the entire story. You may find it a provocative experience to review an EIS in your area and determine how accurate and complete the document is by whatever standards of acceptability you establish. If an ecological assessment does not exist on a proposed local project of some importance, try your hand at drafting a statement about its ecological impact on the urban ecosystem near you.

To give you an idea of what you might discover, we shall use an illustration from an area familiar to the author, San Jose, California. This example of a new subdivision will be an actual case similar to dozens of other project proposals that occur at the fringe of a large urban area. Perhaps the one unique circumstance is that the state of California has

FIGURE 12-15.
Olympic National Park. Blue Glacier on Mount Olympus. (Courtesy of the U.S. National Park Service.)

FIGURE 12-16.
Construction of a pipeline. (Courtesy of the Pacific Gas and Electric Company.)

FIGURE 12-17.
Cooling towers attached to a nuclear power plant. Each atomic plant requires compliance with the Environmental Protection Agency and the filing of an environmental impact statement. (Courtesy of Spencer W. Havlick.)

borrowed the intent and language of NEPA and applied it to *both* public and private projects which come up for approval. Every major construction project that may have a significant effect on the environment must have an environmental impact report prepared which is judged to be accurate and complete by several public review bodies. After approval of the findings, the elected legislative body (e.g., city council or county board of supervisors) assumes the legal and moral responsibility of implementing the policy set forth in the report. In California, the environmental impact report must include measures that will help minimize or mitigate the adverse impacts of a project (Figure 12-21). The mitigative features and possible alternatives to the project are the creative paragraphs in a document that otherwise tends to be traditionally descriptive. Another excellent California variation from NEPA is a section that must state the positive or negative impacts if "no project" occurred.

The case in point is a proposed housing development at the southwestern boundary of San Jose. In the last several decades San Jose has grown from a tiny agricultural and food-processing center in the Santa Clara valley to a sprawling metropolis of approximately 575,000. The class I agricultural lands have been converted to a somewhat monotonous swath of residential structures. The city is approximately 19.3 kilometers (12 miles) wide and 56.3 kilometers (35 miles) long, with a most irregular boundary. Part of the proposed housing project under discussion straddles the city limits, with nearly 40 percent of the parcel extending into the foothills of the Santa Cruz Mountains not far from the San Andreas fault. The housing tract is projected to contain about 800 units on 125.8 hectares (311 acres), thus adding nearly 2400 people. For the last 10 years San Jose has added roughly 21,000 persons to its population each year, so this project represents more than 11 percent of the total annual growth.

A curious feature about this housing project is that almost half of the site would require cut-and-fill and grading excavation on slopes with a

gradient up to 70 percent. Habitat of diverse flora and fauna would be eliminated, including the golden eagle and bobcat. The agricultural area would no longer produce the apricots, prunes, walnuts, and row crops which have been harvested in the past. The elimination of trees and grazing land would be an aesthetic loss to the 28,000 residents who live in viewing distance on the valley below the project. However, these on-site costs to the natural ecosystem are dwarfed by the off-site impacts that would result if the project were to be built.

The consequences of locating new housing units beyond a line where municipal services are available raises a question about the costs of providing new services to new dwellings when present facilities and services are inadequate for a population already in place. About 70 percent of Santa Clara County population increase is attributed to net immigration from other areas. The impact of 2400 "instant" people creates serious problems in a community where growth has been more rapid than schools, roads, fire and police protection, libraries, water, electricity, and other services with which they could keep pace.

It is estimated that each new housing unit of this economic category ($65,000 to $100,000) generates 12 automobile trips per day on the average, which produces 9600 additional auto trips per day on roads and expressways presently jammed with twice their designed capacity. The critical smog conditions would be further aggravated, as would the noise problem along present roadways. The physical impacts are relatively easy to list as the web of influences reaches out to intensify the off-site problems of solid waste disposal, water supply, and wastewater treatment. Usually, other net results include poorer phone service, less responsive police and fire protection, heavily impacted schools, and supersaturated recreation facilities. Despite the seriousness of each

FIGURE 12-18.
Aeration ponds that clean the water used by this Springfield mill of the Weyerhaeuser Company. An environmental impact statement is required, as the effluent is eventually placed in a navigable stream. Navigable streams and waterways are under federal jurisdiction, including the Environmental Protection Agency. (Courtesy of the Weyerhaeuser Company.)

FIGURE 12-19.

View of a highway system in New York State. Here is another example of a development that requires an environmental impact statement. (Courtesy of the U.S. Department of Housing and Urban Development.)

added impact on the biophysical environment, even more critical questions emerge as we consider the total urban ecosystem from the human perspective.

What provision is made for low- or modest-income housing? How does this project accommodate the elderly, the handicapped, the small children, or the working mother? Are immediate, short-term construction jobs created for workers who could never afford to live in or near the units they are building? Who are the direct recipients of the external costs and benefits of a new project of this kind? Do the inner-city poor have to wait another 5 years for a new fire station as the priority gets shifted to the fire station in the elite suburb? Each of these questions is of profound ecological significance if one is to understand the subtle functions of the urban ecosystem. The issue of human equity is always a difficult one to raise, but it must be answered if a realistic assessment of an urban environment is to be made.

Once the social implications are added to the biological and physical considerations of our proposed project, it becomes understandable why the modern metropolitan area is the most complex environment that civilization has created. Of course, man's urban communities do not totally fit the biological parameters of an ecosystem, because their sphere of influence goes beyond the city limits and beyond the national borders to include the total spaceship. Our atmospheric resources, mineral

FIGURE 12-20.
Removing coal from the bench by strip mining. The strip mining of coal from land controlled by the Tennessee Valley Authority or Bureau of Indian Affairs, in national forest, or on other federally controlled land requires an environmental impact statement. (Courtesy of Spencer W. Havlick.)

resources, energy resources, and virtually all others have become common resources with a world urbanizing system which C. A. Doxiadis and others call ecumenopolis. A new subdivision in San Jose literally is affected by what happens in the world biosphere, and in a multitude of ways, the new housing tract has implications, needs, and impacts that affect the farthest corners of the earth.

Let us take a final look at our new housing project and see what modifications or mitigating factors might be incorporated to lessen the severity of the adverse consequences, such as avoiding the archeological sites which are located on the land to be developed. One of the first actions should be to provide tax relief at the state and federal level for those who wish to continue agricultural activities or to maintain open space. The county assessor is forced to tax agricultural land at its highest potential rate. This procedure provides compelling incentives for agricultural landowners near cities to sell part or all of their land to speculators and developers. The rancher or farmer eventually walks away six times a millionaire if 121.4 hectares (300 acres) are sold at $49,423 a hectare ($20,000 an acre), which, incidentally, is about one-half the current market price of land described in our case study. Other alternatives or mitigative possibilities to soften the ecological, social, and economic impacts (Figure 12-22) include:

1. A housing project of substantially smaller size.
2. A residential-service-oriented community with rentals and purchase costs for diverse income levels.

FIGURE 12-21.
An environmental impact report is required in California for private projects such as housing developments. If this tract, located on the San Andreas fault in Daly City, were to have to file an environmental impact statement, serious questions would be raised about slope and seismic problems. (Courtesy of Spencer W. Havlick.)

3. A project to be built over a 10- to 20-year period.
4. Non-automobile-dependent transportation facilities, such as bicycle paths and personalized rapid transit units to link with high-speed arterial systems.
5. Mandatory supplementary solar heating and cooling.
6. Mandatory water conservation practices in plumbing, pricing, and non-water-dependent landscaping.
7. Maximum insulation requirements.
8. Preservation of the natural vegetation and topographic features.
9. No housing development on this site at all. Donate land for addition to the county park system as a tax-deductible item instead of a tax liability.

If the complexity of an urban ecosystem begins to overwhelm you, your research efforts have been thorough. You have discovered the limitless interrelationships, the almost untraceable energy flows, the amazing linkages of competition and cooperation of chemical cycles, political campaigns, power struggles, and the infinite human spirit that seems to make great cities of the world immortal.

It may also occur to the thoughtful observer of urban environments that no one individual can be a practicing urban ecologist. The target is so vast, so dynamic, so experimental, so evasive, and so complex that urban ecological research requires a remarkable team effort made up of sociologists, economists, biologists, engineers, psychologists, architects, geologists, legal researchers, philosophers, and many others. In fact, each of us assumes the multiple roles of guinea pig and research director simultaneously, as we carry out personalized environmental impact reports on our daily automobile use or on our individual water and energy use.

Despite the afflictions of the urban ecosystem or the human settlement (whichever you prefer to call your site of habitation), there seem to be

FIGURE 12-22.
The potential impact of the vegetation and other natural features is very substantial when a huge housing project is proposed on the land shown here. Only by abiding by ecological principles will urban ecosystems ever be able to integrate with the more pervasive ecological imperatives that have proved to be workable over thousands and millions of years. (Courtesy of Spencer W. Havlick.)

ample advantages to continue the urban experiment which began nearly 10,000 years ago. Cities with proper design do provide for an optimum potential for human contact. Improved educational, medical, economic, and cultural opportunities are found in great cities like London, New York, and Tokyo instead of in the rural hinterland. And cities can be built with economies of scale and efficiencies that do not dehumanize the residents. Great towns and cities thrive within the urban ecosystem of the world when the happiness and security of all the residents are goals. Out of community dignity and pride come the great cultural storehouses of civilization, the museums, the libraries, and the great cathedrals (which never would have passed a benefit–cost ratio test in Congress). Dynamic cities, like biological climax communities, have clear clues about their heritage and about their future. Consumers, producers, parasites, saprophytes, and predators are always present.

The unknown variable in the urban ecosystem of the future is you and what niches you decide to live in.[2]

Summary

As one contemplates urbanization in the 1980s and beyond, hope and despair emerge intertwined. The urban problems of the past have been well documented. However, at least one optimistic possibility emerges in future urban development if builders of cities begin to consult and utilize ecological principles that have been operative in biological communities for thousands of years. Human settlements can be built that abide by laws of thermodynamics and ecological imperatives. Future urbanites, once informed about old constraints and new opportunities for creating

[2] Portions of this chapter have been extracted from S. W. Havlick's *The Urban Organism* (New York: Macmillan Publishing Co., Inc., 1974), pp. 25–30, 409–414, 470–475.

humane, balanced communities will be required to ponder the next large decision, which deals with equity of resources among the poor and the rich in the United States and all the other inequities borne by passengers on this spaceship of finite resources and uneven energy supplies.

Humankind as a total species never needed to worry about chronic inequities of goods and services before megalopolitan urbanization became a reality. We have just begun to learn and experiment with new towns and new life-styles and hopefully new "consumer–producer" patterns that will produce a harmonious equilibrium between the urbanite and biological–physical systems that have the survival of a species as an inconsequential by-product. Whether or not human settlement at the national and global scale becomes a benefit or a cost to the majority of people will be a function of how many sacrifices the minority (of predators) are willing to make for the majority.

Perhaps one of the best chances for people making their environment work for them will be in the way people plan, design, and build their cities of the future and recycle or restore their great cities of the past.

Selected References

BERRY, B. J. L., and F. E. HORTON. 1974. *Urban Environmental Management.* Prentice-Hall, Inc., Englewood Cliffs, N.J.

BOUGHEY, A. S. 1975. *Man and the Environment.* Macmillan Publishing Co., Inc., New York.

Council on Environmental Quality. 1974. *Environmental Quality* (The Fifth Annual Report of the Council on Environmental Quality). Government Printing Office, Washington, D.C.

CURTIS, V., ed. 1973. *Land Use and the Environment.* Washington, D.C.

DANTZIG, G. B., and T. L. SAATY. 1973. *Compact City: A Plan for a Liveable Environment.* W. H. Freeman and Company, Publishers, San Francisco.

DETWYLER, T. R. 1971. *Man's Impact on Environment.* McGraw-Hill Book Company, New York.

DETWYLER, T. R., and M. MARCUS, eds. 1972. *Urbanization and Environment.* Duxbury Press, North Scituate, Mass.

EHRLICH, P. R., and R. H. EHRLICH. 1972. *Population, Resources and Environment.* W. H. Freeman and Company, Publishers, San Francisco.

GOTTMAN, J. 1975. The evolution of urban centrality. *J. Ekistics* **39**(223), 220–228.

HAVLICK, S. W. 1974. *The Urban Organism.* Macmillan Publishing Co., Inc., New York.

McHARG, I. L. 1969. *Design with Nature.* Doubleday & Company, Inc., Garden City, N.Y.

MILLER, G. T., JR. 1975. *Living in the Environment—Concepts, Problems, and Alternatives.* Wadsworth Publishing Company, Inc., Belmont, Calif.

MUMFORD, L. 1968. *The Urban Prospect.* Harcourt Brace Jovanovich, Inc., New York.

ODUM, E. P. 1974. *Fundamentals of Ecology.* W. B. Saunders Company, Philadelphia.

Real Estate Research Corporation. 1974. *The Costs of Sprawl.* Report prepared for CEQ, HUD, and EPA. Government Printing Office, Washington, D.C.

SCHMID, J. A. 1974. The environmental impact of plants and animals. *J. Ekistics* **37**(218), 53–61.

SMITH, R. L. 1972. *The Ecology of Man: An Ecosystem Approach.* Harper & Row, Publishers, New York.

STEADMAN, P. 1975. *Energy, Environment and Building* (A Report to the Academy of Natural Sciences of Philadelphia, Cambridge Urban and Architectural Series). Cambridge University Press, New York.

WHITTAKER, R. H. 1975. *Communities and Ecosystems.* Macmillan Publishing Co., Inc., New York.

WHYTE, W. 1968. *The Last Landscape.* Doubleday & Company, Inc., Garden City, N.Y.

13

The Population Problem: Its Implications for People and Plants

George F. Estabrook

Questions for Consideration

1. How to better understand, predict, and control the dynamics of populations of organisms is one of the central questions asked by population biologists. What are some reasons why answers to this question are important?

2. The text mentions green plants as examples of creatures that might be considered friends of people; the text also mentions animals that eat our crops as examples of creatures that might be considered competitors or enemies of people. Make two lists, one called "Friends" and the other called "Enemies," of creatures you think might be friends or enemies of people. Are there any creatures on both lists?

3. The text mentions birth, death, and migration as three factors that can change population size. Consider each of the following populations: the students in your school; the people in Ireland in 1846; the mosquitoes in the Canadian Northwest Territory in the fall of the year; the people in Philadelphia in 1970; the ragweed at the side of a newly constructed road (Figure 13-1); a forest whose trees are selectively cut to provide a continuous supply of timber; the algae in a small lake near which many summer cottages have just been built. What is the relative importance of each of the three factors noted above within each of these populations?

4. The intrinsic rate of natural increase of a kind of organism is the maximum per individual growth rate for population of that kind of organism living under ideal conditions. This concept enters frequently into discussions of population dynamics. Discuss the difficulties that might be encountered in trying to determine its value for people, for the photosynthetic open-sea algae that reoxygenate the air, for other kinds of creatures that might be interesting to you. Can you think of a more practical-to-measure concept that might substitute for intrinsic rate of natural increase?

5. The carrying capacity for a given kind of organism, of a collection of resources, is the maximum number of individuals of that kind that can live indefinitely (i.e., for a long sustained period) on these resources. Clearly, the nonrenewable resources in a collection contribute nothing to the real carrying capacity of the collection. Thus consuming nonrenewable resources permits a population to actually exceed the real carrying capacity of its resources. What other factors might permit a population to exceed the real carrying capacity of its resources? How do these factors apply to people?

FIGURE 13-1.
Common ragweed (Ambrosia artemisiifolia), a weed that is ubiquitous along roadsides in the summer and a pioneer invader species that appears along newly constructed roads and other excavated areas. (Courtesy of Larry Mellichamp.)

6. All the various kinds of creatures came to be the way they are through the process of evolution by natural selection. Why does this fact lead us to expect difficulties in trying to impose any program to control the human population of the world?

Why Study Population Biology?

Of all the people alive today, approximately 143,000 of them will be dead tomorrow.[1]

How to better understand, predict, and control the dynamics of populations of organisms (including human beings) is one of the questions Peter Kaufman of the University of Michigan raises in his course, entitled "Plants, People, and Environment." This question, now

[1] This number is calculated from an estimated world population of 3.6 billion in 1970, an estimated population growth rate of 0.02 individual per individual per year, and an estimated death rate of 0.013 individual per individual per year. These estimates are given by Frejka (1973).

an important subject for research by population biologists, should merit your attention for several reasons.

1. As the resources on which the human population currently depends reveal themselves to be in limited supply, the benefits to be gained by deliberately planning their use become more apparent. Many of these resources, for example, the forests, the fisheries, and the soil, are themselves made up of populations whose dynanics must be better understood if we are to use these resources most effectively.

2. Good planning requires accurate predicting not only of the future availability of resources, but also of the future demand that will be made on them. Thus it is important to be able to *predict* the number of human beings of each age that will be alive at any given time in the future.

3. No kind of creature lives completely independently of every other kind of creature. Human beings are *not* an exception. We have allies and competitors among the other species in the world. For example, the soil microorganisms that maintain the soil as a suitable place to grow our crops, the crop species themselves that provide us with food, all the green plants that capture the sun's energy in a form that we can use and that replenish the oxygen in our atmosphere might be considered among our allies. Herbivorous animals, pathogenic fungi, bacteria, and viruses that attack our crops, our domestic animals, and our own bodies might be considered among our competitors or enemies. Now that modern technology has given us the power to change the world very rapidly, it is even more important that we understand the dynamics of the populations of our friends and enemies so that we do not inadvertently do ourselves a "grave" disservice.

4. The maintenance of the good health of our own bodies depends to a large extent on the preservation of the proper balance among the populations of organisms that inhabit us. Especially important are the microorganisms that live in our respiratory and digestive tracts. When these microecosystems are in balance, it is more difficult for pathogenic organisms to colonize us. A better understanding of this aspect of population biology will help our medical technicians serve us more effectively.

5. The most important and immediate reason why the study of the dynamics of populations of organisms should merit your attention is because organisms are interesting things to think about and fascinating things to look at. Where do the millions of flies and mosquitoes, of which we are made so aware in the summer, go when winter comes? How do these populations arise again the next spring? Thirty years from now, what trees will grow where we now see abandoned farmland dotted with young trees and shrubs (Figure 13-2)? Why are there more eastern bluebirds now than there were 20 years ago? An awareness, and increasing understanding of the creatures with whom we share the earth contributes greatly to the quality and excitement of our lives. We should, therefore, be motivated to learn how to behave ourselves in such a way that the earth will continue to support not only human beings but all creatures.

The most accurate predictions, and the wisest controls, must ultimately come from the best understanding of the factors that influence

FIGURE 13-2.
An old field in Michigan, viewed from abandoned farmland that is being re-colonized by young trees and shrubs. The dark ever-green trees are the common upright juniper, Juniperus virginiana. (Courtesy of Larry Mellichamp.)

population size and structure, and of the mechanisms through which the effects of these factors are mediated. There is much that we do not understand, but some basic concepts have been put forth that may help us to organize our thinking and to ask specific questions that can be the basis for scientific experiments and research. We shall introduce a few of these basic concepts here and then discuss how we might use them to think about human population dynamics.

The Factors That Change Population Size

BIRTH

All organisms that have not yet become extinct reproduce. The rate at which they do so is believed to be influenced by many factors. Any thing or condition that is necessary for an organism's reproduction can be considered a resource. Thus sunlight is a resource for all green plants. Temperatures above freezing would be a resource for some green plants. Various kinds of food and shelter are also resources for animals as well. Space is an important resource for both plants and animals, and especially man.

We can define *birth rate* to be an operationally meaningful concept in many ways. The total number of individual organisms born into a population in a unit of time is an obvious possibility. According to this concept, we would expect larger populations to have higher birth rates. For this reason, most population biologists feel that it is more meaningful to define birth rate on a per individual basis.

DEFINITION:
The birth rate of a population is the number of individuals born into a population during a unit of time, divided by the number of individuals in that population at the beginning of that unit of time.

FIGURE 13-3.
Staghorn sumac (Rhus typhina) in summertime, with shaggy inflorescence and pinnately compound leaves. (Courtesy of Larry Mellichamp.)

In the study of plant populations we often speak of recruitment rather than births. Even this is inadequate in the study of some plant populations, where the concept of an individual is itself elusive. Entire hillsides of big tooth aspen (*Populus grandidentata*), or staghorn sumac (*Rhus typhina*; Figures 13-3 and 13-4) are often made up of genetically the same organism (these are called *clones*).

Often, it is instructive to subdivide a population into age classes. All individuals in the population under study, whose ages are between 5 and 10 years, would be an example of an age class. Populations can also be subdivided into developmental classes. In a mouse population, for example, all prepuberty mice that have been weaned would be an example of a developmental class. Birth rates are defined for such classes just as they are defined for populations. As we might expect, birth rates vary greatly from one class to another within the same population. In many human populations, birth rate remains at zero for prepuberty classes, rises to a maximum in the mid to late twenties, and falls to zero again at menopause after the age of 40.

In many populations, individuals can be distinguished on the basis of sex. Since males and females of a given age class may not contribute equally to the birth rate of that age class, and because it is often difficult to measure the contribution of males to birth rates anyway, birth rates are often determined only for the female segment of a populaton. In this case, the birth rate would reflect only the females born. A typical birth rate for human beings is 0.02 to 0.04 per year.

But birth is only one of the factors that affects population growth.

DEATH

A single organism cannot live forever. It will eventually die, and at what age it does so has an important influence on the growth and age structure of its population.

FIGURE 13-4.
Old abandoned farmland, where one of the invader shrubs is staghorn sumac (Rhus typhina), identified by the shrubs in the lower right corner with the dark-appearing infructescences. (Courtesy of Larry Mellichamp.)

Similar to the concept of birth rate that was defined above, *death rate* will be defined on a per individual basis.

DEFINITION:
The death rate of a population is the number of individuals in that population that die during a unit of time, divided by the number of individuals in the population at the beginning of that unit of time.

Many of the factors that influence birth rate also influence death rate, although not in exactly the same way. For example, if there is a shortage of food, the reproductive activities of some may decrease, while others will starve to death. A contagious disease may cause many more deaths per individual in areas of high density (many individuals per unit area) than in areas where individuals are more widely spaced. Death may also come from cataclysmic factors such as being run over by a snowmobile, or being caught in a natural disaster (e.g., a flood) or an unnatural disaster (e.g., radioactive fallout from military bomb testing). Predation (animal eating) and, to some extent, herbivory (plant eating) are often significant causes of death in natural populations of animals and plants.

Considerations of the causes of death are very important in understanding and predicting the changes in population size. For any particular population under study, the relative importance of various causes of death contributes much to the determination of the ecological role of that population. One very general classification of causes of death that seems to have theoretical validity recognizes two basic kinds: (1) those causes that result in a *higher* death rate (in the sense of our definition above) when the population density (individuals per unit area) is high, and (2) those causes that contribute to the death rate equally at high and low population densities. The first kind is often called *density-dependent.* Examples are limited supplies of food and shelter, faster spread of disease, and increased exposure to predators. The second kind is often called *density-independent.* Examples are old age, earthquakes, or an unusually late spring frost.

MIGRATION

Individuals entering or leaving a population other than by birth or death can cause an increase or decrease in the size of a local population. *Immigration* ("in + migration") and *emigration* ("ex + migration") *rate* will be defined on a per individual basis.

DEFINITION:

The immigration (or emigration) rate of a population is the number of individuals entering (or leaving) that population other than by birth (or death) during a unit of time, divided by the number of individuals in the population at the beginning of that unit of time.

Birth and death rates are always important to an understanding of how the size of a population changes. Unless the population is soon to become extinct, both these numbers are positive. Migration may, or may not, be a factor in the dynamics of a local population. For many kinds of plants and animals (especially those with short life spans and high birth rates) that depend on finding new disturbed habitats to colonize, emigration (often called dispersal in this context) is very important. For populations whose numbers are small relative to the availability of resources, immigration may also be an important factor. Occasionally, immigration and emigration are equal self-canceling effects, or, as in the case of the world human population, not presently considered possible.

POPULATION GROWTH RATE

The population growth rate is the result of the combined effects of the birth rate, the death rate, the immigration rate, and the emigration rate. Thus, it, too, is defined on a per individual basis. It can be observed and measured directly as indicated.

DEFINITION:

The growth rate of a population is the number of individuals in the population at the end of a unit of time minus the number of individuals in the population at the beginning of a unit of time divided by the number of individuals in the population at the beginning of that unit of time.

Population growth rate can also be derived indirectly as: [birth rate] plus [immigration rate] minus [death rate] minus [emigration rate]. Population growth rate, unlike the other rate concepts that we have defined so far, can be negative in the event of a declining population.

Population dynamics means the changes in the size of populations with the passage of time. Numerical quantities, such as the size of a population, that are not necessarily the same at different times, are called *functions of time.* If we use the symbol N to represent the number of individuals in a particular population, it is very helpful to use the notation $N(t)$ to represent the number of individuals in that population at time t. Time is usually given in the number of units of a specified length that have elapsed since a specified absolute time (or date). The notation $N(t)$ reminds us that population size is a function of time.

The concepts of birth rate, death rate, immigration rate, and emigration rate that have been defined above are, often, also functions of time; they may depend on the particular interval of time for which they were calculated. Performing the calculations specified by the definitions may result in different values at different times during the growth of a population.

INTRINSIC RATE OF NATURAL INCREASE

Suppose that a population of mice were being bred and raised in a laboratory. The mice were kept in clean and roomy cages, given plenty of food and water and nesting material, and maintained in the temperature and humidity regime most conducive to their well-being; in short, they were raised under ideal conditions. The population would increase. The growth rate could be calculated with the procedure described above. When every external or environmental condition that can contribute to the growth rate of this mouse population has been provided, what prevents the population growth rate from increasing any more must be something intrinsic to the mouse itself and *independent* of the environment. If we also measured the maximum growth rates under ideal conditions for a population of cattle, and for a population of fruit flies, we would expect to get different maximum population growth rates for fruit flies (high), for mice (not so high), and for cattle (lowest).

DEFINITION:
The intrinsic rate of natural increase of a kind of organism is the maximum (per individual) growth rate for a population of that kind of organism living under ideal conditions.

There are two important things to notice about this definition.
1. This definition would suggest that, for a specified time unit (e.g., per year), to every distinct kind of organism can be associated a number, called the intrinsic rate of natural increase of that kind of organism, and this number depends in no way whatsoever on the environment.
2. This is the first definition we have encountered that does *not* include calculating (or measuring) instructions. It is extremely important that scientific definitions contain measuring or calculating instructions, if possible. Some would argue that if a definition fails in this regard, it is not a scientific definition. This is a very strong point.

One way to measure intrinsic rate of natural increase of a kind of organism is to raise a population of that kind of organism under ideal conditions. How do we know when conditions are ideal? We never know for sure, but sometimes we can be quite confident that our estimates are not too far off. We can usually be reasonably certain that there is no immigration or emigration, and that density-dependent death factors are very low (by providing plenty of space, food, etc.). Density-independent death factors cannot all be eliminated because death from old age is a proper component of the intrinsic rate of natural increase. In laboratory conditions, cataclysmic death factors can usually be controlled. Birth rate may be more difficult to maximize, as in many kinds of organisms there are environmental and behavioral factors in mating, gestation, and child rearing that are unknown to man and very difficult for us to study. A detailed discussion of methods for calculating intrinsic rates of natural increase can be found in the book by Andrewartha and Birch (1954).

The concept of intrinsic rate of natural increase has been an idea in biology for many years. In his book published in 1931, R. N. Chapman proposed the term "biotic potential" to represent this concept. It is safe to say that we virtually never observe populations in nature growing for long at their intrinsic rate of natural increase. Rather, what we observe is a rate less than this (of course), and usually much less, and not uncommonly negative for some populations during some periods. Chapman suggested the term "environmental resistance" to include all the factors that reduce the intrinsic rate of natural increase to the observed growth rate of a population. Although neither of these terms is in common use today, their introduction in 1931 evidences the fact that biologists have been thinking about these ideas for many years.

CARRYING CAPACITY

In measuring intrinsic rate of natural increase, we provided unlimited quantities of consumable resources and adjusted environmental conditions to be optimal for population growth. Things are rarely, if ever, this ideal in nature. Consumable resources are provided at a rate that is more or less independent of population size. Nonconsumable (in the short term) resources, such as space, are fixed in amount, and environmental conditions, such as temperature, are whatever they are. For a particular habitat, with its resources fixed in this way, it is interesting to ask: What is the maximum population size that these fixed resources can support?

DEFINITION:
The carrying capacity for a given kind of organism of a collection of specified resources (such as those found naturally on a designated piece of land, or within a designated pond, etc.) is the maximum number of individuals of that kind that can live on those resources for a sustained period of time.

Similar to the definition for intrinsic rate of natural increase, this definition of carrying capacity has two properties:

1. It would suggest that for a specified set of resources, to every kind of organism can be associated a number, called the *carrying capacity*, of those resources for that kind of organism, and that this number depends in no way on time.

2. The number defined as carrying capacity is usually even more difficult to measure than is intrinsic rate of natural increase. Even if we are generous and say that "everything" is a resource, some resources are not really provided at a rate but are actually used up irreversibly or are renewed so slowly that, on the time scale of lifetimes, they are not renewed. Other "resources," such as climate, have a strong random component that becomes difficult to control or predict. Some organisms have the capacity to "improve" their resources so as to support more individuals. How do we choose the time scale so as to determine how long is a "sustained period of time"? Unlike rate of natural increase, it is possible to encounter in natural populations that exceed (temporarily) good estimates of carrying capacity. Sometimes in laboratory situations, when conditions are held "constant" and "identical" populations are allowed to increase to a maximum so as to measure experimentally carrying capacities, different maxima are achieved.

However, in many cases, we can derive numbers that we might call carrying capacities, with strong predictive properties when used in the theories of population biology, so the concept does often meet one of the important tests of scientific validity. Furthermore, the idea makes sense conceptually, and you will find it useful as a guide for thinking about populations, especially our own.

UNLIMITED GROWTH

Mathematical Models for Calculating and Predicting Changes in Population Size

T. R. Malthus suggested in the early nineteenth century that the population size increases as a geometric series. Exactly what could this suggestion mean? In order to discuss the idea both simply and clearly, let us use the following notation:

$$N(0) = \text{number of individuals in a population at some stated time}$$

and

$$N(t) = \text{number of individuals in that population } t \text{ time units (e.g., days or years) later}$$

Now we can express Malthus's 150-year-old idea as

$$\frac{N(t + 1)}{N(t)} = \lambda \qquad (1)$$

where λ is a positive constant number that is the same for all values of t and λ is the rate of multiplication of the population. If time were measured in years, for example, λ would be the yearly rate of multiplication. In a population that increased as a geometric series, if $N(3) = 100$ and $\lambda = 1.02$, then

$$N(4) = N(3) \times \lambda = 100 \times 1.02 = 102$$

In words this says: If after 3 years the population has 100 individuals, and the yearly rate of multiplication is 1.02, then after 4 years, the population would have 102 individuals.

If we express λ as an exponential,

$$e^r = \lambda (e = 2.718 \ldots , \text{the natural base})$$

then r is often called the instantaneous population growth rate. If we express λ this way, in terms of instantaneous population growth rate, then we may write

$$N(t) = N(0)e^{rt} = N(0)\lambda^t \qquad (2)$$

Expression (2) shows the modern concept of *exponential growth*. It is logically the same idea as Malthusian growth in a geometric series determined by the Malthusian parameter λ. It differs only in its form of expression in terms of population growth rate, r. Figure 13-5 is a graph of an exponentially growing population.

For an exponentially growing population, the growth rate, r, is determined by many things. For this reason, it is often called the *realized growth rate* of the particular population under study, to indicate that it is the growth rate actually observed, and, as such, it reflects all the factors (known and unknown) that influence its magnitude. Usually, the realized growth rate of a population is *not the same* as the intrinsic rate of natural increase for the kind of organism comprising the population. If not equal, realized growth rate is, of course, less than intrinsic rate of natural increase. Wilson and Bossert (1971) and Solomon (1969) are examples of modern expositions of population biology that fail to maintain this distinction. Similar failures are common in popular writings as well.

LIMITED GROWTH

Just as organisms must die eventually, populations cannot go on increasing at the same rate forever. Their growth must be limited when resources run short. There are different ways to express this idea with the

FIGURE 13-5.

Model of unlimited population growth. $N(0) = 10$, $N(20) = 40$, r $= 0.693$ per 10 years, and $\lambda = 2$ per 10 years.

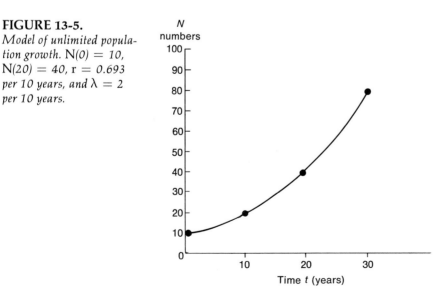

Probing the Nature of Our Environment

precision of mathematics, each leading to a slightly different prediction about population size. Two formulations are common. Each makes more general the idea of exponential growth discussed previously.

The instantaneous *per individual* growth rate in a population that is growing exponentially is r more individuals after a unit of time for every individual present in the population at the beginning of a unit of time. If r were 0.02 and there were 98 individuals at the beginning of a unit of time, there would be very nearly 100 individuals at the end of that unit of time.

Suppose that the carrying capacity of the place under study were K individuals. Suppose also that when $t = 1$, $N(t) \equiv N(1)$ was very small compared with K, so that no density-dependent growth-limiting factors are in effect and the population grows at rate, r. But suppose that density-dependent growth-limiting factors do take effect as the population increases in size, so that only a fraction of the number of individuals are capable of reproducing at rate r and the rest do not reproduce at all. It is reasonable to assume that this fraction is the proportion of unfilled carrying capacity:

$$\frac{K - N(t)}{K - N(1)} \tag{3}$$

Thus, at time t, the population grows at the rate

$$r = \frac{K - N(t)}{K - N(1)} \tag{4}$$

A graph of population growth limited in this way is shown in Figure 13-6.

This approach seems to predict population size well for grazed pasturage and certain agricultural crops. It also predicts the growth of domestic animals. (Brody, 1945).

LOGISTIC GROWTH

To require a decreasing total population growth rate with increasing population size, as does the model just presented, is to place a very strong requirement that is frequently not met by natural populations, especially when the sizes of these populations appear to be well below the carrying capacity of the resources of the place being studied. In these cases, the exponential growth model of expression (2) gives much better predictions. It is only when population size is close to the carrying capacity of the resources that the model shown in Figure 13-6 gives good predictions. A population growth model that is like exponential growth when population sizes are small relative to K, but like restricted growth when population sizes are near K, would make good predictions for many different kinds of populations.

Such a growth model results if we require that the per individual growth rate of individuals in the population decreases with increasing population size. The resulting population growth rate at time t is thus:

$$r = \frac{K - N(t)}{K} \tag{5}$$

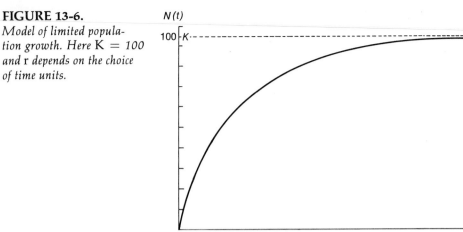

FIGURE 13-6.

Model of limited population growth. Here K = 100 *and* r *depends on the choice of time units.*

Figure 13-7 is a graph of a population growing in accordance with this idea. P. F. Verhulst, a Belgian, is credited (Batschelet, 1971) with suggesting this model for population growth in 1838. The curve of Figure 13-7 is often called a logistic curve.

Predictions made from these mathematical models are rarely *exactly* right, although frequently they are quite close. Natural populations rarely conform to the assumptions on which these models are based, for several reasons. Populations usually show age structure in which only some of the population is reproductively mature. Density-independent death and birth factors are not accounted for. Changes in weather, seasons, and other environmental factors may change enough during the life span of an individual in the population to have an effect, yet may not change enough for these effects to "average out." Yet, the general trends (if not the exact quantitative predictions) of the logistic growth curve are very common in biological populations, and the concepts of growth rate and carrying capacity are useful when thinking about our own population and the limited resources on which it depends.

Human Population Dynamics

EVOLUTION BY NATURAL SELECTION

Your ancestors and mine had at least one thing in common—they each had children. It is extremely unlikely that the desire *not* to have children will evolve. Individuals who *do not* want children generally produce fewer children than individuals who *do* want children (other things being equal). Thus fewer children will inherit the desire not to have children; and more children will inherit the desire to have children. If we expect anything to evolve consistently in all biological populations, it is the ability and apparent desire (as evidenced behaviorly) to have children.

This concept is at the very heart of the phenomenon of evolution by natural selection. An inheritable morphological, physiological, behav-

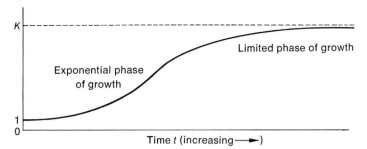

FIGURE 13-7.
Logistic model of population growth.

ioral, and so on, quality evolves in a population of organisms by virtue of the fact that organisms in possession of that quality have more children than organisms that lack that quality.

Most kinds of organisms that have ever lived are extinct. The kinds of organisms we see around us, including ourselves, are those kinds that have not yet gone extinct. There are many reasons why kinds of organisms become extinct. Important among them are (1) changes in their physical environment that occurred *too rapidly* for anogenesis (evolutionary change within a species) to keep pace; and (2) changes in other kinds of organisms (predators, herbivores, competitors). Paleontologists have shown that, in the history of life, phenomena called *adaptive radiations* are common. An adaptive radiation occurs when some inheritable quality evolves that makes its possessors especially good at surviving to reproduce. Creatures with this quality increase and spread rapidly. Mammals, for example, are especially good at regulating their internal temperatures and at feeding their young a truly superior food (milk). About 60 million years ago, during the epoch called the Paleocene at the beginning of the Tertiary period, mammals became very abundant. Shortly after special tissues for the transport of water evolved in plants, plants with this quality (vascular plants) became very abundant about 400 million years ago during the Devonian period. Perhaps half a million years ago, during the Pleistocene epoch, the fantastic central nervous system of modern man evolved to permit the rapid spread of human culture that apparently took place 10,000 to 12,000 years ago.

There is no reason to believe that the possession of such a central nervous system frees man from the pragmatics of the phenomenon of evolution by natural selection. If anything, our intelligence permits us to understand better how we and our world came to be, and to decide more deliberately how we want to behave in it.

THE QUALITY OF LIFE

It is difficult to define and measure the quality of life for many reasons: the personal or subjective factor is enormous; the theoretical foundations for our ideas differ; and, even if we could all agree on a criterion, the difficulties of measurement are likely to be great. Notwithstanding, we suggest three general considerations that we can make in the context of our discussions so far.

1. The quality of life increases as physical pain, discomfort, sickness, death, and so on, decrease (the medical approach).
2. The quality of life increases as availability of food, shelter, pleasure, and other resources increases (the American dream of affluence).
3. The quality of life increases as the number of children you can successfully raise increases (an evolutionary concept).

These considerations warrant some discussion.

DEATH CONTROL

One of the important causes for the increase in world population growth rate in the last few hundred years is increased "death control." Many of us apparently feel that consideration (1) above is quite important. We want no one to die. This was the case in Aldous Huxley's 1932 fantasy *Brave New World*.[2] Of course, we cannot live forever. We must die sometime. In *Brave New World* this is done with great efficiency, so as to waste a minimum of resources. At the appropriate age of death, age 70 say, the citizens of *Brave New World* could cheerfully and willingly step into the furnace. In contrast to this, modern society often spends great resources to prolong the life of an elderly human vegetable by a few months. We spend vast resources searching for a cure for "cancer" as if it were a disease. Perhaps we should call old age a disease, too.

Virtually every kind of organism is capable of producing offspring at a rate so high that, if *all* were to survive, the surface of the earth would become literally covered with that kind of organism. There are other reasons why it is important that not every organism survive to reproduce. The processes of mutation (over very long periods of time) and of recombination (over short periods of time) that produce genetic variability operate more or less at random. Many of these new genetic combinations are not so biologically adequate as the parents that produced them. Natural selection (death before reproductive maturity) has a conservative influence on biological populations by eliminating from the breeding population biologically inferior organisms. The conservative force of natural selection has been an important factor in the evolution of modern man.

The application of modern medical technology to control all kinds of death has undermined nature's conservative mechanism. Biologically inferior organisms with genetically based defects that would formerly have died may now survive to reproduce and spread their genetic deficiencies into the population. To this extent, medicine has had a deleterious effect on the quality of our species. Let us suppose that death control became nearly perfect, so that everyone who was born lived to be 70, and that birth control (a very unlikely phenomenon indeed) was practiced so well that babies were born only often enough to replace the 70-year-olds who step into the furnace. Under these conditions we would predict that the biological quality of human beings would deteriorate rapidly.

[2] You are encouraged to read this book and also Huxley's *Brave New World Revisited* (New York: Harper & Row, Publishers, 1958).

Death is a natural and important part of life. Perhaps controlling death is not what we want to mean by high-quality life. In fact, recent success with death control has contributed much to our population problems.

ÁFFLUENCE

The concept of affluence can also be related to our earlier discussion. Expression (6) has a factor that indicates how much "unused" resource is available:

$$\frac{K - N(t)}{K} \tag{6}$$

This is the fraction of the carrying capacity K that is unfilled by the population of size $N(t)$. When this number is close to 1, there is plenty to go around. When it is close to zero, there is very little to go around. Since our concept of affluence reflects wealth in excess of the bare minimum, in a *general way,* this fraction reflects affluence. Clearly, we maintain a "high quality of life" by maintaining a low value of $N(t)$ (relative to K). But this is in conflict with our desire for death control discussed earlier, and with our desire for *no* birth control (even among genetically inferior organisms). Therefore, $N(t)$ continues to rise.

Modern technology has saved us from death—it has saved us here too. Modern technology simply finds new ways to raise K. Whereas formerly we were limited to people and animal power to do our work, we can now burn fossil fuel (and even smash atoms) to power our lives. We can mine fertilizer and put it on our soils to raise crops, where formerly we had to wait for natural processes to build up the soil's level of natural fertilizer. The air and the sea are receptacles for our wastes. We know how to pave the ground for our dwellings and highways. If we can predict the future on the basis of the past, we should have every confidence that technology (the product of our highly evolved central nervous system) will continue to keep carrying capacity well ahead of population size (especially if we can convince more of the less affluent peoples to practice birth control).

In a sense, the environmental conditions that are necessary for human life have evolved, too. At one time the atmosphere did not contain oxygen but was a reducing atmosphere. Since that time, the high atmospheric concentrations of oxygen on which we depend have come to be. The layer of ozone (O_3) that surrounds the earth and protects us from high-energy radiation was not always there (see Chapter 9). The soils on which plant life (and thereby our own life) depends have come in to being relatively recently. Weather and climate, in part reflecting the thermal activities of the earth, have come to be supportive of human life when formerly they were not.

Modern life lives on the "interest from the capital" that nature has been "investing" for the last thousand million years or so. This natural capital, the resources that make life possible, limits the carrying capacity. Technology can raise the carrying capacity of the land somewhat through wise and careful management of our natural capital. But when technology permits the consumption of resources in such a way or at such a rate that the natural organization on which life depends is randomized, then technology raises the carrying capacity by spending natural capital. We can harvest renewable resources at their growth rates indefinitely. But perhaps some of our technological activities are spending natural capital and our environmental endowment will soon be gone. Mass extinctions, including that of our own species, will result.

LARGE FAMILIES

An understanding of the basic workings of evolution by natural selection teaches us to expect the desire to reproduce to be strong in human beings as well as in other kinds of organisms. We might expect much satisfaction and a sense of fulfillment to attend successful reproduction. Successful reproduction also means raising your children to reproductive maturity, as well as giving birth to them. We sometimes hear adults express the sentiment that they wish to make the world a better place for their children. This sentiment is understandable. They might wish also to improve the chances of reproductive success for their nieces and nephews, and brothers and sisters, and grandchildren as well. We would also expect this wish because of the genetic relatedness.

Birth control runs counter to the very forces that powered the evolution of human beings. That its practice meets with resistance now is only natural. In the name of restoring the quality of life, birth control threatens the consummation of the most basic component in the quality of life for any organism: successful reproduction.

ARE THERE ANY SOLUTIONS?

What about war? The biologically superior organisms exterminate the biologically inferior organisms, making up for the evil of modern medicine by "cleansing the gene pool" and reducing $N(t)$ to well below carrying capacity. This solution has been tried from time to time. This solution is not too popular among those judged to be inferior. It is important for us to understand that much of the relatively defenseless third world has good grounds for fearing this solution as a real possibility. In these days of modern military technology, we all must fear this solution.

If we rule out war, we are faced with other conflicts. Death control without birth control leads to lower affluence unless we raise the carrying capacity of the environment, which can only be done temporarily, and then only at the expense of the environment on which we depend for life. Death control *with* birth control leads to the deterioration of the biological quality of our kind and flies in the face of nature. It would probably require something like a police-state world government to enforce effective birth control. Perhaps some progress in birth control could be made if *extended* family units again became the basic economic unit. Desire for reproductive success might be in part fulfilled by contributing to the upbringing of nieces and brothers. But unless birth control is complete, the nonpracticers will have more children, and the fraction of persons willing to practice birth control will decrease.

Clearly, whatever we *do* about these problems as individuals, as societies, as governments, as a species of wild animal in its natural habitat, moral, political, and emotional as well as rational considerations must be made. A better understanding of the biology of populations—our own as well as those with which we interact—will contribute to the ultimate wisdom of those deeds.

Summary

Population biology undertakes to better understand, predict, and control the dynamics of populations of organisms. This is important to people because it helps us to more deliberately manage the resources on which we depend for our continued existence.

The factors that change population size include birth, death, and migration. These can be measured and combined to predict population growth rate. The intrinsic rate of natural increase of a kind of organism, and the carrying capacity of a collection of resources for that kind of organism, have been used to construct theoretical models of population growth. Although these models do not always predict with complete accuracy future population sizes, they do help us organize and think about the relationships among the factors affecting changes in population size.

Evolution by natural selection helps us understand why and how the creatures that are here came to be as they are. The desire to have children has probably been selected for in all creatures that are not yet extinct. This makes humans predisposed to resist birth control programs. Reproducing more progeny than can survive with the subsequent death before reproductive maturity of the genetically inferior ones is the natural selection mechanism for conserving genetically superior organisms. This process has been important in the evolution of the qualities that make people a superior kind of organism. Indiscriminate death control will contribute to the genetic deterioration of people.

Solving the world's population control and resource management problems is *not* simply a matter of getting people to cooperate. These are value judgments that must be made. Human beings *are* wild animals in their natural habitat. A better understanding of the biology and population dynamics of organisms can help us participate responsibly in making the judgments that affect the quality of life.

Literature Cited

ANDREWARTHA, H. G., and L. C. BIRCH. 1954. *The Distribution and Abundance of Animals.* University of Chicago Press, Chicago.

BATSCHELET, E. 1976. *Introduction to Mathematics for Life Scientists,* 2nd ed. Springer-Verlag, New York.

BRODY, S. 1945. *Bioenergetics and Growth. With Special Reference to the Efficiency Complex in Domestic Animals.* Van Nostrand Reinhold Company, New York.

CHAPMAN, R. N. 1931. *Animal Ecology with a Special Reference to Insects.* McGraw-Hill Book Company, New York.

FREJKA, T. 1973. The prospects for a stationary world population. *Sci. Amer.* **228**(3), 15–23.

SOLOMON, M. E. 1969. *Population Dynamics.* Edward Arnold & Co., London.

WILSON, E. O., and W. H. BOSSERT. 1971. *A Primer of Population Biology.* Sinauer Associates, Inc., Sunderland, Mass.

Suggested Readings

ADLER-KARLSSON, G. 1974. Towards a mutual growth moratorium. *Ambio* **3,** 101–106.

BARNEL, H. R. 1974. *Biology and the Food Industry.* Edward Arnold & Co., London.

BORGSTRÖM, G. 1974. The food population dilemma. *Ambio* **3,** 104–113.

HEER, D. M., ed. 1968. *Readings on Population.* Prentice-Hall, Inc., Englewood Cliffs, N.J.

HUXLEY, A. 1958. *Brave New World Revisited.* Harper & Row, Publishers, New York.

LANGER, W. L. 1972. Checks on population: 1750–1850. *Sci. Amer.* **226** (2), 2–9.

MELLANBY, K. 1972. *The Biology of Pollution.* Edward Arnold & Co., London.

PIRIE, N. W. 1967. Orthodox and unorthodox methods of meeting world food needs. *Sci. Amer.* **216**(2), 27–35.

SCHNEIDER, S. H. 1974. The population explosion: Can it shake the climate? *Ambio* **3,** 150–155.

WYNNE-EDWARDS, V. C. 1964. Population control in animals. *Sci. Amer.* **192** (8), 2–8.

Constructive Action and Solutions to Our Fundamental Problems

Our theme in this book is that people must learn to live with nature and not destroy it, especially as seen through plants. Now that we have dealt with plants and their way of life in Section I and probed into major ecological aspects of our environment (natural, agricultural, and urban) in Section II, we must look at solutions to our environmental problems. These solutions come from an understanding of how plants and animals interact, as we have shown in Section II. The most practical solutions we can invoke to solve our environmental problems are those which involve recycling, using alternative sources of energy, significantly reducing all kinds of pollution in our ecosystems, educating people how best to use and manage our resources, and starting to be responsible stewards of our natural environment. These solutions obviously mean altered life-styles. The time is now at hand when we must stop exploiting nature for our own consumptive appetites and start living with nature by returning to it the materials we use, by keeping our populations in bound, and by following the ways of nature (e.g., by using composting, methane generators, fusion, wind generators, and solar power for energy needs). Otherwise, this society, unless it changes its collision course with nature, will perish from the face of the earth in its own synthetic and exploitive quagmire. Hugh Iltis expresses this very pungently in Chapter 27, "Can One Love a Plastic Tree?".

Human Uses of Plants: The Power of Land, the Spirit in Plants
Richard I. Ford
14

Questions for Consideration

1. Why didn't the Tewa Indians destroy their environment?
2. How successful was subsistence farming for the Tewa?
3. How does Indian religion help to protect the environment?

American society could save itself by listening to tribal people. While this would take a radical reorientation of concepts and values, it would be well worth the effort. The land-use philosophy of Indians is so utterly simple that it seems stupid to repeat it: man must live with other forms of life on the land and not destroy it.
VINE DELORIA, JR./We Talk, You Listen

The air still sings and the rivers laugh but not where they once did. When American Indians walked the land alone, the balance of Nature was intact. They tilled the soil, they gathered the products of the earth, they hunted the land guided by a few simple beliefs. They maintained the harmony of forces in the world.

Although we practice little of the knowledge gained from the Indian, the legacy is immeasurable. Immediately, we think of material benefits— corn and other crops, plant medicines, animal names, mukluks—and not the intrinsic ethic of behavior in the physical and biological world. In these terms our best illustrations are the original Americans' interactions with plants.

Ethnobotanists have explored these relationships. *Ethnobotany* is the study of the direct interrelationships between human populations and plant populations. Once it was restricted to a study of "primitive" man's uses of plants. But it is pejorative to characterize the rich intellectual life of nonliterate cultures as primitive, senseless to ignore the nonmaterial significance of plants, and, as will be demonstrated in Chapter 15, foolish to limit our inquiry to non-Western people when it is our plant environment that is in jeopardy. Ethnobotany, then, is not restricted in time or place, and has a closer affiliation with ecology than with economics.

Since plants enter every activity—technological, social, and ceremonial—to the extent that some tribes use corn as a metaphor for life itself, ethnobotanists have an important responsibility. They must record not only which plant is used but who knows about the plant, how it is collected, how much is needed, and what impact humans have on the plant population. The number of individuals in the human and plant populations is crucial. Several Paiute families in southern Utah are assured a few meals of mariposa lily bulbs found growing in five acres of sagebrush, but they would contribute only one snack for a village of 400 Pueblo Indians. The human population, in other words, has certain definite food needs. Some plants are a tasty treat, others provide scarce vitamins or special nutrients, and a few like corn or acorns are staples. On the other hand, plants have particular requirements that must be met during their life cycle. Some relate to growth and others to reproduction. They cannot be collected indiscriminately if the population is to endure. How an Indian population knows when to gather each useful plant is vital for the perpetuation of the plants and for our understanding of Native American ethnobotany.

An Example of Human–Plant Interaction Among Pueblo Indians

In the arid southwestern United States reside six Tewa-speaking Pueblo Indian populations. Their villages are located along the Rio Grande and its tributaries north of Santa Fe, New Mexico (Figure 14-1). Without the rivers, farming would be impossible, since the parched land receives less than 25.4 centimeters (10 inches) of precipitation each year, and irrigation is mandatory. While the vegetation is adapted to a high evapotranspiration rate in summer, native plants must also survive subzero winters. Complicating the environmental picture are the effects of the 3350-meter (11,000-foot) mountain ranges that bound the 1800-meter (6000-foot) Rio Grande trough on the east and west. The close proximity of the mountains causes localized rain shadows, torrential orographic thunderstorms, and frost-producing cold air drainage. The length of the growing season is unpredictable, and grasshoppers, rodents, and deer are a constant threat to the success of any crop (Ford, 1972).

The population size and architecture are similar from village to village. Traditionally, the population has ranged from 100 to 400, and the people reside together in permanent villages. Each village consists of several rows of one- and two-story contiguous adobe houses built around a plaza. The only other buildings are Kivas (religious structures), a Catholic church, and corrals for their domesticated animals, which were introduced by the Spanish several centuries ago. Except for an infrequent shade tree, the landscape within the village is barren and is the same brown color as the houses. A pueblo, however, uses only local adobe and wood, and blends harmoniously into the landscape. Behind each village are several ash heaps on which grow a few hardy plants. Stretching for several miles north and south of a village are the verdant irrigation banks and cultivated fields.

Drawing upon the important work of Ortiz (1969) and my fieldwork, we learn how intrinsic plants are to Tewa thought. At the time of origin, humans were directed from an underworld by white and blue corn mothers symbolized in the life of every Tewa by two perfect ears of the same colors given to each at birth and possessed by all ceremonialists.

FIGURE 14-1.
Nineteenth-century view of Tesuque Pueblo and plaza with hills and mountains in the background. (Courtesy of John K. Hillers.)

From this underworld emerged the people, good and evil, plants and animals, and knowledge possessed by members of secret societies to protect and to perpetuate the people as a metaphorical expression of the corn plant pushing its way through the moist earth that eventually hardens under the rays of the summer sun.

The Tewa world is a duality. Winter is a time for hunting and is symbolized as cold and masculine. Summer is for planting and gathering and is warm and feminine. Each is characterized by a number of binary opposite games, ornaments, and actions.

The ceremonialists belong to several societies, each with a specific role in regard to the forces of life. Referred to as Made People, they possess more knowledge about the world than the uninitiated but must shoulder more responsibility for the awesome power they possess. Each village possesses a group of Spanish-imposed civil officers who deal with outsiders and are appointed by the ceremonialists. The essential leadership is vested in two caciques or headmen, who guide the community for half the year. The Winter Headman and his society conduct the rituals associated with cold, hunting, and so on. In February, the Summer Headman initiates activities leading to warmth, farming, and related things. Other societies work with these men throughout the year to affect the cyclical changes in nature for the benefit of the people.

Although each society possesses its own plant medicine and special plant lore, all work in unison to bring life to the land. The Tewa believe that everything has a spirit that can be appealed to through proper thoughts. The spirits are fed and derive benefit from the work of humans, and in turn, they benefit mankind. The cycle of life-giving events conceptualized as the life cycle of a plant is maintained by the coordinated retreats of the ceremonial societies. They bring the buds to life, then the leaves, the flowers, rains, and the fruits. Finally, they put Mother Earth to rest with a blanket of snow each fall when the Winter Headman once again leads the village.

But despite the efforts and good intentions of the Made People, the Dry Food People or "Weed People," who are the villagers who do not belong to these societies, are held responsible for influencing nature. If one wears the cold turquoise in summer, frost of hail may come. If another breaks a ceremonial taboo, grasshoppers may destroy his crop.

The plants are organized into a hierarchy of more-inclusive names which are used in various contexts. Although we are using the term "plant" in this chapter, the Tewa do not have a name for plants as a general category. They recognize an unnamed general concept of "things growing on the back of Mother Earth." They do have names for what the ethnobotanists call life forms or the major morphological divisions of the plant world (Berlin, Raven, and Breedlove, 1974). These are trees, cacti, grass, leafy things, and edible things. These are general descriptive terms that partition the landscape, but it is the generic level or individual names that are most often used in day-to-day communication. Some of these are further discriminated by one or more adjectives. For example, corn is one of many edible things, but different varieties are distinguishable according to appearance—sweet corn, and red, yellow, blue, white, and speckled corn. In other Indian cultures the crop plants are quite descriptive and complex to the extent that several additional, highly exclusive levels of plant names are distinguished.

However, although this abstract scheme organizes the plant names, it is merely the source of terms selected for use in actual contexts. The most important of the functional classifications is the basic hot—cold duality which orders all food and medicines into one category or another and dictates what will be prescribed for curing illness or balancing daily intake to maintain good health. In other contexts plants are separated into useful versus weeds, general categories of utility, or ceremonial significance. For understanding the actual consequences of Tewa plant procurement strategies, these classifications are the important determinants of behavior.

The geography of a pueblo's environment is also classified. The village itself, trash deposits, rivers, irrigation ditches, farmland, prairies, hills, and mountains are named areas with specific locations recognized within each category. Since various plant populations grow in each of these areas, they are not used either by the same people in a pueblo or in the same frequency from one month to the next. But one rule is used to obtain any plant, regardless of ownership or location: Take only the part and amount needed and leave the rest.

The sun retraces his ageless path northward guided by the good works of ceremonial leaders and the proper behavior of the people. The days lengthen, the air warms, and the frost leaves the back of Mother Earth. A new cycle of activities begins.

In each habitat defined by the Tewa, interactions defined by shared beliefs about nature determine the structure and composition of the plant communities. With the Made People "bringing the buds to life" through their month-long rituals, the time arrives for the irrigation ditches to be cleaned and repaired.

On the appointed day, all able-bodied men and boys assemble in the village to begin several days of hard work before the water supply can again be available for the crops. Each is assigned a section of the main canal and prepares a U-shaped cross section. The lucky worker may find the tasty small tuber of a wild potato (*Solanum jamesii*) or the bun-shaped rhizome of reed grass (*Phragmites australis*) as he shovels the accumulated silt from the previous year onto the bank. These are eaten on the spot or shared with the ever-present children. March is the only time of year when these plants are consumed.

The new soil on the ditch bank forms a man-made habitat. Here, a mixture of perennial grasses used for brooms, roof thatch, and fodder reemerge to join useful pioneer annuals that grow in the silt. Other important plants, such as medicinal wild roses (*Rosa Fendleri*) and sweet-fruited wild plums (*Prunus americana*), are left to grow even if their presence alters the course of the canal. Anything growing along the ditch belongs to the pueblo and can be collected by any man, woman, or child.

With the ceremonial transfer to the village from the Winter Headman to the Summer Headman, interactions with plants intensify. Except for ridding a field of destructive rodents, hunting is forbidden and cold-producing toys, games, and jewelry are replaced by shinny, kickball, and summer ornaments. Thoughts return to farming and plant collecting.

As ritualists "bring the leaves to life" in March, the villagers can have fresh plants for the first time since October. The size of the population naturally limits which plants will be important food for all and which will provide occasional snacks. The first available in quantity are found in the prairie and hillsides beyond the reach of the irrigation ditches. Groups of women and children comb these areas for the sweet roots of chimaja (*Cymopterus purpureus*) and scurf pea (*Psoralea lanceolata*) and the small bulbs of wild nodding onions (*Allium cernuum*). Several days of digging yield a gunny sack of each for a family. During the remainder of the year this area serves as the rangeland for stock herded by old men and boys and, when occasion demands it, as a source of medicinal plants such as Mormon's tea (*Ephedra Torreyana*), snakeweed (*Gutierrezia sarothrae*), and four o'clock (*Mirabilis multiflora*), and of yucca (*Yucca glauca*), whose root is used for washing blankets and hair, each collected by individual women. For each plant, only the part needed is taken; the rest is left undisturbed or reburied. Overall, despite the importance of this area, it is the vagarious weather, not the exploitation of people, that determines the abundance of each plant.

While the women prepare a repast of the remaining stored corn and wheat supplemented with welcome fresh roots, the men toil to prepare their fields for the new season's crops. Each farmer maintains his lateral ditches from the main canal to his several fields. Each field is cultivated, tended, and harvested by its owner.

Recognizing the environmental basis of germination is not sufficient for a successful crop. Prior to planting, a ceremonial game of shinny is played with a ball filled with seeds from all the crop plants throughout the farmland to bring blessings to the new plants, and the seeds are distributed by the summer cacique to each farmer after the ritual. However, the farmer does not plant them immediately or on the same date from year to year. Field and climactic conditions are quite variable. Thus, one does not plant his corn until the snow has left the sacred mountain peak to the west, lest frost destroy the young plants. However, if it lingers and the cottonwood comes into bud, then it is safe to proceed. The new seed is soaked overnight in a solution containing pulverized wild celery root (*Ligusticum Porteri*), a plant growing in the mountains which protects people from snakes and evil and plants from cut worms.

Since only cornmeal that is pure in color is used in various food dishes, and, more important, in all ceremonials to propitiate the spirits, each color must be planted separately to prevent mixing. A border about 3 meters (9 feet) wide is plowed around each 0.1- to 0.4-hectare ($\frac{1}{4}$- to 1-acre) plot. Then the corn is planted in rows, four kernels to a hill, a pace apart. Beans and squash are often planted later in the same field. White corn is planted in one location, red in another, yellow perhaps across the river in a third field, and blue as far as 4 kilometers (3 miles) downstream in still another field. Although these small, scattered fields may appear inefficient to the Western farmer, they actually have an ecological benefit that goes unrecognized by the Tewa. If we recall that they are subsistence farmers living in a patchy environment of low predictability, dispersing fields prevents total crop loss to grasshoppers, frost, or hail, to name a few of the many dangers to the vulnerable crops. One field may be destroyed, but the others will provide some yield. However, a farmer who does not follow the traditional beliefs and rules for farming, in violation of the power of the earth and the spirits of the plants, courts disaster.

Beside prayers and ritual participation to help the plants, the farmer takes other precautions to aid his plants. Each field is irrigated four times and hoed according to a prescribed binary classification that determines the presence of another nonagricultural plant in the field. All plants, regardless of life form or name, are classified as "useful for man" or as "weeds." Food, medicinal, ritual, and even amusement plants are left to grow while all others are hoed out. As a result, a well-tended field might appear unkempt to a Western farm expert, but to the Tewa, the land yields far more value than what one plants. In fact, the many wild plants cultivated by the Tewa farmer do not belong to him and may be gathered by anyone. Not infrequently, women will walk though the fields picking leaves from the small lambs quarter (*Chenopodium album*) and pigweed (*Amaranthus retroflexus*) in the early spring. The same plants will perhaps be revisited by others at harvest time, this time for the seeds to be winnowed for porrage or for baiting bird traps in winter. Other plants gathered include Indian tea (*Thelesperma megapotamicum*); purslane (*Portulaca oleracea*), another spinach; Rocky Mountain bee weed (*Cleome serrulata*), whose leaves provide a cooked vegetable and, with prolonged boiling, a black pottery paint; ceremonial beads from the fruit of silver nightshade (*Solanum elaeagnifolium*); and ground cherry (*Physalis Fendleri*), which has an explosive bladder that gives children endless hours of pleasure. As with all other plants, only the useful portion is taken and other parts are left undisturbed.

The unplanted border provides sunflower stalks (*Helianthus annuus*), used for bluebird and robin snares, and globe mallow (*Sphaeralcea Fendleri*), a ritual glue and ceremonial paint base.

Ceremonies of thanksgiving accompany various cultigens at harvest time. Following the harvest, community members glean any field for produce either missed or deliberately left behind. The stock are then turned loose in the field until the next spring. These animals are obviously a source of manure for the fields and of seed dispersal.

The winter months will be long and cold while the earth sleeps. In preparation, men must chop ample firewood according to the expected hardness of the weather, as predicted by the all-important corn. During the harvest, the women husk and store the ears. If they perceive that the cob has an extra heavy husk, a harsh winter is foretold and extra wood must be obtained. Woe to the cold wood chopper who fails to heed this forecast just once! He will never doubt the power of corn again.

Firewood and construction timbers are obtained from hills and distant mountains up to 32 kilometers (20 miles) from the village. Women never visit these areas unaccompanied by men, and their knowledge of the vegetation is restricted. Following the harvest, families do camp in the higher hills to collect juniper berries (*Juniperus monosperma*), the staple piñon nuts (*Pinus edulis*), and cactus and yucca fruits, if available. Otherwise, most plants needed from here are brought by men on hunting trips or when cutting wood.

Firewood consists of deadwood, with juniper and piñon preferred. Since these village territories have been occupied for well over 600 years, the amount of wood burned through the years is tremendous. By cutting fallen trees and removing dead limbs, instead of destroying the conifer forests, they have actually benefited the trees through thinning and pruning.

Douglas fir (*Pseudotsuga menzeisii*) and yellow pine (*Pinus ponderosa*) are major construction materials. When they are needed, live trees are cut and hauled to the village. In an arid land, however, decay is retarded, and through annual maintenance, which is enforced as part of the ceremonial pattern, house timbers may last centuries, which means that forest cutting is actually quite rare.

Some of the higher elevation plants serve as snacks for the men [e.g., strawberries (*Fragaria ovalis*)], as a reward for children back in the village [e.g., choke cherries (*Prunus virginiana var. melanocarpa*) and skunk bush (*Rhus trilobata*)], or as a condiment requested by a wife [e.g., horsemint (*Monarda menthaefolia*)]. Oak (*Quercus Gambelii*) and mountain mahogany (*Cercocarpus montanus*) limbs provide axe handles and digging tools.

The highest peaks of the four sacred mountains bordering the Tewa territory can only be approached by the Made People and their guards. They go to these mountains to collect secret ceremonial plants and medicinal plants with particular efficacy. On some of these ritual excursions the plants will be classified according to the demands of the ceremony and called names known only to the participants.

Two other areas in close proximity to the village are less important than those just discussed for the plants they support. Several trash mounds are situated adjacent to the outer rows of houses. On these grow some useful plants, such as prickly pear cactus (*Opuntia sp.*), which again belong to anyone in the village. In contrast to this human-created, vegetatively impoverished area is the river and its bank. Here grows a rich mixture of aquatic plants, tangled willows, and stately cottonwoods. This area is visited in the spring by women collecting mushrooms (*Coprinus atramentarius*) and recently escaped asparagus (*Asparagus officinalis*). Otherwise, old men who collect willow stems for baskets, set snares for ceremonially important songbirds in the willows, and gather parts of water plants, such as cattails (*Typha latifolia*), for other ritual purposes are the main trespassers in this plant community.

Tewa interactions with their plant world are very complex. All share a few simple rules when collecting plants that both prevent the degradation of the plant communities and in some cases favor the occurrence of others. Yet all Tewa do not possess the same knowledge about plants or even have the opportunity to interact with the plants in their territory. Women and children have free movement in the immediate vicinity of the pueblo and can go out a mile or so to the foothills. Within this area, one village has exclusive domain over the local resources, but beyond, Tewa from all six pueblos share the hills and mountains, except that it is the men who actually procure the plant resources from this larger region. At the most sacred mountains, only the initiated ceremonialist learns the plants and their importance. Thus relationships with plants vary along three dimensions—distance from village, sex, and sacred knowledge.

Summary

The Tewa example typifies the harmonious balance Amerinds enjoy with nature. Theirs is a oneness; a recognition that the land possesses the power to benefit humans if proper devotion is followed and the spirit of the plants is mollified when proper respect is extended. Plants are readily used but not wasted. Plant populations that are exploited continue to exist without recognition of the laws of ecology. These are only discovered when nature is waning. The dual ideas of preservation and conservation are alien to American Indian and other cultures that behave respectfully toward the plant world.

Selected References

BERLIN, B., D. E. BREEDLOVE, and P. H. RAVEN. 1974. *Principles of Tzeltal Plant Classification.* Academic Press, Inc., New York.

FORD, R. I. 1972. An Ecological Perspective on the Eastern Pueblos. *In,* A. Ortiz, ed. *New Perspectives on the Pueblos.* University of New Mexico Press, Albuquerque.

ORTIZ, A. 1969. *The Tewa World.* The University of Chicago Press, Chicago.

Human Uses of Plants:
Don't Walk on the Grass!
Ethnobotany in Middle America
Richard I. Ford

15

1. Is American ethnobotany so different from the Indians'?
2. What are Americans doing to our plant world?
3. Why do we tend our yards?

Middle-class Americans are alienated from their plant world. Oh sure, poinsettias are given as Christmas presents, lilies adorn the Easter breakfast table, and corsages are still sent on Mother's Day, but by and large most people are content with store-bought food, hazy air, and a panorama of smokestacks. If progress means fewer trees, and leisure time increases proportionately with weed killer, so be it.

But is this image accurate? Perhaps hyperbole distorts the facts to the detriment of our understanding of our neighbors' actual attitudes and relationships with their environment. The answer to this question is important if we are going to effectively communicate the problems posed by an admittedly deteriorated natural landscape and to change attitudes about the ecology.

The author's next-door neighbor never heard of Linnaeus, cannot pronounce Latin binomials, and plants her peas on Good Friday and carrots by the new moon. She has a good income and a well-kept house. There is no reason why a stereotype or an academic disdain for studying the ethnobotany of literate societies should prevent us from learning from her.

Most of our food and many of our clothes, furnishings, and business or school items are made from plants. They are raised, harvested, and processed by others on a commercial basis. Our contact with these plants is *indirect* at best, and in most instances, we do not even know what they look like. The study of these plants and their usages is handled by economic botanists.

On the other hand, ethnobotanists are concerned with the *direct* contact of any population with its plant environment, and they consider cultural beliefs about plants intrinsic to an understanding of the composition of the local vegetation. Thus the ethnobotany of complex, Western cultures like our own is long overdue.

An unresolved problem confronting the student of the ethnobotany of middle America is the magnitude of the project. To adequately sample just one average-size town and to know the floristics of the region is a considerable undertaking. Without pretending to be thorough, for the past 5 years, students in the author's ethnobotany courses have been interviewing residents in Ann Arbor and other communities in Washtenaw County, Michigan, about their ethnobotany. Even though the project is incomplete, and many of our results are preliminary, the patterns are so consistent and the insights gained are so interesting, especially in contrast to American Indian uses and beliefs about the plant world, that we encourage others to undertake similar projects. The University of Michigan students and their courteous "native" consultants are too numerous to name individually; this chapter is an acknowledgment of appreciation for their dedication and cooperation.

Classification

Virtually everyone interviewed had a biology class in high school or college that exposed them to a scientific scheme for classifying biological organisms into a hierarchy of kingdoms, phyla, classes, orders, families, genera, species, and varieties. Yet, even the botany professors in our sample use a folk taxonomy similar to that described in Chapter 14 when discussing the plants in their yard or home (Figure 15-1). Unlike many Amerindian cultures, there is a term for the most inclusive grouping of living organisms growing in place—"plant."

The next division is that of life forms as it is in other folk cultures (Figure 15-1), and some terms similar to those used in non-Western taxonomies—"tree," "shrub," "grass"—are consistently used. Variation enters depending upon one's interest in growing things. Some add "vine," but others include these with "flowers" or simply lump them in the life-form term "plant," meaning small, broad-leaved, and inedible. "Plant" may be a catch-all category to include everything from vines and flowers to mushrooms, algae, and cacti. Finally, some term such as "food" or "vegetables" is used to distinguish edible plants. There exists

FIGURE 15-1.
Middle America's plant taxonomy.

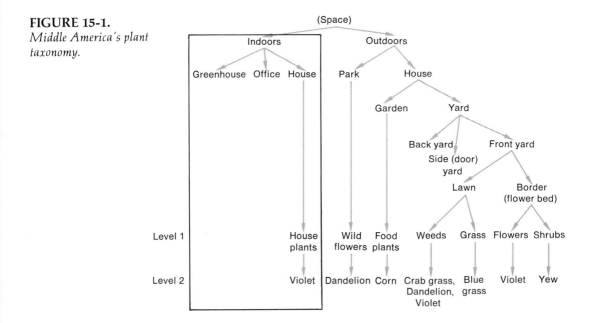

greater variation in this level of nomenclature in our culture than in others, primarily because firsthand experience with smaller plants is so variable. Morphologically, trees, shrubs, and grass are conspicuous even at a distance, and they are used as parts of general descriptions more frequently than are other terms.

Just as in other cultures, middle Americans use single-termed generic names more frequently than any others, and the greatest number of names occurs at this level (Figure 15-1). However, in contrast to the Tewa or other Indian groups, who sometimes group several different plants as determined by Occidental scientists at least into one name, we tend to lump many more plants into one terminal common name. For example, in answer to the questions, "What kind of tree is this?, and this?," maple will, more often than not, be given for both, even though one is a silver maple and the other a sugar maple.

The previous example uses common names that are more exclusive on a specific level (Figure 15-1). Our plant taxonomy is not particularly discriminating and very rarely describes plants with one, no less two, adjectives or other descriptive terms. Again, in subsistence-level cultures, those plants that are grown or used for many culturally important purposes are distinguished by often-subtle criteria into numerous varieties. Even our most experienced gardeners do not do this.

Occurring between the life-form level of taxonomy and the generic are intermediate terms that exhibit a degree of inconsistency (Figure 15-1). For example, trees are sometimes divided into "evergreen" and "deciduous," but "house" and "outdoor" are used only if "plant" is a life form.

This classification for ordering the names of plants with which we have direct contact parallels other folk taxonomies in organization, but our plant terms are not particularly diversified. We should add that few consultants know the scientific names of the plants they so carefully nurture.

Other classifications for using these names are very limited. Few plants that we grow enter into our religious ceremonies, and rarely do they provide the raw materials for our work. We take for granted our economic plants that are familiar only as finished products. Yet our classification of space within our proximate environment is very detailed, and an examination of plant names ordered according to these terms is very interesting.

In Chapter 14, you learned how the Tewa have spatial terms for partitioning their geographical landscape but few terms that apply to the immediate environment of one's house. Anthropologists have found that English has an abundance of words describing the dimensions of space. Figure 15-2 reveals a similar pattern for subdividing our house plots. Here we find tremendous differentiation, which, as we shall discuss, is reflected in our ethnobotany.

Interactions with Plants

Attitudes about plants are best exemplified by the rules people follow to relate to plants. The vegetation growing in each spatial area is treated somewhat differently, and the execution of a single belief about the care of plants will vary from one spatial area to another.

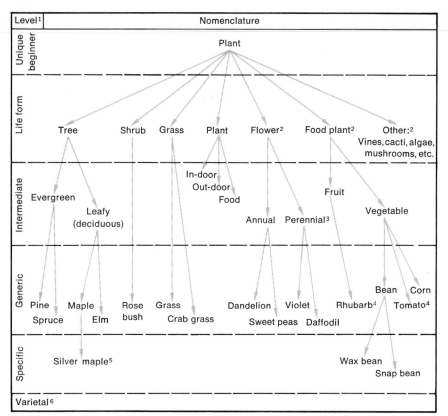

Level[1]	Nomenclature						
Unique beginner	Plant						
Life form	Tree	Shrub	Grass	Plant	Flower[2]	Food plant[2]	Other:[2] Vines, cacti, algae, mushrooms, etc.
Intermediate	Evergreen / Leafy (deciduous)			In-door Out-door Food	Annual Perennial[3]	Fruit / Vegetable	
Generic	Pine Maple / Spruce Elm	Rose bush	Grass / Crab grass		Dandelion Violet / Sweet peas Daffodil	Rhubarb[4] / Bean Corn / Tomato[4]	
Specific	Silver maple[5]					Wax bean / Snap bean	
Varietal[6]							

1. Levels adopted from Berlin, Breedlove, and Raven, 1974. See selected references, Chapter 14.
2. Alternatives to the live form "plant."
3. Most common refinement; an alternative is "from bulbs" vs. "from seeds."
4. Folk usages do not always follow scientific definitions; in this case the folk name and the scientific designation are reversed.
5. Maple trees are usually lumped but when they are distinguished, silver maple is most often recognized.
6. These names occur in other cultures but very infrequently in American folk taxa.

FIGURE 15-2.
Middle America's spatial taxonomy.

The front yard receives the most attention. It is fertilized, watered, mowed, weeded, and raked. This space consists of grass (usually the kinds are not known) and perhaps a few trees and/or shrubs. These are pruned to "look nice"; their leaves are raked in the fall for the same reason and never as a recognized ecological defense against fungus growing under the snow or the destruction of grass by toxic substances.

The grass is fertilized "to keep it green" and watered either at some set interval of time: every other day, on Saturday, or when it "turns brown." Mowing and raking are performed for appearance's sake as well. The prohibition against walking on the grass applies to the front and is for maintaining an even texture and preventing paths or bare spots.

Pity the lowly "weed"! According to the middle American, a weed is any plant growing out of place. In other words, it, too, relates to space. Dandelions are the greatest bane. They are speared with specialized equipment or poisoned with all sorts of concoctions eagerly supplied by any garden shop. We have found, however, that the ecological message against weed killers did affect one consultant, who inquired if her practice of applying an eye-dropper full to each weed would harm the environment. She had previously broad-spread it. Crab grass, classified as a specific type of life-form grass, is a weed when it grows in the yard. However, so is a violet or even a crocus if either wanders from the border or flower bed by chance or through the design of some spiteful squirrel.

The extension, which not everyone has unless a sidewalk is present, receives the same care that the front yard does. Here are located trees planted by the city, which is also expected to care for them. Since it normally ignores them, except in emergencies, the resident reluctantly tends to them. The extension is not a source of particular emotion or commitment.

The side yard or door yard is not as large as the front yard, and its size is even further diminished by driveways, walkways, and borders. Again, this area is mainly planted in grass and carefully maintained if it is highly visible.

And now we move to the back yard. What a contrast. Here a lawn of grass is present along with bare spots, dandelions, and other weeds. The grass is cut, but the look of a fresh manicure is missing. The investment of labor goes into the *public* front yard and not the *private* back yard. For some, a "messy" back yard almost takes the form of cultural rebellion!

Borders are ordinarily found between the front of the house and the yard, often along the side and back of the house, and commonly separating one yard from its neighbors. The choice of plants composing the border varies with its location. Shrubbery adjacent to the house and flowers in front is a typical pattern. Borders along the side of the front yard may be planted in shrubs or flowers or a combination and are a deliberate property and social boundary. In the back, on the other hand, trees or tall shrubs are found providing a curtain of privacy if a fence is not already present. Flowers again add to the color of the environment. Front borders tend to be more orderly and have fewer weeds (in this case escaped grass from the lawn) than in the back. In the past 2 years we have been finding vegetable plants in the back borders.

Flower beds and flower gardens are the same thing. They are areas of larger-than-average size along the perimeters of the yard or as islands within the lawn. These are often "hobby" areas and sources of pride when rose bushes (a shrub) are blooming, and a kaleidoscopic succession of seasonal flowers appears from April until October.

Gardens devoted to edible plants or vegetables are very frequent uses of back-yard space. The square footage varies with family size and the content with taste and interest. Here, again, experience seems to guide behavior. For the gardener of long standing, compost is used for fertilizer, and mulch is a common means to conserve moisture and to retard weeds. The more recent convert tends to use no fertilizer and to devote considerable time to hand weeding. Watering is almost a daily ritual for the novice, and insect repellents are mandatory. This is another area of "pride" for everyone.

Without going into detail, the insides of houses and offices are decorated with certain plants. Office plants are chosen for their ever-green qualities and for their ability to withstand a frequently changing environment, including the forgetfulness of their owners. House plants are more varied, because of the greater opportunity to care for them and to provide them with preferred habitats. Older residents generally have more house plants than younger people do, and in offices, women nurture many more plants than do men.

Naturally, middle-American residents designate spatial areas beyond their proximate residential and occupational environments. To mention only one, parks must have trees for shade and at least some of the vegetation cut to permit comfortable walking and romping. Beyond these two requirements, concern for the floral composition is minimal. This can be appreciated by noting that here dandelions are called flowers or wild flowers, and a bouquet of them may actually be brought home!

Cultural Values and Plants

The preceding description has evaluated only a portion of the importance of plants to Midwestern residents. An even greater significance is seen when we consider how plants actually reflect social and economic transactions.

The structure of the spatial arrangement of plants is culturally determined. First, there is the legal codification and enforcement of dominant cultural values. In Ann Arbor and other municipalities, rank vegetation cannot grow in public areas, and fences cannot exceed maximum heights. Second, there is a social dimension. Those areas more frequently viewed by nonresidents receive much more attention than those areas that are less visible or where social contacts are limited to close friends or relatives. The difference between front yards and back yards is a self-evident example. Third, there is a very important economic constraint. Property values must be maintained. The lawn is maintained not for its cooling effects or for dust control in ecological terms but because it is expected. The resident beleaguered by yard work will entertain the idea of replacing the grass with a synthetic turf and readily dismisses it as inappropriate. For people of all incomes, much is invested in a house and they do not want it depreciated by laziness or thoughtless neighbors. Fourth is an unconscious cultural trait for order. Everything has its place, usually in some variation on a grid. Neatness in yard care and the very concept of "weed" reflect that not only is there an expected spatial arrangement but that everything has its place. Finally, three subjective psychological values enter. Horizontal lines cannot meet vertical planes abruptly. When they do, a sense of visual perspective disappears and aesthetic qualities are lost to uneasiness. Borders along houses solve this by smoothing the transition from ground to height. Green is a cool or relaxing color to Americans. It is pastoral and its dominance in the grass, trees, and shrubs chosen for the home creates a welcome atmosphere. But this is not to say that the list of structural constraints completely stifles individual expression. Obviously, it does not when we see how identical tract-built houses look so different after the yard is planted.

Within these culturally prescribed bounds much choice does exist for the selection of individual or groups of plants, and the results are the idiosyncratic variations distinguishing one house from the next. For long-time residents of the same house, the selection of plantings is made for them by gifts of plants from kinsmen and neighbors. The flowers and shrubs in their yard are the social history of their adult life. A lilac bush from a neighbor who moved to California, an azalea by the front steps from a son for a birthday, and tulip bulbs from a close friend are examples. To think about these plants is to recall happy moments; to tend them lovingly is to preserve these memories forever. For others, judgments result from a lifetime of learning about plants in informal contexts. Folk terms are learned before formal education begins and are used to discuss botany. Children watch their parents, they see homes with their orderly yards on television, and read about them in school books. The newspaper provides garden columns, and the county extension agent has leaflets and telephone advice. Any florist or garden shop will have an "authority" who can answer questions about your lawn or suggest an appropriate grass seed or decorative plant. The sources of information are numerous and actual decisions may be based on cost, effort, children, dogs, previous plantings, and aesthetic ideals.

Individual expression comes in the arrangement of the selections. Some place them for functional benefit such as to shade a corner of a porch or to hide a gas meter. Others seek rows of gay flowers distinctively organized with an artist's eye. Although few express it, working with plants in middle America is about the only opportunity one has to be individualistic and creative on a grand scale!

Since plants permeate our lives to the extent just discussed, they must reveal other values of a more controversial nature, and they do. For one, students have discovered a strong sexual division of labor in the upkeep of yards. With married couples, men do the heavy labor, cut the grass, and prune the larger trees. They are active in the vegetable garden but rarely in the flower beds. Their wives may turn on the sprinkler but most of their outdoor labor goes into the borders and flower beds. The association of flowers with females is very acute. When members of only one sex or the other are able to work, they do all the chores but with a caveat: Women sometimes apologize for not keeping the lawn as nice as a man might, and men occasionally excuse themselves for not having more flowers.

Judgments by other people are prominent in our ethnobotany. Except for the produce from our garden, plants from our immediate environment are rarely used. They are not handled and seldom selected for their fragrance. Plants are for viewing and, as such, are the first impression we have of strangers. Negative ideas about the occupants of a house that does conform to preconceptions are conjured on the spot. Conversely, a resident who wants to alienate himself from his neighbors need only violate the cultural values defining our plant relationships to succeed.

Conclusions

An ethnobotany on American culture has been started, with promising results. Far from the idea that Americans are divorced from plants, we have outlined an intimacy so pervasive that they can reveal the social, economic, and personality characteristics of each residence. In other words, although we no longer "use" the plants we interact with directly, they are important in our lives. They signal information about individuals, they symbolize social relationships, they reveal the importance of economic values and order in our culture, and they permit self-expression.

However, the difference in the relative importance of plants in our society compared to non-Western cultures does not mean we are that much different in our understanding of the plant world. In point of fact, the similarities are quite striking. The manner for classifying plants is identical, although the utilitarian value of plants results in a richer nomenclature in other societies. Our extensive scientific vocabulary is known to only a few and actually used to name plants by fewer still. Furthermore, just as non-Westerners do not express the ecological consequences of their behavior toward plants, so, too, Americans do not give ecological explanations for their labors and, considering the time devoted to yard work, they have an appalling lack of scientific knowledge. The tremendous effort in yard care is for social, economic, or personal reasons, not to control dust, temperature, or humidity.

Summary

Once we understand the ethnobotany of middle America, environmental education is made easier. A folk vocabulary rather than technical jargon must be used. It is primary for all communication about plants. Furthermore, in certain respects, there is no imperative to teaching a new rationale for certain forms of behavior as long as their latent or unexpected result is ecologically beneficial. This allows us to focus on areas of unsound practice, but by knowing why nature is manipulated in certain ways permits development of a strategy that will not be rejected by a cultural reflex. Similarly, by recognizing which values are important, new approaches to conservation can be attempted. For example, in an aerial view of the Midwest, one quickly spots the forests and most complex plant assemblages—the city streets. By propagating and distributing endangered species of plants for inclusion in flower beds, numerous refuges could be available for preserving these plants and even some plant communities. Moreover, we can understand why it is so difficult to introduce native plants into our diets. Most plants we deal with are distinguishable by one or two conspicuous characteristics, but to recognize edible wild plants requires a much greater knowledge of plant morphology than most people have the background to handle, and more important is the negative connotation that accompanies the concept of weeds. If plants are classified in one domain, contradictory information that would place plants in another category is difficult to convey. To change attitudes and to introduce new ideas about ecology requires an understanding of the principles of middle America's plant relationships.

Human Uses of Plants: Indoor Gardening, Organic Gardening, and Home and Urban Landscapes
Peter B. Kaufman

16

Questions for Consideration

1. How can you carry out indoor gardening throughout the year?
2. How are bulbs forced indoors so that they will flower in winter?
3. What kinds of house plants are best adapted, in homes and apartments, to shaded conditions? Sunny exposures? High humidity? Low humidity?
4. What are the basic steps in creating a home landscape, starting with a piece of land devoid of any vegetation (recently graded) and with a new home on it?
5. What are some ways by which you can save labor in maintaining a home landscape?
6. What are some of the kinds of trees and shrubs that can provide you with food in your garden? How can they be used in a landscape scheme?
7. What kinds of shade trees can be used along city streets that are relatively pollution-resistant?
8. What are some of the most effective ways to use landscaping around urban malls? Urban apartments? Urban greenbelts? City buildings? City parking lots?

Indoor Gardening

Indoor gardening has become a dominant pastime for people who live in dormitories, apartments, and homes; for people who work in offices; for children in schools. Basically, we can attribute this craving for plants as our attempt to have a little bit of nature—a bit of the real world—in our synthetic, concrete jungle environment. It is indeed an attempt on our part to be part of nature and to live with nature—the central theme of *Plants, People, and Environment.*

One of the great attributes of indoor gardening is that it can be carried on throughout the year no matter what the weather is like. It can be done in a single window or in a home greenhouse, under indoor lighting systems or in deep shade, and in a refrigerator (when you force bulbs) or in a terrarium.

Our approach to indoor gardening will focus on the kinds of gardening activities you can pursue in order to garden throughout the year. The first activity is that of forcing bulbs in wintertime.

FORCING BULBS IN WINTERTIME

Forcing bulbs in wintertime can provide one with colorful hyacinths, tulips, crocuses, and narcissus in the house, apartment, or office right at a time when very few flowers are blooming. This winter "season" can last from November through March. The technique is quite simple. It is as follows:

1. In October, obtain large bulbs of the desired type. If you are a novice, try paper white narcissus. Later, try tulips, grape and Roman hyacinths, yellow narcissus, and crocuses.
2. The bulbs can be planted in regular potting soil in clay pots or, alternatively, placed in ceramic watertight containers buried halfway into pebbles and small stones mixed together, or in special hourglass-type glass containers in water [the latter are especially good for Roman hyacinths (Figure 16-1), large tulip bulbs, or narcissus].
3. The planted bulbs should be watered well and then placed in a refrigerator or cold room where the temperature is about 7.5°C

FIGURE 16-1.
Forced Roman hyacinth growing in water in an hourglass-shaped vessel.

(45°F). Hardy bulbs can be placed outdoors in a coldframe, buried in the soil, and covered with straw. However, it is essential that the soil around them can be loosened and not frozen so that pots can be extricated and brought into the house or apartment. Paper white and yellow narcissus are *not* hardy, and so cannot withstand outside temperatures below freezing. These can only be subjected to cold of about 7.5°C (45°F).

4. The cold-treated bulbs should remain in cold storage until the roots are well developed and the shoots are beginning to grow. Usually, the time is as short as 6 weeks and can continue until 3 months (especially for bulbs held outdoors). The containers are now ready to bring into a warmer room. Some people bring them to warmer temperatures gradually through succeedingly higher temperatures up to room temperature. This does not take long—a matter of 10 to 15 days. This is the "forcing" treatment. The shoots will develop rapidly, and when buds appear, the bulbs should be placed on display. The flowers will last longer if the room in which they are displayed is kept cool (15.6 to 18°C; 60 to 65°F).

5. After blooming, cut the flowering shoots off, but retain the leaves so as to assure food storage in the bulbs. If your bulbs are large, you may get additional blooms from lateral flowering buds in the bulbs; this is especially true of Roman hyacinths.

After the flowers and foliage have withered, plant the bulbs outside after frost is out of the soil, or store the bulbs in a cool place until they can be planted outdoors. Paper white narcissus will not survive northern temperate winters, so should be recycled into a compost pile after the flowers and leaves have withered.

FORCING TWIGS IN WINTERTIME

In addition to forcing bulbs in wintertime, one can force the buds of many trees and shrubs into leaf and flower at this time. In Europe, it is a common practice to force birch twigs because the catkins and small green leaves are so attractive inside the home or apartment.

Many people force the buds of pussy willows (*Salix* species) in late winter. Other plants, more showy than the pussy willow, are easily forced, including quince, pear, apple, cherry, hawthorn, peach, and forsythia. The twigs are cut off in ½- to 1-meter lengths. The twigs are placed in water in a vase or pail and are forced into bloom at room temperature in a greenhouse or in a south window. The forcing time usually lasts one to several weeks, depending on the species and length of previous cold treatment outdoors. The closer it is to spring, the less time it takes to force the buds into bloom. Forcing does not work well in October and December inasmuch as the buds on the plants must be exposed to 1 to 2 months of cold temperatures before they will respond to forcing.

After the leaves and flowers have withered on the forced twigs, one can root the forsythia and willow branches relatively easily and thus have new plants for the garden. The other species mentioned above do not root easily and can be burned in a fireplace, shredded and placed in a compost pile, or placed in the garden as stakes for peas.

STARTING SEEDLINGS INDOORS IN WINTERTIME TO USE AS FOOD AND SPICE

Many plants can be started indoors in the winter months to use as herbs, spices, salad greens, and protein supplements. In the herb category, this can include coriander for coffee, rosemary for meat flavoring (Figure 16-2), mushrooms from spores for salads and casseroles or meat dishes, and mint for tea. Mung bean and alfalfa sprouts can be grown for salads. Soybeans can be started for high-protein supplements. The seedlings often contain higher levels of protein and soluble amino acids than the seeds. The best way to start these seeds is by the method outlined for starting flower and vegetable seeds. The plants can be grown in window sills of south-facing windows. Sprouts such as those of alfalfa and mung bean can be started in the dark on moist cheesecloth in a jar with a small amount of water in the bottom (Figure 16-3). Special seed germinators have been developed for this purpose. One can simply use a jar with water in it (half-filled) and cheesecloth suspended in the jar above the water containing the seeds. With the high cost of lettuce and other greens, such a procedure for producing masses of seedlings for salads can help a lot financially and provide you with special foods with high levels of vitamins and proteins that are especially necessary in the winter months.

FIGURE 16-2.
Pot culture of rosemary (Rosmarinus offici-nalis), useful for indoor gardening during the winter months.

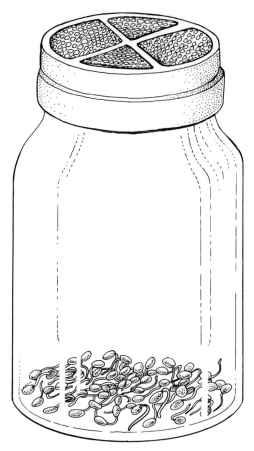

FIGURE 16-3.
Mung bean sprouts grow-ing in a bottle-type germi-nator. The perforated top is for aeration.

STARTING FLOWERS AND VEGETABLE SEEDLINGS INDOORS IN LATE WINTER OR EARLY SPRING

The main objective in starting plants indoors in the late winter or early spring is to have plants of fairly good size ready for transplanting outdoors after the last frost of spring. This is essential in latitudes where the summer growing season is very short. It saves considerable money and gives the indoor gardener something to do during the wintertime. First, one needs viable seeds. Old seeds cause a lot of agony and should be suspect. Vegetable or flower seeds older than 3 years are generally no longer viable. Second, one needs to prepare a good potting soil for starting the seeds. A good mixture is one consisting of 1 part sand, 2 parts loam, and 1 part vermiculite and Perlite mixed together. This mixture should be sterilized dry in an oven or other suitable place (even a fireplace). Thirty minutes at 100°C (212°F) will suffice. The purpose in doing this is to kill microorganisms, such as fungi, which cause "damping-off" disease of the germinating seedlings. Third, one plants the seeds in the soil mix in peat pots or plastic containers or clay pots or, better still, cardboard egg cartons. The seeds should be planted to a depth of twice the diameter of the seed and covered with sand to aid seedling emergence and lessen the danger of seedlings getting too wet and damping off. Good sturdy seedlings can be obtained by starting them under lights with a balanced light spectrum (the entire visible range excluding ultraviolet but including far-red light). Incandescent lamps give off too much heat. Cool white fluorescent lamps are now available which do emit the far-red light that is necessary for normal plant development. New lamps, such as Gro-lux, provide a balanced light spectrum. If lights are not available, seedlings should be started in a south-facing window. Last, one needs to transplant the seedlings, especially when they become crowded in the pot or other container used to germinate them. Pots should be bottom-watered in a pail or sink until the soil surface gets wet; this assures more thorough watering throughout the soil and does not wash out the seed. Pots with germinating seeds can be covered with plastic wrap or glass plates (Figure 16-4), open slightly for air exchange. This is done to prevent the soil surface from drying out too quickly and creates, in effect, a minigreenhouse.

Timing is very important when starting seeds indoors. Seeds of perennial flowers need to be started in late summer or early fall to get plants of any size the next year. This can be done in a coldframe and the plants covered with mulch to overwinter. However, they can also be started in winter, but they will not flower, in most cases, until the next year. Annual plants such as peppers, begonias, and petunias must be started early, as the plants take a long time to reach transplanting size. Others, such as squash and cucumbers, grow very rapidly and can be started later. Tomatoes fall between these extremes. Plants such as these can be started indoors, then transplanted and grown in a coldframe with bottom heat (using a heating cable) and covered with window sash or polyethylene until safe to transplant into the vegetable garden. Many plants, such as peas, beans, and corn, are best started directly in the

FIGURE 16-4.
Pot of germinating brussels sprout seedlings. The glass plate is removed when the seedlings have reached the pot rim.

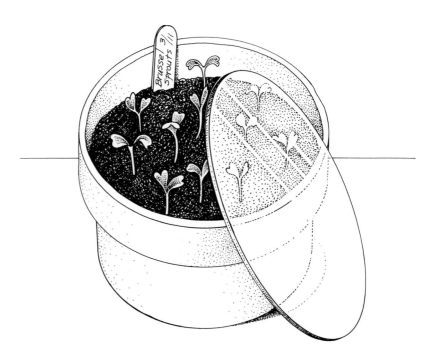

garden. Most herbs are best started indoors, as they develop slowly after they germinate. A head start with herbs ensures that they will have more foliage, flowers, and fruits during short growing seasons. Some herbs, such as rosemary (a tropical shrub) and sage, can be grown indoors as houseplants for use all year.

GROWING HOUSE PLANTS

House plants need care continually (Figure 16-5). Every species usually has its own special requirements for water, fertilizer, light, pruning, and propagation. Here, we shall take up a few different groups of house plants and how to maintain them.

Ferns (Figure 16-6) are distributed in several families, such as the *Polypodiaceae*, which includes most of our common ferns; the *Osmundaceae*, the native royal ferns; and the *Cyatheaceae*, comprising most of the tree

Ferns

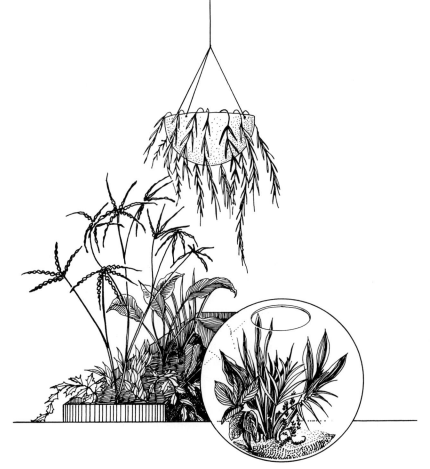

FIGURE 16-5.
House plants grown in hanging pots, in a planter box, and in terraria such as the bottle type shown need care continuously.

FIGURE 16-6.
Button fern (Pellaea ro-
tundifolia).

ferns. Ferns are perennial plants, widely spread in temperate and tropical regions. They include perhaps 10,800 species.

GENERAL CULTURAL REQUIREMENTS

1. Humidity. Avoid too high humidity, as well as a dry atmosphere.
   On warm, sunny days, it is useful to mist-spray the foliage with water.
2. Water. Provide evenly moist soil; standing water is to be avoided.
3. Temperature. Night temperatures should be 13 to 16°C (55 to 60°F)
   with a rise of 6 to 8°C (10 to 15°F) by day in bright weather.
4. Soil. Use fibrous soil (fine peat or leaf mold may be mixed with it).
   Ample drainage should be provided in the bottom of the pot.
5. Fertilizer. Bonemeal in the potting soil is useful. Commercial fertilizers should be used sparingly on ferns because ferns are very
   sensitive to overfertilization.
6. Light. The amount of light to be provided depends on the type of fern.
   Winter sun is not too bright for most ferns, but it is best to avoid
   direct exposure as in south windows. In summer, keep ferns in the
   shade.
7. Propagation. Ferns can be propagated by division of buds or rhizomes
   from the crown or offsets from the fronds of some species. Ferns can
   also be propagated from spores.

CULTIVATED GENERA OF POLYPODIACEAE
USEFUL AS HOUSE PLANTS

1. *Nephrolepis* (Boston fern)	This fast-growing plant with narrow leaves grows best in loose garden soil with rocks. Keep moist. Larger plants should be transferred to hanging baskets.
2. *Platycerium* (Staghorn fern)	Plant this fern on bark. It grows best in a humid atmosphere.
3. *Adiantum* (Maidenhair fern)	Select loose, moist soil with added broken limestone, and provide subdued or filtered sunlight for short periods.

CULTIVATED GENERA OF THE CYATHEACEAE
USEFUL AS HOUSE PLANTS

1. *Cyathea*	This is a tall tree fern with large fronds that will stand full sun.

Araceae: The Arum Family (Aroids)

The aroids are largely tropical herbs with fleshy or semiwoody stems; some are root climbers. They have varied and variable leaf forms and are characterized by inflorescences composed of densely flowered spadix and subtended by a spathe or bract that is often brightly colored and showy. Some are very easy to grow as houseplants because they can tolerate a wide range of growing conditions.

GENERAL CULTURAL REQUIREMENTS

1. Relative humidity. Most prefer at least 30 to 40 percent humidity, but some will survive at lower values.
2. Water. Most like evenly moist soil, except as noted below.
3. Temperature. Maintain average house temperature; do not allow to get below 13°C (55°F).
4. Soil. Use normal potting soil except as noted.
5. Fertilizer. Fertilize flowering plants every 2 to 4 weeks when plants are actively growing; fertilize foliage plants every 3 to 4 months.
6. Light. Only one genus (*Zantedeschia*) likes direct sunlight; most other genera do best in bright indirect light, but several can tolerate lower light intensities.
7. Propagation. Propagate by division, stem-segment cuttings, or air layering for the larger species; stem-tip cutting for the smaller species.

FIGURE 16-7.
*Leaves of dumb cane
(Dieffenbachia).*

CULTIVATED GENERA OF THE ARACEAE

1. *Aglaonema*
(Chinese evergreen)

This is a very tough plant that grows best in shade with barely moist soil. It can stand low humidity.

2. *Anthurium*

This plant has showy inflorescences. Requires high humidity (over 50 percent) and a loose soil (e.g., fir bark and moss); grows best in bright indirect sunlight and a thoroughly moist soil.

3. *Caladium*

This plant has colorful foliage. Prefers a rich soil, fairly bright light, and good humidity; likes evenly moist soil and requires a 4-month dormant period.

4. *Colocasia*
(Elephant's ear)

Elephant's ear requires high humidity and a thoroughly moist soil.

5. *Dieffenbachia* (Figure 16-7) (Dumb cane)

This plant likes to have its soil on the dry side and to have bright light (but it can tolerate low light).

6. *Monstera* (Swiss cheese plant or split-leaf philodendron) (Figure 16-8)

It is best to grow this plant in bright, indirect sunlight and to keep the soil barely moist.

7. *Philodendron*

Philodendron will survive in shade and low humidity; for cutleaf climbing types, provide a support.

8. *Spathiphyllum*

Do not put this plant in direct sunlight. It likes thoroughly moist soil.

9. *Scindapsus* (Pothos)

This plant will survive in low light but may lose its colorful markings.

10. *Syngonium*

This plant will survive in low light.

11. *Xanthosoma* (Elephant's ear)

Provide fresh air; likes high humidity and evenly moist soil.

12. *Zantedeschia* (Calla lily)

Grow in the sun and keep very moist when in active growth.

Marantaceae and Commelinaceae: Arrowroots and Spiderworts

Arrowroots and spiderworts represent two families of tropical herbaceous monocots, many members of which are easily grown as house plants. Although the flowers of these plants are not showy, the foliage is commonly patterned and colored.

GENERAL CULTURAL REQUIREMENTS

1. Humidity. Most can survive in average house humidity but prefer at least 30 percent relative humidity. Low humidity will cause browning of the leaf edges.
2. Water. Most species prefer to be kept evenly moist; those that like to be kept on the dry side are noted below.
3. Temperature. Average house temperature is sufficient but not above 24°C (75°F) in winter.
4. Soil. Normal potting soil should be provided for spiderworts. Arrowroots prefer a richer soil, so add 1 part leaf mold to your potting recipe.
5. Fertilizer. Fertilize every 4 to 6 weeks when the plants are growing.
6. Light. Most spiderworts will tolerate a wide range of light conditions but prefer bright indirect light. Arrowroots prefer semishady light.
7. Propagation. Spiderworts: propagate by stem-tip cutting or layering for trailing types; division for *Rhoeo*. Arrowroots: division at the beginning of the growing season.

MARANTACEAE (ARROWROOTS)

Arrowroots are tropical American herbs, many having tuberous roots; the leaves typically large, with a patterned feather design and sheathing bases. The flowers are surrounded by bracts. These are good plants for east or west windows.

1. *Calathea*	This plant needs high humidity (50 percent or more) and is good for large terraria.
2. *Maranta* (Prayer plant Rabbit track plant)	The leaves fold upward at night.

COMMELINACEAE (SPIDERWORTS)

Spiderworts are semisucculent herbs and creepers. The leaves are alternate, parallel-veined, and have sheathing bases. The trailing types make nice, full hanging baskets if kept pinched back.

1. *Cyanotis* (Teddy bear plant Pussy ears plant)	The leaves are covered with brown or white hairs. Provide bright light and keep on the dry side

2. *Rhoeo*
   (Moses-in-the-cradle
   Moses-in-the-boat)

This is an upright herb with leaves dark green above, purple below. It can tolerate neglect.

3. *Setcreasea*
   (Purple heart)

The leaves are purple. Keep this plant on the dry side.

4. *Tradescantia*
   navicularia
   (Inch plant)

Allow the soil to dry between waterings. There is a wide variety of attractive foliage colors.

5. *Zebrina*
   (Wandering Jew, Figure 16-9)

The foliage is variously colored, with bands of green, red, purple, and silver; bright light intensifies the colors.

Cacti and Succulents

Succulents (Figure 16-10) are plants with thickened leaves and/or stems that contain water-storing tissues. This type of growth habit is found primarily in desert-inhabiting species and occurs in more than 20 plant families. In general, these plants are relatively easy to care for, and many will survive even when grown under less-than-ideal conditions.

GENERAL CULTURAL REQUIREMENTS

1. Relative humidity. Most succulents survive well in low humidity.
2. Water. Most succulents should be allowed to dry out well between waterings; water them less in winter if the plants stop growing.
3. Temperature. An average house temperature is sufficient.

FIGURE 16-9.
Wandering Jew (Zebrina pendula), a house plant that is a rampantly growing ground cover that tolerates deep shade.

FIGURE 16-10.
Cactus (Echinopsis).

4. Soil. Good drainage is important and can usually be achieved by adding extra sand.
5. Fertilizer. Fertilize the plants every 1 to 2 months when they are actively growing, not at all when they are inactive.
6. Light. Some succulents may require at least 6 hours of direct sun per day for best growth; however, others can be scorched if placed in the direct sun but they do need bright, diffuse light.
7. Propagation. Use stem-tip cuttings or seeds to propagate most succulents; some can be propagated by leaf cuttings and some can be divided.

MAJOR GROUPS OF SUCCULENTS

1. Cacti: (Cactaceae can be divided into two types:
   a. Desert cacti: *Mammillaria, Opuntia, Echinocereus, Echinocactus, Gymnocalycium,* and many others — Follow general cultural requirements.

   b. Epiphytic cacti: *Epiphyllum, Rhipsalis, Rhipsalidopsis, Schlumbergera* (Christmas cactus) — These plants need cool nights when setting buds. Night temperatures above 65°F inhibit a maximum bloom. Christmas cactus needs short days (less than 12 hours of light) from

2. Carpet weeds (Aizoaceae): *Lithops* (stone plants), *Faucaria* (tiger jaws), *Conophytum*, *Fenestraria*

October to December and night temperatures from 7.5 to 10°C (45 to 50°F) in order to bloom.

3. Succulent "lilies" (Liliaceae): agave (century plant) *Haworthia*, aloe (*Aloë*), *Gasteria*

The seedlings should be kept moist; mature plants should be watered little during the dormant period.

These plants are rosettelike in habit.

4. Milkweeds (Asclepiadaceae): *Stapelia*, *Hoya*, (wax plant), *Ceropegia woodii* (string of hearts)

Milkweeds are found as vines and herbs. Some have very large and/or unusual flowers.

5. Stonecrops (Crassulaceae): *Crassula*, *Sedum*, *Kalanchoë* *Echeveria*

Many stonecrops can be propagated by leaf cuttings.

6. Composites (Asteraceae): *Kleinia*, (some species)

Most bromeliads are tropical epiphytes with a water-catching rosette of stiff, overlapping leaves. Because of this growth form, bromeliads are often referred to as "tank plants." Other members of the family are terrestrial and live either on the forest floor or on well-drained hillsides. The family is restricted in nature to the American tropics. Many are easy to grow as house plants.

Bromeliaceae: Bromeliad Family

GENERAL CULTURAL REQUIREMENTS

1. Humidity. Average house humidity is tolerated by many bromeliads, but 30 percent or more is best. Good ventilation should be provided.
2. Water. Tank forms: keep water in the tank and occasionally moisten growing medium. Terrestrials: keep the soil barely moist. Water less during the winter.
3. Temperature. Average house temperature is sufficient.
4. Soil. Epipyhtes: use fir bark or osmunda fiber (good drainage is essential). Terrestrials: use normal potting soil.
5. Fertilizer. Fertilize with dilute organic fertilizer during the growing season (usually the summer).
6. Light. Bright diffuse light is best for most bromeliads; some can thrive under less light.
7. Propagation. Propagate by division of offsets or seeds.

CULTIVATED GENERA OF THE BROMELIACEAE

1. *Aechmea*

These plants are epiphytes with flowers surrounded by showy, spiny pink bracts that last for months. The foliage is often banded.

2. *Ananas*
   (Pineapple)

The pineapple is terrestrial and grows best in high humidity and sunlight. To propagate, slice a rosette off the top of the fruit, allow to callus, and plant in normal potting soil.

3. *Billbergia*

These are very durable epiphytes with leaves often variegated and usually forming tubular or urn-shaped rosettes. The inflorescence is drooping, with brightly colored bracts.

4. *Cryptanthus*
   (Earthstars)

These plants are dwarf, low-growing terrestrials that do well in terraria. The leaves are often banded or striped with silver, bronze, and/or red. High humidity is preferable.

5. *Neoregilia*

These plants are mostly epiphytes with spiny-edged leaves. They are very easy to grow.

FIGURE 16-11.
One of the bromeliads (Tillandsia).

| 6. *Tillandsia* (Figure 16-11) | These plants are small epiphytes that need bright light, high humidity, and good ventilation. They are somewhat difficult to grow. |
| 7. *Vriesia* | These plants are epiphytes, some having brilliantly colored, sword-shaped inflorescences. They need warmth, high humidity, and bright diffuse light. Keep the tank filled with water, the soil barely moist. |

The orchid family is one of the largest of all flowering plant families, containing about 600 genera and over 20,000 species. In addition, there are over 35,000 horticulturally produced hybrids registered with the American Orchid Society. In nature, the vast majority of orchids are tropical epiphytes; there are some, however, which are terrestrial, and some do grow in temperate areas (e.g., our lady-slipper). Orchids are generally considered to be difficult to grow, but some genera can be grown successfully as house plants. Many of the popular horticultural cultivars have flowers that will last for weeks.

GENERAL CULTURAL REQUIREMENTS

1. Relative humidity. Requirements vary with the plant, but in general 50 percent humidity is sufficient. Good ventilation is also very important, but do not place plants in hot or cold drafts.
2. Water. Epiphytes: allow the growing medium to become almost dry between waterings. Terrestrials: keep moist but not wet.
3. Temperature. The proper temperature depends on the plant (see below). A drop in temperature of 5.5 to 8.5°C (10 to 15°F) from day to night is very desirable.
4. Soil. Epiphytes: use fir bark or osmunda fiber (should be very porous and allow air circulation). Terrestrials: use finely chopped fir bark, soft brown osmunda, or any mixture that drains well but retains moisture.
5. Fertilizer. The proper fertilizer depends on the plant (see below).
6. Light. The proper light depends on the plant, but in general provide as much light as possible without injuring the plant: too much light will cause yellowing; too little will cause foliage to turn dark green and plant to become stunted.
7. Propagation. Propagate by division or by seeds started in culture; mericloning can be used.

1. *Cattleya*

This plant is an epiphyte. It is a common corsage flower and quite tough. It needs bright light but not summer sun and likes warm temperatures [days: 18 to 24°C (65 to 75°F); nights: 13 to 16°C (55 to 60°F)]. It should be fed a dilute solution of fertilizer every 2 to 4 weeks.

2. *Cymbidium*

This plant is epiphytic with very long-lasting flowers (up to 3 months). Likes bright light and a growing medium that is moist at all times. These plants are heavy feeders and should be fertilized lightly every 1 to 2 weeks.

3. *Oncidium*

These plants are epiphytes, many producing large sprays of small to medium-sized flowers. Grow them under the same conditions as *Cattleya*, except give them more water.

4. *Paphiopedilum*

These terrestrial plants are related to our lady-slipper. They are easy to grow, with long-lasting flowers. They need lower light intensity than most other orchids (best in bright, indirect sunlight). The soil should be kept moist and the plants should seldom be fertilized. The best temperatures are 21 to 27°C (70 to 80°F). In the day and at night, 10 to 13°C (50 to 55°F) for green-leaved types, 16 to 18°C (60 to 65°F) for mottled-leaved types.

5. *Phalaenopsis*

These plants are epiphytes with long arching sprays of flowers. They may stay in perpetual bloom if given the proper treatment. They grow best in temperatures of 24 to 27°C (75 to 80°F) in the day and 18 to 21°C (65 to 70°F) at night, and like shaded conditions, constantly moist "soil," and dilute fertil-

izer every 2 to 4 weeks during
the growing season.

Gesneriads are primarily tropical plants with handsome, often velvety leaves and bell-shaped, tubular, or bell-shaped flowers.

*Gesneriaceae: The
Gesneriad Family*

GENERAL CULTURAL REQUIREMENTS

1. Relative humidity. Provide 30 to 50 percent humidity.
2. Water. Use room-temperature water from top or bottom; keep the soil slightly moist. Cold water or sunshine on water droplets may spot the foliage.
3. Temperature. Provide daytime temperature of 21.5 to 27°C (70 to 80°F) with a 5.5°C (10°F) drop at night.
4. Growing medium. Use a basic potting soil mixture or special commercial African violet mix.
5. Fertilizer. Use a weak solution every 2 to 4 weeks when plants are actively growing.
6. Light. Most gesneriads need plenty of bright light. Direct summer sunlight is not recommended, but diffuse light through a south window is ideal. Many thrive under 14 to 16 hours of artificial light per day.
7. Propagation. Propagate by stem cuttings, divisions, rhizomes, or tubers, depending on the type of plant. Many genera can be propagated by leaf cuttings. Can also be started from seed.

GESNERIADS CAN BE DIVIDED INTO THREE GROUPS:

1. Fibrous-rooted: no thickened underground storage organs.

a. *Aeschynanthus* (Lipstick plant)	Lipstick plants make showy hanging baskets. They may be pinched or pruned to any desired shape.
b. *Chirita*	These plants prefer moderate temperature, light shade, and high humidity.
c. *Columnea*	Most plants are pendent or trailing and prefer to be slightly potbound. They like the sunniest window (except in summer) and high humidity for bud formation.
d. *Episcia*	Many plants have leaves with contrasting veins and ever-blooming flowers, which make attractive hanging baskets. Propagate from runners.

e. *Hypocyrta*
(Goldfish plant)

The goldfish plant is everblooming under good conditions and makes nice hanging baskets.

f. *Saintpaulia* (Figure 16-12)
(African violet)

Eleven described species and countless cultivars are available. They are not related to the garden violet. They must have good light to flower and prefer fresh air (but not drafts).

g. *Streptocarpus*
(Cape primrose)

These plants tolerate lower temperatures and stronger light than African violets. Some need a 2- to 3-month dormancy period after flowering.

2. Rhizomatous: scaly rhizomes are pinecone-shaped, food-storing structures formed from underground stems. Single scales may be planted like seeds.

a. *Achimines*

These plants require medium light and evenly moist soil. They need 4-month dormancy period after flowering.

b. *Kohleria*

Noted for their attractive foliage, these plants need staking for upright growth. Cut back after flowering to encourage new growth.

c. *Smithiantha*
(Temple bells)

Temple bells need a 3-month dormancy period after blooming.

FIGURE 16-12.
Top view of African violet (Saintpaulia ionantha).

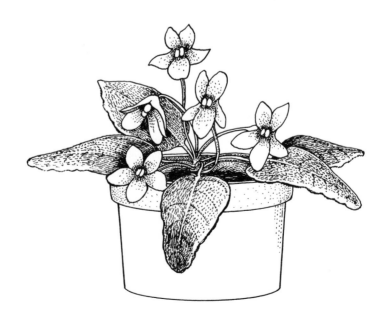

3. Tuberous: tubers are enlarged underground stems.

a. *Rechsteineria* These plants need a 3- to 4-month dormancy period after blooming.

b. *Sinningia* (Gloxinia) After blooming, decrease the water and store in a dark place until new growth appears (check frequently).

Large House Plants

The following are miscellaneous larger plants that are often grown in the house. Unless otherwise stated, these plants will tolerate average house humidity and temperature and should be potted in normal potting soil. Other cultural requirements are also noted.

1. *Araucaria*
   A. excelsa: Norfolk Island pine
   A. bidwillii: Monkey puzzle tree
   Provide bright indirect light and keep the humus-rich soil evenly moist. Propagate *Araucaria* by seed.

2. *Citrus*
   (Lemon, lime, orange, grapefruit)
   C. mitis: Calamondin orange
   Bright light is necessary, as is keeping the soil evenly moist (especially when fruits are forming). Citrus plants like cool temperatures, particularly during the winter rest period. Propagate by seeds or cuttings of young growth.

3. *Coffea arabica*
   (Arabian coffee tree)
   Provide bright indirect light and keep the humus-rich soil barely moist. Propagate by seeds or cuttings of young growth.

4. *Dracaena*
   (Cornstalk plant)
   Provide bright indirect light and keep the soil moist. Propagate by air layering.

5. *Fatsia*
   (Japanese aralia)
   Provide bright light and cool temperatures and keep the humus-rich soil evenly moist. Propagate by rooting young shoots.

6. *Ficus*
   F. benjamina: Weeping fig
   F. carica: Edible fig
   F. elastica: Rubber plant
   F. lyrata: Fiddle-leaf fig
   Provide bright indirect light and keep the soil barely moist. Avoid drafts that might cause leaf drop. Propagate by air layering.

7. *Grevillea*
   (Silk oak)
   Provide direct sunlight and cool temperatures. Allow to dry between waterings.

8. *Phoenix*
   (Miniature date palm)
   Provide bright indirect light and keep the soil evenly moist. Propagate from seeds. Other

		palms have similar cultural requirements.
9.	*Podocarpus*	Bright light and cool temperatures are best. Keep the soil barely moist. Propagate from stem cuttings in the fall.
10.	*Punica* *P. granatum:* Pomegranate	Provide bright light and keep the soil evenly moist. Propagate by cuttings or air layering.
11.	*Schefflera* (Umbrella tree)	Direct sunlight or bright indirect light, is recommended. Allow the soil to become dry between waterings. Propagate from seeds.

Some Helpful Hints About Growing House Plants

See Table 16-1 for help with plant problems. See the following lists for plants that will grow well under various problem conditions.

PLANTS THAT WILL GROW IN WATER

Elodea (*Anacharis*), parrot feather, duckweed, water hyacinth, *Salvinia* (a fern), *Marsilea* (a fern), *Azolla* (a fern).

PLANTS THAT WILL WITHSTAND MOST ADVERSE
HOUSE CONDITIONS AND ABUSE

Chinese evergreen, cast-iron plant, grape ivy, jade plant, dumb cane, *Dracaena*, crown-of-thorns, India rubber, fiddle-leaf fig, *Hemigraphis*, Kentia palm, arrowhead, screwpine, paradise lily, *peperomia*, *philodendron*, *Sansevieria*, devil's ivy.

PLANTS THAT DO WELL UNDER AVERAGE HOME CONDITIONS

Mountain acanthus, *Aechmea*, Norfolk Island pine, Sprenger asparagus, *begonia*, *caladium*, shrimp plant, kangaroo vine, grape ivy, dracena, *Cryptanthus*, holly fern, dumb cane, Japanese fatsia, botanical wonder, banyan fig, ivory fig, botree fig, silk oak, English ivy, arrowhead, *peperomia*, *philodendron*, artillery plant, black pepper, Celebes pepper, *aralia*, Moses-in-the-cradle, *Sansevieria*, strawberry begonia, painted tongue, white anthurium.

PLANTS THAT WITHSTAND DRY, WARM LOCATIONS

Bromeliads, cacti, crown-of-thorns, slipper flower, *peperomia*, leafy cactus, Moses-in-the-cradle, devil's ivy, *Sansevieria*, wandering Jew.

PLANTS WELL SUITED FOR LARGE-TUBBED DECORATIVE SPECIMENS

Acanthus, giant caladium, Australian tree fern, croton, dumb cane, Japanese fatsia, botanical wonder, ivory fig, variegated India rubber plant, fiddle-leaf fig, *philodendron*, screwpine, *aralia*, *schefflera*, bird-of-paradise.

TABLE 16-1 / Listing of Symptoms of Problems with House Plants and Their Probable Causes

Symptoms	Probable Causes
Tips or margins of leaves turn brown	Too much fertilizer
	Soil has been allowed to dry out
	Cold injury
	Wind burn
Wilting, yellowing of leaves, and soft growth	Too much heat
	Damage to root system by disease
	Too much or too little moisture
Small leaves and long internodes	Temperature too high
	Insufficient light
Weak growth, light green-yellow leaves	Lack of fertilizer
	Damage to the root system
	Light too intense
Yellowing and dropping of leaves from the base up	Overwatering
	Poor drainage
	Poor soil aeration
	Gas fumes
	Not enough light
Smaller leaves than normal	Lack of fertilizer
	Soil mix too heavy
	Not enough moisture

PLANTS FOR TOTEM POLES OR TRAINING

Kangaroo vine, *begonia*, grape ivy, bleeding heart, glory-bower, creeping fig, *Hemigraphis*, wax plant, *Pellionia*, philodendron, arrowhead, black pepper, Celebes pepper, devil's ivy, *Stephanotis*, painted devil's ivy.

LOW CREEPING PLANTS FOR GROUND COVERS IN INTERIOR PLANTING BOXES

Episcia, creeping fig, climbing fig, silver fittonia, English ivy, baby tears, *Hemigraphis*, *Pellionia*.

PLANTS REQUIRING BELOW 15.6°C (60°F) AT NIGHT

Austrian laurel, azalea, baby tears, *Calceolaria*, *camellia*, cast-iron plant, Christmas begonia, *Cineraria*, cyclamen, English ivy, *fatshedera*, geranium, German ivy, gold-dust plant, mother-of-thousands, Norfolk Island pine, spindletree, *vinca*.

PLANTS REQUIRING 18.3 TO 23.9°C (65 TO 75°F) AT NIGHT

African violet, Australian umbrella tree, banded maranta, cacti, succulents, *caladium*, Chinese evergreen, *philodendron*, *pothos*, *peperomia*, snake plant.

PLANTS REQUIRING MEDIUM LIGHT

Aluminum plant, *Chlorophytum*, Boston fern, *philodendron*, devil's ivy, holly fern, kentia, *peperomia*, *begonia*, dumb cane.

Fatshedera, fiddle-leaf fig, grape ivy, piggyback plant, *pothos,* silver marble, variegated India rubber plant, English ivy.

Organic Gardening

Organic gardening is not new. It has been practiced for centuries in China, India, and central Europe. Farmers and gardeners have relied on organic fertilizers (mainly manure) and compost for improving the growth of their plants, and on natural predators to control pests that attack them. Organic gardening has recently become very popular as an alternative to corporate agribusiness and the type of gardening that has relied so heavily on extensive use of commercial inorganic fertilizers and environmentally polluting and food contaminating pesticides. We shall discuss alternative methods of pest control in detail in Chapter 18. Here we simply list some of the practices of organic gardening:

1. Organic gardening relies on heavy use of organic fertilizers such as manure, sterilized sludge from secondary treatment of sewage, and blood meal. These "organic" fertilizers contain most of the inorganic nutrients essential for plant growth.
2. Organic gardening involves use of composting to recycle vegetation and return it to the land. Composting is discussed in detail in Chapter 20.
3. Organic gardening utilizes mulches, such as straw, rice hulls, cocoa bean hulls, buckwheat hulls, and newspaper to control weeds.
4. Organic gardening practices for control of insects and pests rely on the use of companion plants (such as marigold, chives, and many herbs) that produce their own "insecticides." Only mild insecticides, such as rotenone and pyrethrum (both plant products), and dormant oils are used to control pests. Natural predators, such as certain wasps, lady bugs, and praying mantis, are used to control crop-eating insects.
5. Organic gardening relies on crop rotation (which includes legumes that fix nitrogen) and cover crops, such as annual rye grass or legumes.

What are the benefits accrued from practicing organic gardening? One of the primary ones is that the crops and soil are not polluted by pesticides. A second one is that soil management with organic gardening is vastly superior; soils under this regime are not impoverished of nutrients, have better water-holding capacity, contain more organic matter, and thus are less likely to blow away in high winds. A third is that even though yields may be slightly lower than when there is heavy use of inorganic fertilizers and pesticides, the costs of growing crops is less, so that total costs balance out. The organic gardener wins in the long run because of the better quality of crops and flowers produced, far less soil loss due to wind erosion, better water retention by the soil, no pesticide residues, lower production costs, and less waste of plant residues. What better way is there to work with nature than through the sensible, nature-compatible practices of organic gardening?

FIGURE 16-13.
Formal garden at entrance to the Climatron at the Missouri Botanical Garden. (Photograph by Peter Kaufman.)

FIGURE 16-14.
Formal garden and pool at the Frankfort Botanical Gardens in Germany. (Photograph by Peter Kaufman.)

Home and Urban Landscapes

Home landscapes are basically of two types: *formal* and *informal.* Formal landscapes involve strong geometry (i.e., the use of circles, rectangles, squares, straight lines, and even balance). They reached their epitomy in many early European gardens around palaces and manor houses, in public parks, and in botanical gardens. Formal landscapes are costly, and more important, require much labor to maintain them. Many home owners use some formal landscaping in their house foundation plantings and front gardens. Some examples of formal landscapes are illustrated at the Missouri Botanical Garden in St. Louis (Figure 16-13), the Frankfurt Botanical Gardens in Germany (Figure 16-14), and a city park in Lund, Sweden (Figure 16-15). Examples of informal gardens are illustrated at Cypress Tropical Gardens in Florida (Figure 16-16), a city park in Lund, Sweden (Figure 16-17), and a demonstration garden terrace at Narviken Gardens in Sweden (Figure 16-18).

Many homes are not landscaped at all, and the effect is disastrous! A home without foundation plantings and without trees and shrubs in the yard "sticks out like a sore thumb" (Figure 16-19). One that is landscaped is a pleasure to behold (Figures 16-20 and 16-21). The plants soften up harsh corners and add more interest to the house. Even the magnificent old farmhouse shown in Figure 16-19 would be much more handsome if properly landscaped, either formally, informally, or both.

To landscape a property, one can follow a fairly simple set of procedures. These are illustrated in Figure 16-22, steps A to E. This particular scheme utilizes both formal and informal landscaping. It is a type of landscape which is aimed at ease of maintenance, makes provision for recycling vegetation in a compost, includes food plants, and gives one a variety of functions in the different unit areas. In more

FIGURE 16-15.
Formal garden in the city park in Lund, Sweden. (Photograph by Peter Kaufman.)

FIGURE 16-16.
*Informal garden at Cypress
Gardens, Florida. (Photo-
graph by Peter Kaufman.)*

FIGURE 16-17.
*Informal garden in a small
city park in Lund, Sweden.
(Photograph by Peter
Kaufman.)*

crowded situations, one or more of the unit areas would have to be made smaller or eliminated. The use of espalier-type training of fruit trees or ornamentals on fences or against the sides of buildings or homes conserves a great deal of space. This treatment is widely used in Europe with pear and apple trees. Espalier basically involves pruning a tree so that its trunk and branches lie more or less in one plane against a building, and the branches are oriented in horizontal tiers, affixed to wires. "Living fences" as used in Figure 16-22, steps D and E, can be small shrubs with edible fruits (e.g., elderberry, currant, gooseberry, highbush cranberry), grapes, or cane berries (e.g., raspberries, boysenberries, youngberries). See also Table 16-2, which presents labor-saving devices for use in home landscaping.

Urban landscaping is probably one of the most crying needs in our urban ghettos and rundown central business districts. Spencer Havlick alluded to this in detail in Chapter 12. One of the best places to start is with an urban mall that is not just blocked off from traffic and is full of concrete walks and steps. It must have plants of all types: hanging baskets, planter boxes, shade trees, beds with bulbs and summer annuals, and beds with prostrate and upright evergreens (both coniferous and broadleaf evergreens). Figures 16-23 and 16-24 illustrate a very successful urban mall in Rotterdam in the Netherlands that utilizes many of these features.

The urban ghetto needs more parks and places for recreation, replete with water in ponds, aquatic plants, shade trees that are well adapted for urban areas (Elias and Irwin, 1976, p. 114), community vegetable gardens, and nurseries where people can raise plants for their gardens. A solar-heated greenhouse coupled to a wind generator should also be included for raising winter vegetables, flowers, and vegetable plants to

FIGURE 16-19.
Unlandscaped home. (Courtesy of Larry Mellichamp.)

FIGURE 16-20.
Informal landscape surrounding an urban home in Ann Arbor, Michigan. (Photograph by Peter Kaufman.)

FIGURE 16-21.
Close-up view of home shown in Figure 16-20, illustrating Cotoneaster divaricata, a shrub being trained to grow flat on the side of the house. (Photograph by Peter Kaufman.)

put in the garden. Finally, it should include a compost pile, where leaves are placed and the cut tops of vegetables and flowers are recycled. The humus from such a compost would then be placed in the community vegetable gardens to improve the soil. Such projects are low in cost,

FIGURE 16-22.
Basic steps in how to make a landscape plan.

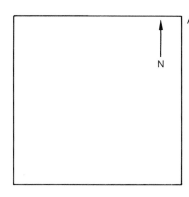

A. Draw in the plot of land, showing its dimensions.

LANDSCAPING
How Do You Make a Landscape Plan?

B. Put in the buildings.

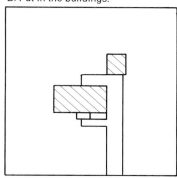

C. Block out the unit areas you want. (Basic uses should be kept in mind.)

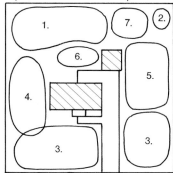

UNIT AREAS

1. Informal garden 2. Compost 3. Formal garden
4. Recreation 5. Vegetable garden 6. Terrace
7. Herb garden

Key to items in unit areas

- ✳ Conifers
- ⊛ Deciduous trees
- ∿ Living fences
- ⌇ Flowers (annuals)
- ═ Grape arbor
- ✿ Shrubs
- ∘∘∘∘ Stone pathway
- ⊙ Fountain
- ✱✱✱ Flower garden
- ▫ Bird feeder
- ▤ Picnic table

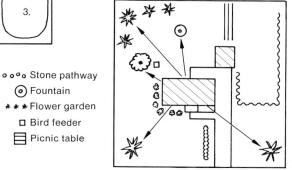

D. Put in focal points and axis lines.

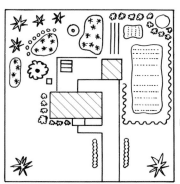

E. Landscape plan for a formal-informal garden.

stimulate people to work together to improve their community, and provide both beauty and food for those who live there.

TABLE 16–2 / Some Ways to Save Labor in Maintaining a Home Landscape

Labor-Saving Device	*Example*
Use of ground covers in beds around shrubs, under trees.	*Ajuga, Pachysandra,* English ivy (*Hedera helix*), lily-of-the-valley (*Convallaria majalis*), myrtle (*Vinca minor*), stonecrops (*Sedum*), *Dicondra,* sandwort (*Arenaria verna*), and *Gazania*.
Earth-made saucers around individual plants to retain water, conserve water, and lessen frequency of watering.	Grapes, shrubs on hillsides; tomatoes, young squash, cucumbers, and pumpkins; newly planted trees.
Use of mulches (leaves, straw, cocoa bean hulls, rice hulls, newspaper, and plastic film) to control weeds and conserve soil moisture. In some urban areas, gravel or crushed rock (with a layer of plastic film underneath) is used for these purposes.	Newspaper under squash, cucumber, and pumpkins. Straw or black plastic film under and between strawberries; and organic mulches around shrubs and perennials.
Composting to recycle vegetable scraps from the kitchen, trimmings, weeds, etc., to improve soil humus and improve water-holding capacity of the soil.	Old sod, soil from excavations, hedge trimmings, grass clippings, leaves, ashes from fireplace, eggshells, old vegetables, citrus rinds, coffee grounds, and the like.
Planting annuals and perennials in masses to curtail weed growth in individual flower beds.	Zinnias, marigolds, salvia, chrysanthemums, asters, dahlias, and many other garden flowers.
Keeping functions in different unit areas distinctly separate from each other.	Vegetable garden lawn area, area for pets, areas of high recreation use, and compost pile all in distinctly separate unit areas.
Landscaping the foundation of a house with plants that are of the *right size* (initially and years later), *spaced properly,* and *pruned as needed*.	To prevent foundation planting from covering windows, growing up to roofline, and becoming overcrowded. Many such plantings are *initially* overcrowded, and the plants in the scheme get overgrown in 5 to 10 years.

FIGURE 16-23.
Urban mall in Rotterdam, The Netherlands. (Photograph by Peter Kaufman.)

FIGURE 16-24.
Hanging baskets in a tree in an urban mall in Rotterdam, The Netherlands. (Photograph by Peter Kaufman.)

Plants, grown indoors or outdoors, can provide us with a bit of the real world of nature in our urban, synthetic environment. But, to enjoy them, we must learn the art and science of growing plants and arranging them in the proper niches in our landscapes or plant collections indoors. We must learn where and how they grow in their native habitats and take our cues from this knowledge to grow them at their best in our homes, apartments, offices, and gardens. In addition, great patience, eternal vigilance, and continual maintenance are needed. In this chapter we have provided some practical hints on indoor gardening and on landscaping and maintaining an outdoor garden with the least amount of work possible.

Summary

CARPENTER, P. L., T. D. WALKER, and F. O. LANPHEAR. 1975. *Plants in the Landscape.* W. H. Freeman and Company, Publishers, San Francisco.

ELIAS, T. S., and H. S. IRWIN. 1976. Urban trees. *Sci. Amer.* **235**(5), 100–118.

New York Botanic Garden. 1965. Creative ideas in garden design. *Plants and Gardens* **21**(3), 1–65.

TAYLOR, J. L. 1977. *Growing Plants Indoors.* Burgess Publishing Company, Minneapolis.

WOLF, R. 1975. *Organic Farming Yearbook of Agriculture.* Rodale Press, Inc., Emmaus, Pa.

Selected References

17

Edible Wild Plants[1]
Ellen Elliot Weatherbee

Questions for Consideration

1. Is it possible to collect wild plants to be used as a source of food?
2. What precautions are needed before attempting to utilize wild foods?
3. What type of historical knowledge has been collected about these wild edibles?
4. Do these wild plants contain any significant nutritional levels?
5. Which types of habitats are most likely to contain wild edibles?

The study of edible wild plants is a bright spot in the often-disappointing ecological picture. Even on a cold winter day, within a few miles of a metropolitan area, a feast is made easily after a few hours spent gathering and processing. A sample February menu might include such delicacies as wild leek and watercress soup, broiled cattail shoots, the underground tubers of groundnut sautéed with velvet-foot mushrooms, and boiled Jerusalem artichokes garnished with wild chives. Dessert specialties could be highbush cranberry sherbert, wild ginger candy, and a choice of bergamont tea made from the leaves of the plant, or lemonade made from sumac fruits.

The best collecting areas are those with diversified habitats. Some plants are adaptable and will grow in a variety of terrains; others require precise locations with proper temperature, humidity, and acidity ranges. A woodlot, bordered by an old field on one side, a garden plot in need of weeding on another, and with a stream in a low area potentially incorporates many of these plants. Often ideal looking places contain nothing, while the unkempt backyard of an apartment house could be loaded with edibles.

With all the edibles available, why has man not fully used these resources? Probable reasons are (1) the ease of buying already processed food; (2) the lack of knowledge of plant identification; and (3) a rejection of the traditions of immigrants, who collected wild plants both from necessity and for enjoyment.

What leads people to devote time to the collecting of these edibles?

[1] The photographs illustrating this chapter were arranged and, for the most part, made by T. Lawrence Mellichamp, Department of Biology, University of North Carolina, Charlotte, North Carolina 28223.

1. Wild food can cause a significant reduction of the food bill, especially when plants are dried, canned, or frozen for other times of the year.
2. A knowledge of plants enables man to appreciate his natural environment, to increase the possibilities of surviving in a crisis situation, or to add fresh food to a hiking trip.
3. Some wild plants can be used year round; others, especially the weedy types, can be used earlier than their commercial counterparts.
4. Nutritional values run very high in some of the wild edibles (see the nutrition table, Table 17-1).
5. Wild plants bring many new tastes, with excellent flavors, textures, and their own special appeal. Some are truly gourmet adventures and please almost everyone who tries them.

There are five general habitats that interest the wild food addict, and among these will be found most of the prominent edibles. Some of the areas are natural habitats, some are disturbed sites whose existence and perpetuation is largely determined by man's activities, and some are intermediate phases several years after man's influence ceases. These areas may be divided into dry and wet domains.

TABLE 17-1 / Abbreviated Nutritional Table for Selected Wild Plants[a]

Plant	Calories	Vitamin C (mg/100 g)	Vitamin A (I.U./100 g)	Calcium (mg/100 g)
Asparagus	20	26	900	21
Dandelions	33	18	11,700	140
Highbush cranberry	—	100	2,105	—
Jerusalem artichoke	7–75	4	20	14
Lamb's quarters	32	37	9,700	258
Mints	—	68	8,575	—
Nettles	—	76	6,566	—
Pokeweed	20	82	8,700	53
Purslane	15	12	2,100	86
Strawberry				
Berries	37	59	60	21
Leaves	—	229	—	—
Violets				
Blossoms	—	150	—	—
Leaves	—	210	8,258	—
Watercress	19	79	4,900	151
Domestic Values for Comparison				
Milk	65	1	140	118
Oranges	49	50	200	41
Spinach	23	51	8,100	93

[a] Some edible wild plants are high in nutritional values. The figures shown here are taken from the U.S. Department of Agriculture (1963) and from Gibbons (1966). Vitamin C values indicate the number of milligrams (mg), vitamin A values indicate the number of international units. (I.U.), and calcium levels indicate the number of milligrams (mg). All values are given for each 100 grams of edible portion of food. A dash (—) indicates that that particular value is not available. Three domestic foods are given for comparative purposes.

FIGURE 17-1.
Typical roadside and ditch bank. Shown surveying the vegetation is Professor Warren H. Wagner, Jr., of the University of Michigan, well-known botanist and expert on wild edible plants and poisonous plants.

FIGURE 17-2.
Old field with junipers, sumac, young wild cherry trees, and many herbaceous plants.

FIGURE 17-3.
*Recently abandoned culti-
vated field, now rich in
weedy species.*

FIGURE 17-4.
*Rich, mature beech—maple
woods with carpet of
spring wildflowers. Shown
is Professor Warren H.
Wagner, Jr., fern specialist
at the University of
Michigan.*

1. Roadsides (Figure 17-1) and railroad banks. These are man-produced and are maintained by occasional mowings and fire. Railroad embankments frequently contain native prairie species and thus form some of the last remnants of the once vast prairie that characterized much of the Midwest.
2. Old fields (Figure 17-2) and meadows, unweeded fields (Figure 17-3), and old orchards. These areas were man-produced and have now begun to revert to natural conditions.
3. Oak—hickory and beech—maple woods (Figures 17-4 and 17-5). These mature woods are the result of the natural progression of growth after the extensive logging operations at the turn of the century.

Dry Habitats

FIGURE 17-5.
*Disturbed areas in beech—
maple woods with under-
growth of shrubs.*

Wet Habitats

1. These naturally existing areas that include shallow marshes (Figure 17-6) and swamps (Figures 17-7 and 17-8) and stream and river banks (Figure 17-9), deeper streams, lakes, and ponds form further special-ized habitats.
2. Bogs (Figure 17-10). Fortunately, some of these spectacular bogs have escaped draining and contain their unique plants in a very acid environment.

From a large assortment of edible plants, the plants mentioned are chosen for the following reasons: they are readily found, are not easily confused with poisonous species, exist in sufficient abundance to justify usage, are not too tedious to prepare, and are good to eat.

Roadsides and Railroad Banks

These easily accessible areas (Figure 17-1) contain a wealth of edibles. Avoid sprayed or dusty localities.

FIGURE 17-6.
Extensive shallow marsh with cattails, sedges, and shrubs.

ASPARAGUS (*ASPARAGUS OFFICINALIS; LILIACEAE*)

Asparagus (Figure 17-11) is a common perennial of roadsides, farm fences, old orchards and railroad embankments. It is especially abundant where seeds have spread from cultivated fields to nearby areas. The plants produce young shoots which come up the end of April; if kept cut every 5 days, the plants will bear through June.

Wild asparagus looks and tastes like store-bought asparagus. The plants are the easiest to locate in the fall, when the brightly colored foliage makes yellow, waist-high feathery splotches. This is the best time to note localities, for in the spring, the dead stalks are broken down, although they can occasionally be spotted, especially if they are caught in a fence.

Wild asparagus can be prepared in any recipe using domestic asparagus. The easiest method is to steam the shoots until tender, about 10 minutes. Asparagus is excellent any way it is prepared—cooked and chilled for salads, hot in omelets and souffles, boiled, or sautéed. It is always a treat.

Asparagus is found in many areas of the world, including Europe and Russia. The Romans made extensive use of the young shoots as early as 200 B.C. (Sturtevant, 1919). There are many varied and exciting references to its enjoyment.

JERUSALEM ARTICHOKE (*HELIANTHUS TUBEROSUS; ASTERACEAE*)

Jerusalem artichokes (Figure 17-12) occur along roadsides and in abandoned garden sites and borders. The underground tubers are filled with inulin[2] from September through April. Cultivation of the plants is

[2] Inulin is a storage polysaccharide made up of the sugar, fructose; starch, another storage polysaccharide, is made up of the sugar, glucose.

FIGURE 17-7.
Hardwood swamp forest with characteristic woody species such as black ash, basswood, and red maple.

FIGURE 17-7 *(cont.)*

FIGURE 17-8.
Old elm swamp.

FIGURE 17-9.
Riverbank and floodplain forest.

FIGURE 17-10.
Typical sphagnum bog in Michigan's Upper Peninsula. Open water is being encroached by mat of sphagnum moss and shrubs. Behind this zone, on more solid ground, develops spruce and tamarack swamp forest.

FIGURE 17-11.
Asparagus officinalis, young shoots.

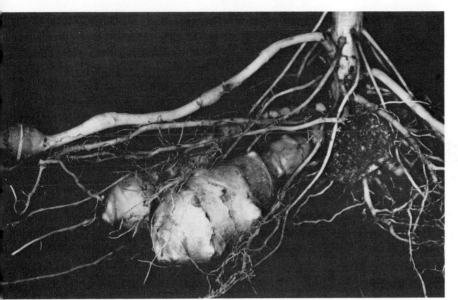

FIGURE 17-12.
Helianthus tuberosus,
Jerusalem artichoke, tubers.

simple, as each of the paired sprouts along a tuber grows well. Gardens are often overwhelmed by these easily spreading plants, so, if you cultivate them intentionally, keep them away from the main area of the garden.

Jerusalem artichokes are tall [1.8 to 2.4 meters (6 to 8 feet)] plants with many miniature [2.5 to 5 centimeters (1 to 2 inches)] sunflowers near and at the top. The stem is hairy almost to a raspy texture. They are the only sunflower that forms large tubers. These tubers, found 7.6 to 30.4 centimeters (3 to 12 inches) below the plant, range in size from thin, pencil shapes to more commonly, those the size of an elongate potato.

Slice the tubers thinly and rub with lemon juice to prevent browning. They can be boiled until tender (10 minutes) and served with butter or fried in a little oil. They are good raw in salads and with a sour cream dip.

The name is a corruption of the Italian "Girasoli articocco," which means "sunflower artichoke." The plants may have originated from a sunflower native to the Mississippi River valley (Sturtevant, 1919). It has been domesticated in the United States since the mid-1700s.

Nutritional values (Table 17-1) (given only for the raw tubers) show an interesting phenomenon: when first harvested, they contain only 7 calories/100 grams of edible portion. After long storage, the calorie count jumps to 75 for the same amount.

MILKWEED (*ASCLEPIAS SYRIACA*; ASCLEPIADACEAE)

Milkweed (Figure 17-13) is found along roadsides, railroad embankments, and old fields. Various portions of the plant are useful from the end of May to mid-July. Four parts of this plant are used—the young shoots to 20.3 centimeters (8 inches) in height (Figure 17-13), the unopened buds, the young pods (Figure 17-14), and the young leaves.

This plant has milky juice and opposite leaves. The young leaves have a white hairiness on their underside, and there will often be found the familiar pods of the previous year. Milkweed can be confused with dogbane (*Apocynum* spp.), which has smooth sides to its leaf and branches more quickly. Dogbane has been known to cause cattle poisoning (Hardin and Arena, 1974). According to a student who inadvertently cooked dogbane, the product remained extremely bitter.

Milkweed is one of the few wild edibles that require special processing to make it palatable. After a thorough washing, immerse the milkweed parts in a roaring boiling water bath; boil 1 minute, drain, and throw out the water. Repeat twice, using three waters in all. Taste for bitterness; by this time the vegetable should taste good. Serve as a boiled green—it goes especially well with a cheese sauce.

In 1772, the French Canadians were reported to use the young shoots, and in 1845, the Sioux Indians were known to have eaten the boiled young pods with buffalo meat (Sturtevant, 1919). In the southern Appalachians, the natives use milkweed in the spring as a tonic for "what ails you" (Wigginton, 1973).

FIGURE 17-13.
Asclepias syriaca, common milkweed, young shoots.

FIGURE 17-14a.
*Asclepias syriaca, young
pods.*

FIGURE 17-14b.
*Asclepias syriaca, milk-
weed, bowl full of young
pods.*

FIGURE 17-15.
*Phytolacca americana,
pokeweed, young plant
1 meter (3 feet) tall.*

POKEWEED (*PHYTOLACCA AMERICANA*; PHYTOLACCACEAE)

Pokeweed (Figures 17-15, 17-16) is found in old fields, along railroad tracks, and especially in disturbed areas in those woods where some light can reach the plants. In the southern states, it is found abundantly along roadsides.

Pokeweed is best identified by the stalks of the previous year. The 1.8-meter (6-foot) stalks will still show their branching and the chambered pith of the interior of the stalk, even when bent over. The young shoots may or may not have a reddish tinge to the stalk. Use young shoots up to 25.4 centimeters (10 inches) in height. The berries are probably poisonous, and the root is very poisonous and should never be eaten.

Pokeweed is another plant that must be boiled in at least two waters, preferably three, in order to remove the bitter and perhaps poisonous compounds. Traditionally poke, also called poke salad, is served with bacon bits and vinegar.

Many records of Indian edibles mention pokeweed. Poke has been exported from North America to Europe, Mexico, Brazil, and to the Mediterranean as a garden vegetable (Sturtevant, 1919). Poke is commonly used in the mountains of the southeastern United States (Wigginton, 1973).

Poke is low in calories and high in vitamins A and C (Table 17-1).

FIGURE 17-16.
Phytolacca americana, young edible shoot.

SUMAC (*RHUS* SPP; ANACARDIACEAE)

Sumacs are found in dry, weedy habitats, in or outside of town, and they especially like old fields. Most species of sumac form large clones with interconnected roots. Several species have edible fruits. Any of the species that have red berries forming erect heads or clusters and which are found growing on dry soil are all right to use. Poison sumac (*Toxicodendron vernix*), with its skin-irritating oils, grows in wet areas (such as bogs and swamps) and has white berries that hang down. The edible sumac is available for year round use; pick the freshest looking, totally red heads.

Place the washed red heads in cold water. Bring almost to a boil and steep until a dusky pink in color. Pour the mixture through a piece of cheesecloth or a clean pillowcase to extract the hairs and debris that gets caught among the berries. The juice needs sweetening with honey or sugar and is especially good when served chilled.

Seeds of sumac have been found in archeological sites dating from 2000 B.C., and remains as early as A.D. 500 are known from the Midwest.

WILD FRUITS AND NUTS

There are many types of wild berries, which, conveniently, ripen at spaced intervals throughout the summer. Among these are raspberries (*Rubus* spp.) (Figure 17-17), blackberries (*Rubus allegheniensis*), gooseberries (*Ribes* spp.), elderberries (*Sambucus canadensis*), mulberries (*Morus* spp.) (Figure 17-18), strawberries (*Fragaria virginiana*), and the blueberries (*Vaccinium* spp.).

FIGURE 17-17.
Rubus occidentalis, black raspberry.

FIGURE 17-18a.
Morus rubra, red mulberry, ripe and unripe multiple fruits.

FIGURE 17-18b.
Morus rubra, red mulberry, fruits.

Fruit trees are found in abandoned orchards, as escapes due to animal activity, and in other deserted areas. In low-yield years, the unsprayed fruits tend to be wormy and, when processed, the fruit blemishes need some cutting or ignoring. In years of exuberant fruiting, the worms cannot compete, and the fruit is properly presentable even to the most fastidious palates. Pears, apples, and grape vines are commonly found in these areas.

Nuts are abundant in the fall and are easily recognized: walnuts (*Juglans nigra*) (Figure 17-19), butternuts (*J. cinerea*), hickories (*Corya ovata* and *C. glabra*), and acorns from the many oak species (*Quercus* spp.).

FIGURE 17-19.
Juglans nigra, black walnut.

Old Fields and Meadows, Unweeded Fields, and Old Orchards

Plants found in old fields (Figure 17-2), unweeded fields (Figure 17-3), and old orchards are "camp followers" whose existence and distribution are due to the influence of man. These plants are usually weeds, and they thrive and reproduce rapidly. Their seed production and dispersal continue throughout an extensive period of time, and the plants adapt to a variety of disturbed conditions. Some plants need periodic disturbance (e.g., a sometimes-weeded-garden, a lethargically worked construction site).

DANDELION (*TARAXACUM OFFICINALE*; ASTERACEAE)

Dandelions (Figure 17-20) are found in grassy areas, meadows, old fields, wood's openings, and vacant lots. Collect the new leaves, the crown, and the unopened flower buds. Dandelion leaves are easiest to dig by slicing the crown [the white, underground part to which the leaves are attached about 2.5 centimeters (1 inch) below the soil surface]. Wash thoroughly by vigorously dipping up and down in a big pot of water to remove all possible grit. Do not remove the leaves from the crown until the cleaning is finished.

Boil the washed leaves until tender, 10 to 12 minutes. The slight bitterness is enhanced by being served on bread with bits of bacon and a cruet of vinegar.

Dandelions have been extensively used in Europe and sometimes in the United States. The name comes from a corruption of "dent de lion" because it was thought that the indentations of the leaf looked like a lion's tooth.

Dandelions are low in calories and high in vitamin A (Table 17-1).

JUNIPERS (*JUNIPERUS VIRGINIANA* AND *J. COMMUNIS*; CUPRESSACEAE)

Junipers (Figure 17-21) are found year round in open, sunny areas. They are also found in woods where there are open spots; however, as the tree canopy matures, the junipers will die for lack of light.

The upright juniper (*J. virginiana*) has a small trunk reaching to 9.1 meters (30 feet) and has both pointed and scaly leaves, while the low, 0.9 to 1.2 meters (3 to 4 feet) spreading juniper (*J. communis*) has only sharply pointed leaves. Both have blue berrylike cones.

The needles are steeped for tea, and the berries are used to flavor meats, especially wild game, and in the manufacture of gin. They are commonly available commercially for flavoring meats. The ripe berries have been used in Europe for preserves, a beer, and as a coffee ingredient (Sturtevant, 1919).

LAMB'S-QUARTERS (*CHENOPODIUM ALBUM*; CHENOPODIACEAE)

Lamb's-quarters (Figure 17-22) are found in highly disturbed areas from the end of April through the end of September. The plant especially likes unweeded gardens and construction sites.

FIGURE 17-21.
Juniperus communis, spreading juniper, female plant with fleshy cones.

FIGURE 17-22.
Chenopodium album, lamb's-quarters (right in picture). Pigweed, (Amaranthus sp.) on left, also has edible leaves when young.

The young shoots appear before most garden vegetables are ready for harvest. If the ground is kept in a disturbed condition, the young plants will continue to be produced until September. If the plants are over 15.2 centimeters (6 inches), take only the top tender leaves and discard the stem, which is now stringy. The leaves have a ruffled arrow shape with a mealy quality underside when young and edible.

Lamb's-quarters (also called pigweed), known since the Bronze Age, were used both fresh and dried by many tribes of Indians (Yarnell, 1964). Sometimes the plants were "double-harvested," meaning that some plants were used as shoots for spring greens and others were left to go to seed and ground into a meal. Fernald and Kinsey (1958) refer to Napoleon's memoirs, in which he mentions times of living solely on lamb's-quarters.

Lamb's-quarters is fairly high in vitamin C, very high in vitamin A, and extremely high in calcium content (Table 17-1).

NEW JERSEY TEA (*CEANOTHUS AMERICANUS;* RHAMNACEAE)

New Jersey tea (Figure 17-23), a low shrub found in old fields, prairies, and barrens, is usable any time there are leaves, from midspring until late fall. New Jersey tea becomes well established when left undisturbed.

FIGURE 17-23.
Ceanothus americanus, New Jersey tea, leaves and old fruiting pedicles.

The plant often grows abundantly as a clone. The small fruits form tiny saucer-shaped receptacles, which remain conspicuous throughout the winter. The leaves have three prominent veins.

The dried leaves make one of the best wild teas. Dry the leaves indoors on newspaper until crinkly dry; use a small handful of dried leaves per liter of water. Bring to a simmer and steep until the color of regular tea.

Sturtevant (1919) reports that the leaves were used extensively for a tea during the American Revolution.

PURSLANE (*PORTULACA OLERACEA*; PORTULACEAE)

Purslane (Figures 17-24 and 17-25) occurs in flower boxes, sidewalk cracks, construction sites, and gardens, where it has a special affinity for corn. It grows from July to September and quickly recovers after cutting if the root is left in the ground.

Purslane is a succulent plant with prostrate, branching, fast-growing stems, small yellow flowers, and flat, paddle-shaped leaves. No matter how young the plant is collected, it manages to fruit with small, dot-sized black seeds by the time it is cooked.

Any part of the aboveground portions can be used raw in salads, boiled for 10 minutes and served with butter, or chopped in soups and casseroles. The stems make a refreshing pickle. This is a good plant to know, as it is very tasty and often can be gathered in great quantities. It freezes well for winter use.

FIGURE 17-24.
Portulaca oleracea, purslane, succulent, edible shoot.

FIGURE 17-25.
*Portulaca oleracea, pur-
slane, flowering plant.*

Purslane is a well-known edible used in many parts of the world. French gardeners raise three varieties. Champlain in 1605 found the Indians along the Maine coast using the prostrate shoots. The plant originated in tropical and subtropical regions and is especially common around the Mediterranean (Sturtevant, 1919).

Purslane is low in both calories and nutritional value (Table 17-1).

COMMON SORREL (*RUMEX ACETOSELLA*; POLYGONACEAE)

Sorrel (Figure 17-26) also known as sheep's sorrel is commonly found in lawns, old orchards, old fields, and along fences. The leaves are found year round, although they will be much larger and more flavorful in the summer.

Children are especially fond of sorrel and think that the leaf looks like an arrowhead when held up straight and like a fish when held on its side. In summer, the leaves are about 7.6 centimeters (3 inches) long and occur on a 0.3-meter (1-foot) high plant. When sorrel flowers, the reddish inflorescences color whole fields with their exuberant color.

Sorrel is good raw as a nibbling food, chopped raw in salads, boiled alone, or as an ingredient in soups or casseroles. Collect more than seems needed, as the leaves cook down considerably. Sorrel adds a pleasant, sour-lemon taste to food.

The French cultivate sorrel, and in this country, the larger food markets occasionally sell the leaves. The seed can be obtained from seed companies.

FIGURE 17-26.
Rumex Acetosella, common sorrel, young leaves (edible).

The leaves are low in calories and very high in vitamin A, with fairly high values for vitamin C.

STRAWBERRIES (*FRAGARIA VIRGINIANA*; ROSACEAE)

Strawberries occur in old fields and orchards, in vacant lots, along railroad tracks, and in sunny borders of woods. The berries, which reach $\frac{1}{2}$ inch in size, ripen from late May to mid-June. In good years, many quarts can be collected. It is best to clean them while collecting; snip off the stem end to avoid tedious cleaning during later preparation.

Strawberries are perennial herbs identified by their lobed, three-times-divided fuzzy leaves. They are low, nonvining plants with white flowers. Spreading runners often connect these sun-loving plants.

Wild strawberries are prepared exactly as the domestic ones, although the untamed varieties are sweeter, smaller, and more flavorful. They are always a delight, whether eaten raw, in pies, in jams, or soaked overnight in white wine and added to champagne for festive occasions.

The leaves of strawberries are used for one of the finest of wild teas. They are available any time of year, including the winter, but they are best when the plant is in flower.

Do not wash the leaves before drying them. Clean them with a vegetable brush and place upon a layer of newspaper indoors; leave until crinkly dry, about 48 hours. Store the dried leaves in covered jars. To make the tea, add a small handful of dry leaves to 1 liter of cold water. Bring to a simmer and steep until a robust flavor is obtained, about 15 minutes.

There has been extensive worldwide use of strawberries, with such authors as Pliny, Ovid, and Virgil mentioning their excellence. The Great Lakes Indian tribes ate great quantities of the fresh fruits and dried them as well for winter use (Yarnell, 1964). First records of cultivation are in 1536, and white-fruited types, as well as various colorations of red, yellow, and white ripe berries are known (Sturtevant, 1919).

Wild strawberries are high in vitamin C, and the leaves are extremely high in vitamin C. (Table 17-1).

WILD CHIVES AND WILD ONIONS (*ALLIUM* SPP.; LILIACEAE)

Many species of onions and chives occur commonly, and the easiest to find are those in lawns and fields, where they resemble plump grass. Some species prefer railroad embankments, some like meadows, and the wild leek (*A. tricoccum;* Figure 17–27) requires rich woods (see p. 351).

The flower is a small white lilylike flower and may or may not be present at the time the leaf is seen. The distinctive bulb, with its unmistakable onion smell, is the identifying characteristic. The various species are often difficult to differentiate, but all can be used in the same way.

Use the bulbs, and leaves if present, like regular onions and chives; if too strong, use only the finely chopped tops or parboil the bulbs until comfortably jolting. Onions are sumptuous in stews, casseroles, and soup (especially the French onion type).

Onions have been cultivated for hundreds of years. The explorer Marquette and his party in 1674 were forced to use wild onions as a large portion of their food source on a journey from Green Bay to Chicago (Fernald and Kinsey, 1958). There is evidence of extensive use of the plants by the Great Lakes Indians as fresh vegetables and dried for winter use (Yarnell, 1964).

Onions are low in calories and have insignificant nutritional value.

Both oak–hickory and beech–maple associations (Figures 17-4 and 17–5) contain a number of good edibles. Many of the early spring edibles are found on the wood's floor before the trees leaf out. The best collecting woods are those which combine several habitats. Some gentle disturbance by people seems to encourage the growth of additional interesting species. The best local areas for edibles sometimes compete with motor bikes, snowmobiles, and log cutters—not a pleasant thing

Oak–Hickory and Beech–Maple Woods

FIGURE 17-27.
*Allium tricoccum, wild
leek, bulbs and leaves.*

for the woods but more productive for plant diversity. Where ecotones
exist, the potential for edibles increases. The direction of exposure determines the time when plants are available (e.g., a south-facing slope
will warm the quickest and have the first plants). North-facing slopes
can still be harvested long after the south-facing plants have become too
old to collect. Plants taken from these woods should be carefully selected,
so as not to destroy in any way their quality and abundance.

MAPLE (*ACER* SPP.; ACERACEAE)

Maple trees are found abundantly in beech–maple woods. They are
the predominant tree in "sugar bushes," which were planted and/or
encouraged to grow by early settlers in the area. Maples have often been
planted in rows along country roads.

The bark of sugar maple is plated near the ground and becomes gray and smooth as it nears the upper canopy. The leaves and branches are opposite.

Any species of maple can be used for syrup making, including box elder (*A. negundo;* Figure 17-28). Sap begins to flow in late January and will continue throughout most of April, when the weather is freezing at night and reaches 5 to 10°C (40 to 50°F) during the day. Although there are elaborate syrup-making methods, it is easy to make only a few liters of syrup in a makeshift fashion. A spile, which acts like a spigot and has a place to hang a pail, is inserted into the tree. A hole is first drilled in about 5 centimeters (2 inches) into the tree and approximately 3 feet from the ground. Empty the buckets every day (twice if the sap is running well) and boil down with a roaring fire until the syrup reaches 103.8°C (219°F). Since the volume of the sap must be reduced roughly 40:1, the initial boiling should take place outdoors; traditionally this is done over a wood fire. As the color of the sap darkens, bring indoors, put through a filter, and boil gently to prevent burning; longer cooking results in maple sugar.

The first reliable report of maple sugar making is mentioned in the late 1700s in Canada (Sturtevant, 1919). Often, whole Indian tribes would move near the sugar bush during tapping season (Yarnell, 1964). Some authorities feel that the Indians did not make syrup until the early settlers taught them to boil down the sap. The making of syrup continues to be an income-producing job for residents of the northeastern states and parts of Canada, and there are maple syrup festivals held each year in Vermont and Michigan.

FIGURE 17-28.
Acer negundo, boxelder, buckets collecting sugary sap for sugar and syrup.

MAYAPPLE (*PODOPHYLLUM PELTATUM*; BERBERIDACEAE)

Mayapples prefer low areas in rich, open woods, especially around wood's pools and damp areas. Mayapples also occur along roadsides where protected by trees. The fruit is ripe from mid-August through mid-September. Good collecting localities should be spotted in the spring because the leaves die and lie on the ground when the fruit is ripe. If the fruit is not quite ripe when collected, place on newspaper indoors in a dry area until soft and yellow.

Mayapples are clonal and come up with either one or two umbrella-shaped tops. The solitary, waxy flower hangs down from the fork of the two-topped plants. Do not eat the vegetative parts or the unripe berry, as they contain a poisonous resinoid, podophyllin, which has been used by herb and witch doctors as well as in home remedies (Hardin and Arena, 1974). The fruit is yellow-green and soft when ripe; it is a good-tasting fleshy berry about the size of a chicken egg with a pleasant fruity odor. The fruit can be eaten raw or made into marmalade or juice. A mayapple chiffon pie, made with brown sugar and gelatin, is delicious.

Other names for this plant are "raccoonberry" and "wild lemon." There are numerous records of Indian usage. In the southern mountain regions, a drink is made from Madeira, sugar, and juice from mayapples (Fernald and Kinsey, 1958).

SASSAFRAS (*SASSAFRAS ALBIDUM*; LAURACEAE)

Sassafras is found along roadsides and in woods. Since it is clonal, the plants are interconnected by a complex network of roots. Where one tree is found, there are almost certain to be others, varying in size from small shoots to full-size trees. The pleasant sassafras aroma is found in all parts of the plant, with the best concentration in the roots (Figure 17-29). The bark progresses from mottled brown near the base of the tree to salmon pink in the upper branches. The tip of the twigs is smooth and olive-green.

An excellent tea is made from steeping the roots in cold water. Bring almost to a boil and steep until a musky pink color results. A small handful of roots per quart of water is usually the proper portion. The young twigs can also be used for tea, but they are inferior in taste and strength.

Sassafras tea is a well-known spring tonic, even though it is available year round. The leaves (Figure 17-30) are the basis for the gumbo filé of Creole cookery.

Sassafras became famous during the colonization of the United States. For a time, the roots were extensively exported to Europe for their reported medicinal values and for their superior flavor. There is good evidence indicating much use of the roots by the Indians. *There has been some suggestion that sassafras tea may be carcinogenic. We are therefore recommending that it no longer be used.*

FIGURE 17-29.
*Sassafras albidum
sassafras, roots.*

FIGURE 17-30.
*Sassafras albidum, char-
acteristically lobed leaves.*

VIOLETS (*VIOLA* SPP.; VIOLACEAE)

Violets (Figure 17-31) are found in woods, thickets, shaded roads, and lawns. The plants bloom from the third week in April to late May, depending on the species. Both leaves and flowers are used. Any color of

FIGURE 17-31.
Viola sp., violets, edible flowers and leaves.

wild violet is edible, but the dark blue varieties have the most flavor. When the blossoms are collected, more grow to take their places.

Violets are low plants, rarely exceeding 10.1 to 15.2 centimeters (4 to 6 inches) in height. There are many different colorations of the flowers—blue, rose, yellow, white, and white on top with blue or pink underneath. The shape of the leaf also varies—heart-shaped, "bird-foot"-shaped, rounded, and pointed.

The buds and blossoms are eaten raw in salads. A good and nutritious syrup is made by pouring boiling water over the blossoms; allow to stand 24 hours, then sweeten with sugar and add a squirt of lemon juice. Bring the mixture back to a boil to dissolve the sugar.

Fernald and Kinsey (1958) mention that violets were used to extend food supplies during wartime. More peaceful uses are in the making of candied violets.

Violets contain a high amount of vitamin C, both in the leaves and in the blossoms. In addition, violet leaves are high in vitamin A (Table 17-1).

WILD GINGER (*ASARUM CANADENSE*; ARISTOLOCHIACEAE)

Wild ginger (Figure 17-32) is found in rich woods, always in colonies. The surface rhizomes remain viable year round and look like green twigs. Wild ginger smells exactly like the unrelated commercial ginger, which comes from the West Indies.

The hairy stems and the hairy, heart-shaped leaves, combined with the very pleasant smell of ginger, identify this plant. The rhizomes can be transplanted to rich, shady soil.

Wild ginger is potent and flavorful, requiring only a few 5-centimeter (2-inch) pieces to flavor tea. Place the rhizomes in cold water and bring almost to a boil; steep until the water is a light brown. The pieces can be candied in a sugar syrup or used without sugar to flavor vegetable and poultry dishes.

The Ojibwa and Potawatomi Indians used the rhizomes (Yarnell, 1964).

WILD LEEKS (*ALLIUM TRICOCCUM*; LILIACEAE)

Wild leeks (Figure 17-27) are found in rich woods, often in groupings of several thousand individuals. The young leaves come up in early April and last until mid-May. The flower stalk comes up separately in mid-July. The seeds on the 20.3 to 25.4-centimeter (8 to 10-inch) stalk often last into the winter. The stalk greatly aids in spotting the plant when the leaves are not present.

The tasty, young leaves look like tulip leaves and are 15.2 to 20.3 centimeters (6 to 8 inches) long and 2.5 to 10.1 centimeters (1 to 4 inches) broad. The white bulb and the leaves have the unmistakable odor of onions.

Wild leeks can be eaten raw, but the taste is extremely strong and many leave a burning sensation in the mouth. The best use is to cook the tops (when in season) and the bulbs in several waters until the desired strength is reached. They are excellent sautéed in butter and served with a cream sauce, and they make worthy additions to soups and casseroles. Be sure to use only a small handful of bulbs so as not to overpower the other ingredients.

FIGURE 17-32.
Asarum canadense, wild ginger, flower and heart-shaped leaves.

Wild onions have been extensively used throughout many parts of the world. Fernald and Kinsey (1958) report that the city of Chicago was the locality for many wild leeks; Chicago is derived from the Menomini Indian word for "skunk"—"Pikwu'te sikaku'shia."

WINTERGREEN (*GAULTHERIA PROCUMBENS*; ERICACEAE)

Wintergreen is a small, aromatic plant found year round in cool, sandy woods. This low, creeping plant has waxy, dark-green leaves and 1.9-centimeter ($\frac{3}{4}$-inch) berries which turn bright red in the late fall. The leaves remain on the plant year round, and both leaves and berries have a very pleasant smell and refreshing taste.

The best use of wintergreen is for a tea and a sherbert. Steep the fresh leaves and berries until a good smell and flavor exudes. Since the aroma is quite delicate, the flavor is strengthened by bruising the leaves or by letting the tea stand for several hours or even overnight. Sherbert is made using a strong tea mixture.

Ojibwa and Iroquois Indians used wintergreen as a beverage and a flavoring (Yarnell, 1964), and the berries were sold on the Boston markets in the late 1800s (Sturtevant, 1919).

Marshes, Swamps, and Streambanks

Wet areas have their own set of edible plants. Most of the plants found in such areas need a fairly constant source of water to survive, although many cannot flourish in water over a foot deep for any extended period of time. These are the plants of stream and river banks (Figure 17-9), low-lying ditches, swamps (Figures 17-7 and 17-8), and marshes (Figure 17-6).

CATTAILS (*TYPHA LATIFOLIA* AND *T. ANGUSTIFOLIA*; TYPHACEAE)

Cattails (Figure 17-33) are found in almost any low-lying area that contains shallow water. They especially like ditches and swamps, where they are a familiar sight, with the overwintering brown spikes rising 0.9 to 1.2 meters (3 to 4 feet) above the water level. Even in the most severe winters, these telltale spikes and linear leaves persist. Do not confuse the young shoots with the leaves of the iris family. Iris members have equitant leaves that form tight Vs rather than the circular arrangement of the cattail leaves.

This sun-loving plant is usable throughout the year. A flour is made from the roots; the young shoots (which are formed late in the fall) can be used throughout the winter and into the spring until their growth reaches about 38.1 centimeters (15 inches). The young male spike, which is positioned on top of the female spike, is delicious in June (Figure 17-33). A few days later, the yellow pollen is ready to collect for use in bread.

Flour is extracted from the root by stripping off the bark and soaking the remaining starchy fibers in water. After a thorough mixing, remove

FIGURE 17-33.
*Typha latifolia, cattail,
male inflorescence shedding
pollen above.*

the fibers and allow the starch to settle to the bottom of the bowl. Repeat, then use with part regular flour in breads. The young shoots are used raw or boiled; they are bland and so mix well with more-flavorful foods. Boil the young male spikes for 8 to 10 minutes and eat with butter. The male pollen is gathered as an ingredient for bread.

Cattails are known and used in many parts of North America and Europe. There are detailed accounts of their use in Russia during the 1700s (Fernald and Kinsey, 1958) and many examples of use by Indians and early settlers in the United States.

GROUNDNUTS (*APIOS AMERICANA*; FABACEAE)

Groundnuts (Figure 17-34) can be found year round, but the tubers (Figure 17-35) are in the best shape for edibility from November through March. The plants prefer to grow in moist soil close to a source

FIGURE 17-34.
Apios americana,
groundnut, flowers and
leaves on the smooth-
stemmed vine.

FIGURE 17-35.
Apios americana,
groundnut, tubers, usually
occur in chains under-
ground.

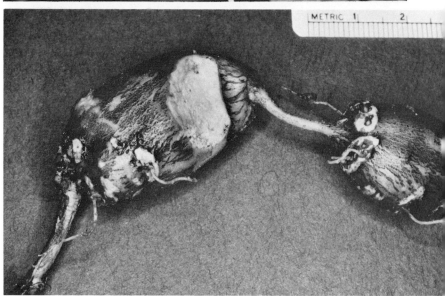

of water. As the twining vine needs support, it soon climbs on nearby, more rigid vegetation. In late winter, the foliage forces easily and makes an attractive indoor house plant.

The vine, which twines counter-clockwise, has leaves of five to seven leaflets, and the flowers are dusky pink. When in fruit, there are many miniature bean pods on the vines.

The underground tubers, which grow as a sequence of swollen regions along the roots, are often the size of small eggs. The tubers are prepared like potatoes—thinly sliced and fried, or boiled and served with butter.

Fernald and Kinsey (1958) state that the groundnut was one of the most extensively used wild edibles by the Indians and the early settlers. The Pilgrims evidently made much use of the tubers, especially during the first year before their gardens were planted.

MINTS (*MENTHA SPP., MONARDA FISTULOSA,* AND *NEPETA CATARIA;* LAMIACEAE)

Mints (Figure 17-36) occur in quite diverse habitats, each species having its own requirements. Some of the best in flavor are found in low-lying swampy areas, covered with water in the spring. A few species are available year round.

The mints are easily recognized by the following characteristics: square stems, opposite leaves, and by their distinctively pleasant smell. The individual species are sometimes hard to differentiate, but they can all be used in the same way.

Mint leaves are used either dry or fresh: a small handful of leaves flavors 1.4 liters ($1\frac{1}{2}$ quarts) of water. Bring the infusion to a simmer and steep 10 minutes; serve with honey. Add 125 milliliters ($\frac{1}{2}$ cup) crushed leaves to an apple jelly recipe, or serve raw as a garnish or in mint julips.

Mints have an exalted place in history and are mentioned in such diverse literature as the writings of Dioscorides and Pliny and in the *Bible,* where mint was "strewn in the temple to make a sweet smell before the Lord." Mint has been famous worldwide in the processing of liquors and candies, and the Indians made extensive use of its aromatic oils.

Mints contain quite high levels of vitamin C and vitamin A (Table 17–1).

NETTLES (*URTICA DIOICA* AND *LAPORTEA CANADENSIS;* URTICACEAE)

Nettles are found in swamps and other damp places. The best time to collect the shoots is in the spring, during their early growth. By summer, the plants are too tough to use, except for the top, tender leaves, which can be utilized before the plant flowers. Fall often brings a new, young crop. The plant overwinters as 5- and 7.6-centimeter (2- and 3-inch) offshoots of long, underground rhizomes.

The opposite leaves of *Urtica* (Figure 17-37) and the alternate leaves of *Laportea* (Figure 17-38) are armed with stinging bristles which act like

FIGURE 17-36.
*Mentha arvensis, field
mint, characterized by
square stems and oppo-
site leaves.*

mosquito bites on any part of the body that touches them. The effect lasts
from 30 minutes to 3 days, depending upon an individual's sensitivity to
it. For protection, pick the shoots with rubber gloves. Once cooked, the
prickly effect vanishes entirely.

Boil these excellent-flavored greens in salted water for 8 to 10 minutes
and serve with butter or a cream sauce. Nettles are excellent in soups and
casseroles.

There has been much use of nettles in Europe. In France, there are
seven main types of dishes made with nettles; in England, nettles are
used for such diversified consumables as nettle beer and nettle pudding.

Nutritional analyses show fairly high amounts of vitamin C and
vitamin A (Table 17-1).

FIGURE 17-37.
Urtica dioica, stinging nettle, older female plant with young fruits. Note the opposite narrow leaves.

OSTRICH FERN (*MATTEUCCIA STRUTHIOPTERIS*; POLYPODIACEAE)

Ostrich ferns occur along the banks of rivers and streams as well as in many gardens. In northern Michigan, Maine, and some parts of Canada, it is one of the more common inhabitants of the river banks and lakesides. Its favorite home is on the alluvial plains of rivers. In the lower Great Lakes area, the ostrich fern is less common, although it can be abundant locally. Cut only one or two fiddleheads per plant and return at 5-day intervals for additional harvesting. The season varies depending on location, but the first 2 weeks in May usually find the plant at its prime. Take great care to protect this lovely fern.

FIGURE 17-38.
*Laportea canadensis,
wood nettle, mature plant
with flowers. Note the
alternate broad leaves.*

Ostrich fern is one of two common ferns to retain a separate, brown fertile frond year-round. The uncurled fronds, the mature fronds, and the fertile spikes have a definite, easy-to-spot groove running down the entire length of the leaf stalk. These characteristics distinguish the ostrich fern; the "other" fern with the brown fertile stalk has a rounded leaf stalk.

Wash the young croziers and recut the stalks just before tossing them into boiling water. Cook about 10 to 12 minutes, until just tender. Serve with butter and lemon or vinegar. This is an excellent plant.

There is a long history of "fiddlehead" collecting by the natives of Nova Scotia and the state of Maine. Thousands of pounds are collected each year, both for personal use and for selling at expensive prices in specialty food stores.

FIGURE 17-39.
*Asimina triloba, pawpaw,
small tree with broad,
entire leaves.*

PAWPAW (*ASIMINA TRILOBA;* ANNONACEAE)

Pawpaws are the only representative of the tropical custard apple family growing in the Great Lakes area. These small trees are found in rich, damp woods, around swamps, and along river banks (Figure 17-39). Since they are clonal, they are usually found in large groups. The fruit ripens in September and is best collected after the first frost (Figure 17-40).

The alternate leaves, with their peculiar, rank odor, reach 25.4 to 38.1 centimeters (10 to 15 inches) in length. The knobby bark of the older trees [to 76.2 centimeters (30 inches)] and the smooth brown bark with gray blotches of the young, plus the unusual elongate fruit, helps to identify this plant. The green fruits grow in clusters of one to seven and are between 5 and 17.7 centimeters (2 and 7 inches) long when mature. When completely ripe, the fruit turns brownish green and has orange pulp with flat, dark brown seeds (Figure 17-41).

FIGURE 17-40.
Asimina triloba, pawpaw, large cluster of unripe fruits.

FIGURE 17-41.
Asimina triloba, pawpaw, mature fruit cut open to show pulp and seeds. Scale in centimeters.

The fruits can be eaten raw or made into jelly, juice, or chiffon pie. Often it is difficult to find enough ripe fruit, as the raccoons also enjoy pawpaw pulp. It is, however, possible in ideal localities to collect as much as a bushel of fruit.

Fernald and Kinsey (1958) claim that the fruits have "long been famous among those who have the fortune to live where it occurs."

Pawpaws make a poor showing nutritionally.

FIGURE 17-42.
*Nasturtium officinale,
watercress, young shoot
showing lobed leaves.*

Plants in the very wet habitats of streams, lakes, and ponds include those which have standing water year around. These plants cannot tolerate periods of extended drought.

*Streams, Lakes,
and Ponds*

WATERCRESS (*NASTURTIUM OFFICINALE*; BRASSICACEAE)

Watercress (Figure 17-42) is typically found in moving water, rivers, fast-moving streams, and by bubbling springs. Watercress has also been found in vast quantities in roadside ditches. The leaves are available year-round, although they are very ragged looking in winter. Spring and early summer are the best times to harvest the new young growth (Figure 17-42).

Watercress is a perennial aquatic plant with underwater or partly floating stems. The many white rootlets facilitate colonization anywhere a small piece is placed. The leaves have three to nine rounded segments, with the terminal segment larger and the smaller, lateral segments paired.

All watercress should be soaked at least 20 minutes in a solution of Clorox (1 teaspoon per quart of water) or water-purifying tablets. This precaution is necesary as it is safe to assume that all water is polluted. After soaking, rinse the greens well with several immersions in clean water. Watercress is used raw in salads and sandwiches, and it is also delicious cooked in a chicken or potato soup.

Watercress is a native of Eurasia and is quite common in the United States and southern Canada. Sturtevant (1919) claims that "the young shoots and leaves of watercress have been used as a salad from time immemorial." Both the Romans and the Greeks made mention of watercress. The Latin name comes from *nasus tortus* meaning "convulsed nose," to describe the pungent taste (Fernald and Kinsey, 1958).

The raw leaves are low in calories, average in vitamin A, fairly high in vitamin C, and high in calcium (Table 17–1).

WILD RICE (*ZIZANIA AQUATICA;* POACEAE)

Wild rice (Figure 17-43) is a robust, annual grass that occurs in small bays along river banks, mixed with cat-tails in marshes, or standing forlornly in isolated muddy wood's pools. The best localities are those found in shallow lakes, where there are almost pure stands of the wild rice, involving hundreds of acres. The grains ripen in late August and early September, when they are eagerly sought by water fowl and people.

This tall grass [3 to 3.6 meters (10 to 12 feet) high] has terminal spikelets, with the female portion on top and the male section directly beneath. The leaves are long and narrow [0.9 meters $\times$ 5 centimeters (3 feet by 2 inches)] (Figure 17-43). Tenacious bracts house the green grains, which turn brown when ripe.

Wild rice is fun to gather. A canoe is used to edge through the tangle of wild rice stalks because the lake bottom is much too mucky for wading. Place a sheet in the bottom of the boat to collect the grains; bend the stalks over the boat and shake the heads. If the stalks have been previously tied together like a shock of wheat, the collecting time is shorter. Because it is difficult to remove the protective covering from the grains, the theory is to dry the grains well outdoors in the sun and to finish up the parching in an oven with the door partly open. At that point, the chaff can be winnowed out. However, it is tedious to extract the grains without mechanical help. The rice is so tasty and expensive that it is worth any time spent processing it.

Study of pollen deposited in lake beds in Minnesota and South Ontario has shown that wild rice spread into little glacial lakes, perhaps as early as 1000 B.C. The grains are spread now by ducks and geese. As early as A.D., 600 to 700, there is evidence of the Indians going to ricing camps in northern Michigan (Yarnell, 1964).

Acid bogs (Figure 17-10) are among the most specialized of habitats. The acidic conditions of the cool, wet sphagnum bogs create a physiologically dry (xerophytic) condition for the plants. The bog is the fascinating home for many beautiful species of orchids, pitcher plants,

Bogs

FIGURE 17-44.
Vaccinium corymbo-sum, highbush blueberry, fruits.

sundews, leather-leaf, and tamaracks. Blueberries and cranberries can be abundant in these localities.

BLUEBERRIES (*VACCINIUM* SPP.; ERICACEAE)

Blueberries come in various shapes, sizes, and flavors. The best blueberries are the highbush type (*V. corymbosum;* Figure 17-44), which requires bog habitats, where they prefer very moist sphagnum moss. The berries may be very abundant, and two people can collect as many as 8 gallons in 4 hours. Frequently, poison sumac (*Toxicodendron vernix*) of equal height and exuberance occurs in the middle of the blueberry bush. Avoid contact with the sumac (see p. 365).

Highbush blueberries are tall shrubs, often reaching 3 to 3.6 meters (10 to 12 feet); sometimes there are several main stems. The short, narrow leaves are numerous, as are the berries.

The berries are excellent raw, cooked in pies, in muffins, and as jam, or made into juice.

Blueberries have been used extensively by Indians and modern man. People of varying economic and social backgrounds can all be fanatic collectors of this delicious fruit.

Blueberries are not especially high in nutritional value.

CRANBERRIES (*VACCINIUM MACROCARPON* AND *V. OXYCOCCUS*; ERICACEAE)

Cranberries (Figure 17-45), while they prefer the true northern bogs with the floating sphagnum mat around an open lake, also occur in acidic areas along big lakes, and occasionally in damp, acidic habitats not associated with a lake. Fruitings vary considerably in abundance from

year to year. When fertile, the fruit is ready for use after the first hard frost, and it may persist on the plant until March.

Cranberries are borne on small, evergreen, trailing shrubs which look impossibly small to support the ¾-inch berries. Prepare wild cranberries any way that domestic ones are used—in sauces, pies, relish, and juice.

There has been extensive use of both the wild berries and the semidomesticated berries throughout the northeastern United States, especially in the New England states.

Poisons

When collecting plants for human consumption, there is always a chance that a poisonous compound occurs in the plant. Some plants are very poisonous in all parts, and a small piece included in a meal could kill the eater (e.g., water hemlock, *Cicuta maculata*). Other plants are poisonous at certain times of the year (e.g., mayapple, *Podophyllum peltatum*); some plants have both edible and poisonous parts (e.g., pokeweed, *Phytolacca americana*). Certain plants (e.g., milkweed, *Asclepias syriaca*) require special processing. Each plant should be correctly identified and checked that the right part is being used during the corresponding time of year. There are also plants in the grocery stores that contain poison, such as the soluble oxalate found in rhubarb leaves (*Rheum rhaponticum*) and the cyanide found in the pits and seeds of cherries, plums, and apples.

Some plants contain contact poisons that cause painful itching of the skin. Two of these commonly found plants are poison ivy and poison sumac (*Toxicodendron radicans* and *T. vernix*). Some people contract poison sumac from water in which the plant grows. The oil of the plant should be immediately removed from the skin with soap and water. The itching usually does not appear until 24 hours after contact and lasts for 5 to 10 days. The itching of nettles, both the common nettle and the wood nettle (*Urtica dioica* and *Laportea canadensis*), begins almost immediately and lasts from 30 minutes to 3 days. Soaking in water helps relieve the stinging.

Pollutants in rivers and streams are a hazard to water plants, such as watercress (*Nasturtium officinale*). Be sure to soak these plants in a Clorox or other water-purifying solution.

Plants that are found in gardens and lawns which have been sprayed with poisonous weed killers should not be used. Roadsides and railroad tracks, both of which offer rich collecting areas, have the possibility of lead poisoning as well as poisoning from the weed-killing chemicals. Some areas are sprayed with salt solutions, and others are merely mowed. Check with the county road commission and the railroad personnel before collecting in these areas. State trunk lines will have different spraying policies than will the county roads.

Possibilities of Environmental Damage

In the process of collecting edible plants, there exists a possibility of damage to both the plant populations and to the sometimes very delicate habitats.

There is no opposition to collecting quantities of weedy plants. Given average weed tenacity and the great powers of seed production, it would be almost impossible to eradicate them. The weed group fortunately includes some of the best edibles—lamb's-quarters, sorrel, and purslane.

Seeds of wild fruits often grow where expelled by birds, animals, and man, so the utilization of these fruits and berries often serves to spread the plant.

Some small trees and shrubs are clonal. These plants are usually connected by an intricate root system. Sassafras, for example, can be dug between two trees without distorting the trunks themselves. The collecting of a few of the trees seems to stimulate the branching points on the subterranean shoots.

Certain plants have renewable resources. Examples of this are pokeweed, asparagus, and the ostrich fern. If all the stalks are cut in the pokeweed and asparagus, and several at a time in the fern, it is possible to return every 5 days for several weeks during the growing season to obtain a new crop. After the last collection, allow the plant to mature.

Maple trees and their sap are another renewable resource, as they can be tapped year after year if they are properly drilled and cared for. Always remove the spiles (taps) after the tapping season, and apply tree wound paint to combat infection.

Moderation in the collection of any plant should be the primary goal in foraging. Always leave a fruitful supply behind and follow the practice of taking only what can be used. Cognizance of animal and bird usage prohibits overgathering of important food supplies that are needed by birds and animals. Persistent seeking and conscientious collecting enable the wild food connoisseur to make sensitive use of the environment.

Summary

Wild foods are abundant, tasty, and nutritious if collected with skill in identification and prepared with patience. Many edibles are found in a variety of habitats, while others are very demanding as to specific time of year, exact temperature and rainfall, and detailed habitat requirements. Many of the edibles have been used throughout the world for thousands of years; local plants were used abundantly by the various Indian tribes. Care must be taken to avoid poisonous look-alike species and to ascertain the proper plant part and the correct time of year, carefully noting any preparation techniques required to make the plant palatable and nonpoisonous.

The seeking out and preparation of these wild edibles, while taking care not to damage the future of any of these plants or to interrupt an important wild life chain, is an interesting, tasty, nutritional, free, and addicting pursuit.

COFFEE CAKE

Mix 178 ml ($\frac{3}{4}$ cup) of sugar with 59 ml ($\frac{1}{4}$ cup) of soft shortening. Add 1 egg and 118 ml ($\frac{1}{2}$ cup) of milk. Sift together 355 ml ($1\frac{1}{2}$ cups) of flour, 10 ml (2 tsp) baking powder and 1.25 ml ($\frac{1}{4}$ tsp) of salt. Stir in 474 ml (2 cups) of wild berries (e.g., blueberries) that have been combined with 118 ml ($\frac{1}{2}$ cup) of flour. Place the cake mixture in a buttered cake pan. Prepare a crumb mixture by combining 118 ml ($\frac{1}{2}$ cup) of sugar, 79 ml ($\frac{1}{3}$ cup) of sifted flour, 5 ml (1 tsp) of cinnamon, and 118 ml ($\frac{1}{2}$ cup) of butter or oleomargarine. Then sprinkle the top of the cake with the crumb mixture and bake at 191°C (375°F) for 50 minutes.

JELLIES AND JAMS

Follow the recipe on the folders of commercial pectin that seems to most resemble the wild fruit. Add an extra 118 ml ($\frac{1}{2}$ cup) sugar and 118 ml ($\frac{1}{2}$ cup) of water if the fruit has a strong flavor. Sumac heads can be added to the cooking fruit for a lemon flavor. Many wild fruits contain high amounts of pectin, but for the beginner, commercial pectin gives excellent, uniform results.

JUICES

Wash and crush the fruits and berries. Place in a pan and add only enough water to keep the fruit from burning. Cover and cook until the juice flows freely; add more water if necessary. When the fruit is tender, strain it through a damp jelly bag or several layers of cheese cloth. Add a squeeze of lemon and sugar or honey to taste. Reheat to dissolve the sugar. Serve well chilled.

PIES

Fruit Pies

Prepare pastry for a two-crust, 23 centimeter (9-inch) pie. Combine 947 ml (4 cups) of fresh berries with up to 237 ml (1 cup) of sugar, 44 ml (3 tbl) of all-purpose flour, a squirt of lemon, and a dash of salt. Place in a pastry shell and top with bits of butter. Cover with a lattice or plain crust topping and bake at 204°C (400°F) for 50 minutes.

Chiffon Pies

Use May apples, cranberries, highbush blueberries, raspberries, or other fruit. Cover 1.90 l (2 quarts) of cleaned fruit with water, cover, and simmer for 30 minutes. Put through a sieve or food mill. Combine 79 ml ($\frac{1}{3}$ cup) of sugar, 1 envelope of unflavored gelatin, 1.25 ml ($\frac{1}{4}$ tsp) of salt, and stir well. Beat the pulp with 4 egg yolks and stir into the sugar–gelatin mixture; cook until it boils, remove from the heat, and chill until slightly thickened. Beat 4 egg whites until soft peaks form; add 1.25 ml ($\frac{1}{4}$ cup) of sugar and beat until stiff. Fold the fruit pulp mixture gently into the egg whites. Place in a graham cracker crust and chill.

SOUFFLES

This is a good recipe for small amounts of collected wild food: Use asparagus, nettles, lamb's-quarters, groundnuts, wild leeks, or other food. Boil 118 to 474 ml ($\frac{1}{2}$ to 2 cups) of wild food in salted water until almost tender. Drain well and set aside. Melt 79 ml ($\frac{1}{3}$ cup) of butter in a sauce pan or frying pan. Add 44 ml (3 tbl) of flour and stir until smooth. Add slowly 237 ml (1 cup) of milk, scalded, and stir constantly until the mixture thickens and boils. Cool for 5 minutes. Add 3 well-beaten egg yolks, 2.5 ml ($\frac{1}{2}$ tsp) of salt, a dash of pepper, and 44 ml (3 tbl) of grated cheese. Fold in 3 stiffly beaten egg whites and bake at 191°C (375°F) for 45 minutes.

SOUPS

Wild Bouillabaisse

Cook 118 ml ($\frac{1}{2}$ cup) of wild leeks slowly in 118 ml ($\frac{1}{2}$ cup) of olive oil until tender. Stir in 4 cloves of garlic and 1 l (1 qt) of ripe tomatoes and cook 5 minutes. Add 2.375 l ($2\frac{1}{2}$ qt) of water, 2 wild bayberry leaves, 2 pinches of saffron, 15 ml (1 tbl) of salt, and several pounds of fish (e.g., bluegills, perch). Boil gently for 1 hour, then strain, pressing out the juice. Add collected wild greens and boil 20 minutes.

Leek and Potato Soup

Simmer 1 l (4 cups) of peeled potatoes, 0.5 l (2 cups) of parboiled wild leeks, and 15 ml (1 tbl) of salt in 1.90 l (2 qt) of water for 1 hour.

Force through a food mill or sieve and add the collected wild greens. Cook 20 more minutes, add 44 ml (3 tbl) of whipping cream, and season to taste.

TEAS

Dry the leaves and blossoms for at least 24 hours on newspaper, indoors, away from any draft. The plants should be in one layer only and be completely dry before storing in jars. When making tea, use a small handful of dried material for each 1.425 l ($1\frac{1}{2}$ qt) of cold water. Bring slowly to just under a boil and steep until the desired strength is reached. Serve with honey.

Selected References

ANGIER, B. 1974. *Field Guide to Edible Wild Plants.* Stackpole Books, Harrisburg, Pa.

BERGLUND, B., and C. E. BOLSBY. 1971. *The Edible Wild.* Charles Scribner's Sons, New York.

BILLINGTON, C. 1968. *Shrubs of Michigan,* 2nd ed. Cranbrook Institute of Science, Bloomfield Hills, Mich.

BROCKMAN, C. F. 1968. *Trees of North America.* Western Publishing Co., Inc. (Golden Press), Racine, Wis.

CROWHURST, A. 1972. *The Weed Cookbook.* Lancer Books, New York.

FERNALD, M. L., and A. C. KINSEY (revised by R. C. ROLLINS). 1958, *Edible Wild Plants of Eastern North America.* Harper & Row, Publishers, New York.

GIBBONS, E. 1962. *Stalking the Wild Asparagus.* David McKay Company, Inc., New York.

GIBBONS, E. 1966. *Stalking the Healthful Herbs.* David McKay Company, Inc., New York.

GLEASON, H. A., and A. CRONQUIST. 1963. *Manual of Vascular Plants of Northeastern United States and Adjacent Canada.* Van Nostrand Reinhold Company, New York.

HALL, A. 1973. *The Wild Food Trailguide.* Holt, Rinehart and Winston, New York.

HARDIN, J. W., and J. M. ARENA. 1974. *Human Poisoning from Native and Cultivated Plants,* 2nd ed. Duke University Press, Durham, N.C.

HARRINGTON, H. D. 1967. *Edible Native Plants of the Rocky Mountains.* University of New Mexico Press, Albuquerque, N.M.

HATFIELD, A. W. 1971. *How to Enjoy Your Weeds.* Macmillan Publishing Co., Inc. (Collier Books), New York.

Ipswich River Wildlife Sanctuary. 1971. *Eating Wild.* Massachusetts Audubon Society, Boston.

KINGSBURY, J. M. 1965. *Poisonous Plants of the United States and Canada.* Prentice-Hall, Inc., Englewood Cliffs, N.J.

KIRK, D. R. 1970. *Wild Edible Plants of the Western United States.* Nature-graph Publishers, Happy Camp, Calif.

KROCHMAL, A., and C. KROCHMAL. 1973. *A Guide to Medicinal Plants.* The New York Times Book Co., New York.

MARTIN, A. C., H. S. ZIM, and A. L. NELSON. 1951. *American Wildlife and Plants: A Guide to Wildlife Food Habits.* Dover Publications, Inc., New York.

MEDSGER, O. P. 1966. *Edible Wild Plants,* 2nd ed. Macmillan Publishing Co., Inc., New York.

PETERSON, R. T., and M. McKENNY. 1968. *A Field Guide to Wildflowers.* Houghton Mifflin Company, Boston.

STURTEVANT, E. L. 1919. *Sturtevant's Notes on Edible Plants.* J. B. Lyon Co., Albany, N.Y. Also published as a Dover paperback.

U.S. Department of Agriculture. 1963. *Composition of Foods,* rev. ed. Government Printing Office, Washington, D.C.

U.S. Department of Agriculture. 1971. *Common Weeds of the United States.* Dover Publications, Inc., New York.

WIGGINTON, E. 1973. *Foxfire 2.* Doubleday & Company, Inc. (Anchor Books), New York.

YARNELL, R. A. 1964. *Aboriginal Relationships Between Culture and Plant Life in the Upper Great Lakes Region* (Museum of Anthropology, No. 23). University of Michigan, Ann Arbor, Mich.

Alternative Means of Pest Control
J. Donald LaCroix

18

Questions for Consideration

1. What is meant by biological magnification of pesticides?
2. Discuss the pros and cons of integrated pest control. What is the likelihood that such a means of control will replace the chemical means of control?
3. What is the probability of a complete ban of pesticides and would it be desirable?
4. What is organic farming?
5. How does one account for the disease resistance naturally found in some plants but lacking in others?

Pests in our environment, in our homes, our forests, and our crops, including those in storage as well as those in the field, have long been controlled with pesticides. There is little doubt among pesticide consumers that their continued use is essential. Indeed, it is highly unlikely that constant increases in crop production, acreage harvested, and yield per acre could have kept pace with a burgeoning world population had it not been for the use of pesticides.

Until very recently, people have been so elated with the benefits derived from the use of pesticides that they gave little thought to their possible harmful effects. Even more alarming, however, is the fact that too many consumers are totally unconcerned or ignorant of the potential hazards associated with the use of pesticides. It is now apparent that many of these chemicals produce some short-term, damaging effects, while the long-term effects remain relatively unknown. Widespread use of pesticides during the last 30 years has resulted in the distribution of these chemicals in the air, soil, water, and all kinds of living organisms.

Misuses of Pesticides

Agriculturists are becoming increasingly aware of the phytotoxity to some species of plants following the indiscriminate use of fungicides. Symptoms of damage include stunted growth, foliage injury, blossom drop, lower-quality fruit, and drought damage. Mercury is an effective fungicide, but it can also be a deadly pollutant. Dangerously high levels of mercury have been discovered in fish, including those in the Great Lakes, in birds, in the eggs of birds, and in farm animals who had ingested mercury-treated grain. The consumption of mercury-containing food has resulted in serious health problems, including crippling, brain damage, and death in many parts of the world. The Michigan Department of Natural Resources recently issued a report concerning the widespread presence of several contaminants in Great Lakes fish. According to the report, levels of dieldrin, an extremely toxic insecticide, are increasing in lake trout and chubs. Some popular fish and their contaminants included in the report were walleye pike (mercury, DDT, and PCB or polychlorinated biphenyl), salmon (mercury and PCB), lake trout (mercury, DDT, PCB in addition to dieldrin), rainbow trout, carp and catfish (PCB), and largemouth and smallmouth bass, sheepshead, and muskellunge (mercury).

Today, and in view of the final outcome of the war in Vietnam, one must question whether or not it was sound strategy for the military to employ high concentrations of the herbicides 2,4-D, 2,4,5-T, and picloram, which is quite persistent, to defoliate forests and croplands. This action damaged or destroyed crops, mangrove and hardwood forests, adjoining vegetation, and natural ecosystems. The long-term ecological ramifications and the effects on the Vietnamese people are unknown.

Unlike many pesticides, 2,4-D is believed to be harmless to humans because it is biodegradable, and, as a result, its effects are temporary. Such may not be the case with 2,4,5-T, since the National Institutes of Health have reported that even very low levels of it are capable of producing tumors and birth defects in animals. Additional concern arises from the fact that 2,4,5-T and perhaps other herbicides and numerous chemicals are often contaminated with a deadly toxin called dioxin (2,3,7,8-tetrachlorodibenzo-p-dioxin or TCDD). It has been reported that this industrial chemical kills animals at extremely low concentrations (in the parts per trillion range), is showing up in mother's milk, and may be cumulative.

After 30 years of widespread and indiscriminate use of the various chemical insecticides, people are becoming more and more conscious of their harmful side effects, and it is all too obvious that these chemicals are not the total answer to our pest control problems.

Unlike selective herbicides, nonselective insecticides are capable of eliminating natural controls such as predator insects and birds, along with the target organism. As a result, the insect pest multiplies and the problem is far more serious than before. Related to this is the use of broad-spectrum insecticides, employed to control a single pest, which also eradicates several natural enemies of still another pest, resulting in a much larger population of an heretofore naturally controlled organism. Another extremely serious problem developing from the repeated use of an insecticide over a period of years is that a pest may become tolerant to the insecticide and produce genetically resistant offspring. The resultant insect-resistant and now impotent insecticide suggests to the manufacturer of chemical sprays that a new, more toxic insecticide must be produced and, of course, applied at higher concentrations.

The accumulation and concentration or *biological magnification* of persistent insecticides, such as the chlorinated hydrocarbons, in the environment, presents still another argument against the widespread and careless use of many insecticides. Any one of the broad-spectrum insecticides (aldrin, chlordane, DDT, dieldrin, lindane, toxaphene) are capable of entering food chains, increasing in concentration as they move through them (Figure 18–1). The second major group of insecticides, the organophosphates (malathion, parathion), although nonpersistent, are the most toxic of the chemical sprays.

One recent disaster, now referred to as the "Kepone Disaster," resulted from the unlawful dumping of an extremely stable chlorinated hydrocarbon called kepone, a powdery pesticide used to control banana and potato insects, ants, and roaches, into Virginia's James River, long one of the country's most productive bodies of water. Fishermen, whose livelihood depended on their catch of crab, oysters, scallops, striped bass, bluefish, shad, and trout, took scant consolation from the fact that the chemical company was fined in excess of $13 million.

Rather Than Chemicals . . .

With the ever-increasing need for better crop production to feed the expanding world population and the never-ending need for controlling pests, the search for more efficient and safer means of pest management must continue. In spite of the many harmful effects of chemical pesticides on organisms, including people, they will continue to be used, and probably must be, but in moderation and judiciously.

With evidence mounting that the use of many of the chemical pesticides is a "mixed blessing," alternatives that are less harmful to people and their environment, yet effective in the control of pest populations, are currently being utilized or investigated. Among these are methods whereby the normal reproductive and growth cycles of the pests are interrupted, the use of natural enemies, the development of more resistant crop varieties, improved farming methods, the use of mechanical devices, and more stringent quarantine regulations.

Normal reproductive and growth cycles of pests can be interrupted through the use of sterilization techniques; attractants such as sex (pheromones), sound and light, and hormones. Sterilization, by irradiation, of the male screwworm fly pest of cattle, their release and mating with normal females, who mate only once in their lifetime, ultimately resulted in complete eradication of the fly population. Chemosterilants, or chemical sterilizing agents, have also been released in a natural environment. These sterilants are, however, dangerous chemicals that can strike nontarget species.

Pheromones, released by some female insect species, are capable of attracting male members to traps that are baited with insecticides. Control of yellow jacket wasps and the gypsy moth has been achieved using this method. Sound and light attractants and chemical repellents have, thus far, had little effect on the control of insects.

The juvenile hormone, a naturally occurring compound, when applied to insects at a specific stage of development, prevents metamorphosis, and, as a result, the organism seldom develops into an adult. These chemicals are biodegradable, very specific for insects, but are toxic to all insects. Another hormone, ecdysone, accelerates metamorphosis or ecdysis of the larva so rapidly that the organism dies. It has recently been

FIGURE 18-1.
Biological magnification of DDT (parts per million). (G. T. Miller, Jr., Living in the Environment: Concepts, Problems, and Alternatives, Wadsworth Publishing Company, Inc., Belmont, Calif., 1975, p. E39. **Data** *from Woodwell, Craig, and Johnson, 1971, and Edwards, 1970.)*

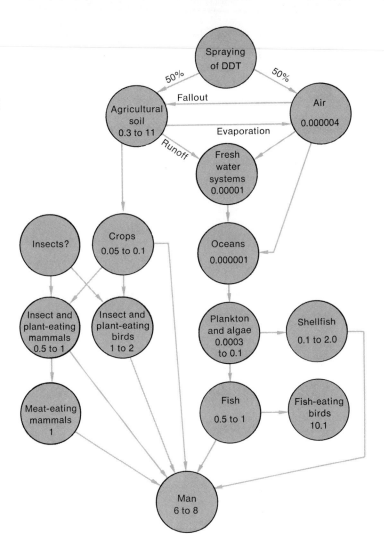

reported that two chemicals, precocene I and II, isolated from the common mistflower (*Ageratum grandiflorum*), prevent certain insects from developing normally by blocking the synthesis of juvenile hormone. Indications are that these antihormone substances are nontoxic, biodegradable, and effective against several insect species.

Great success in controlling insects and undesirable plants has been achieved through the introduction of natural predaceous enemies such as other insects (e.g., ladybugs, green lacewing, praying mantis, and certain species of wasps), bacteria (e.g., milky bacteria, *Bacillus thuringiensis*), and viruses (e.g., polyhedra virus). This method of biological control is more permanent than other control measures and chemicals are not involved (Figure 18-2). An example illustrating this means of control is the discovery that the fungal spores of *Coelomomyces psorophorae*, which first infect an intermediate host, the copepod (*Cyclops vernalis*), are capable of controlling the larvae of certain mosquitoes. Another possible means of controlling mosquitoes is through the use of the protozoan *Hydra* in certain freshwater environments.

FIGURE 18-2.
Before (top) and after (bottom) control of alligator weed by its natural enemy, the flea beetle. (Courtesy of the U.S. Department of Agriculture.)

The development of crop varieties that are resistant to pests can be an extremely valuable method of biological control. However, this resistance, which may take many years to develop (e.g., rust-resistant wheat cultivars), can be temporary, since the pest is capable of adapting to the new situation.

Improved farming methods to change the agricultural environment has proved to be an effective alternative to the liberal use of chemical insecticides. Crop rotation, cultivation of soil to dislodge insects, scheduling planting times to avoid specific insects, spacing of crops, separating crops with uncultivated land, erecting barriers between crops, early harvest of some crops, use of "trap" crops (e.g., marigold, soybeans, and barley) and companion plants (e.g., chives, marigolds, tansy, nasturtium, and radish), destruction of crop refuse, and eradication of weeds (but not with pesticides) are all good farming practices that have resulted in higher crop yields.

Organic farming is another approach to be considered. Eliminating the use of pesticides and synthetic fertilizers could very well be feasible for the organic gardener but unrealistic for the large-scale conventional farmer. On the one hand, chemical companies claim that herbicides help reduce erosion, conserve water, and improve wildlife habitat; on the other, U.S. Department of Agriculture scientists report more soil erosion, more water loss, and lower yield per acre in chemically treated plots. Perhaps a compromise between the two extremes, organic farming versus traditional agriculture, where the farmer believes that pesticides and fertilizers are sometimes necessary, is a solution if their utilization is kept to a minimum.

Mechanical means of control are sometimes employed to limit pests that might infest stored crops. Providing spacious, clean storage areas and high- and low-temperature treatments of grain are feasible, but hardly on a large scale.

Continued, strict enforcement of quarantine regulations among the nations will do much to control the widespread distribution of serious pests.

Because the worldwide production of food is so vital to the survival of mankind, and because the use of toxic pesticides to control insects has proved to be less efficient and more hazardous to people and their environment than was previously believed, a panel of experts was convened under the auspices of the U.S. Department of Agriculture, the Environmental Protection Agency, and the Ford Foundation to investigate the total area of the utilization of pesticides and alternatives to their use.

Undoubtedly, the most ideal method of pest control is an integrated one that combines biological, chemical, cultural, and mechanical means of control and is, at the same time, effective, nonhazardous, and economical (Figure 18–3). Integrated control measures are ecologically and financially advantageous in the long run and, although requiring long-range planning, time for development of techniques, intelligence on the part of users, and indirect approaches, must be expanded. The research necessary to implement these measures must be state and federally supported, since it is highly unlikely that much financial backing will be forthcoming from the chemical companies.

Safety Precautions   Because pesticides are used extensively, it is essential that the Environmental Protection Agency regulate their application. Since its formation several years ago, the Agency has banned aldrin, dieldrin, and DDT and suspended chlordane and heptachlor. Despite the necessity for these measures, the chemical industry, agricultural groups, farmers, and

FIGURE 18.3.
The cotton boll weevil, one of the pests controlled by integrated control measures. (Courtesy of U.S. Department of Agriculture.)

politicians have not wholeheartedly supported the Environmental Protection Agency's pesticide control program. As a result, the EPA just recently appointed an advisory board to assist it in decisions relating to control measures.

In the meantime the individual consumer should be familiar with the hazards as well as the benefits inherent in the use of pesticides. Users should always:

1. Read the label before purchasing and using any pesticide.
2. Use only according to the directions, including warnings and cautions.
3. Store materials in a safe place (a locked cabinet) in the original container out of the reach of children.
4. Avoid inhaling materials and the treated area temporarily.
5. Wash thoroughly immediately after using the chemicals.

Summary

In our effort to protect ourselves and our environment from the ravages of undesirable pests, we have increasingly turned to the use of chemical pesticides. Until fairly recently, we have accepted the beneficial aspects of these chemicals with little regard for their hazardous side effects. In the light of this mounting evidence, alternatives that are less harmful to people and their environment, yet effective in the control of pest populations, must be sought. It is becoming more and more apparent that the solution to pest control lies in the use of an integrated approach that combines biological, chemical, cultural, and mechanical means of control, and at the same time is nonhazardous to people and their environment.

Selected References

CARSON, R. 1962. *Silent Spring.* Houghton Mifflin Company, Boston.
CROSBY, D. G. 1973. The fate of pesticides in the environment. *Ann. Rev. Plant Physiol.* **24,** 467–492.

DeBach, P., ed. 1964. *Biological Control of Insect Pests and Weeds.* Van Nostrand Reinhold Company, New York.

DeBach, P. 1974. *Biological Control by Natural Enemies.* Cambridge University Press, New York.

Edwards, C. A. 1970. *Persistent Pesticides in the Environment.* CRC Press, Inc., Cleveland, Ohio.

Gould, R. F. 1966. *Organic Pesticides in the Environment* (Advances in Chemistry Series No. 60). American Chemical Society, Washington, D.C.

Graham, F. 1970. *Since Silent Spring.* Houghton Mifflin Company, Boston.

Holcomb, R. W. 1970. Insect control: Alternatives to the use of conventional pesticides. *Science* **168**, 456–468.

Huffaker, C. B. 1974. *Biological Control.* Plenum Publishing Corporation, New York.

Irving, G. W. 1970. Agricultural pest control and the environment. *Science* **168**, 1419–1424.

Kraybill, H. F., ed. 1969. Biological effects of pesticides in mammalian systems. *Ann. N.Y. Acad. Sci.* **160**, 1–422.

Martin, E. C. 1972. *Modern Concepts of Pest Management* (Entomology Series, No. 1). Gadjah Mada University, Jogjakarta, Indonesia.

Marx, J. L. 1977. Applied ecology: Showing the way to better insect control. *Science* **195**, 860–862.

Matsumura, F. 1975. *Toxicology of Insecticides.* Plenum Publishing Corporation, New York.

Miller, G. T., Jr. 1975. *Living in the Environment: Concepts, Problems, and Alternatives.* Wadsworth Publishing Company, Inc., Belmont, Calif.

Pest Control: *An Assessment of Present and Alternative Technologies.* 1976. National Academy of Sciences, Washington, D.C.

Pimentel, D. 1971. *Ecological Effects of Pesticides on Non-Target Species.* Office of Science and Technology, Washington, D.C.

Rudd, R. L. 1964. *Pesticides and the Living Landscape.* University of Wisconsin Press, Madison, Wis.

Schwartz, P. H., Jr. 1975. *Control of Insects on Deciduous Fruits and Tree Nuts in the Home Orchard—Without Insecticides* (Home and Garden Bulletin No. 211). U.S. Department of Agriculture, Washington, D.C.

Shea, K. P. 1977. Profile of a deadly pesticide. *Environment* **19**, 6–12.

Woodwell, G. M., P. P. Craig, and H. A. Johnson. 1971. DDT in the biosphere: Where does it go? *Science* **174**, 1101–1107.

Yepsen, R. B., Jr., ed. 1977. *Organic Plant Protection.* Rodale Press, Inc., Emmaus, Pa.

Controlling Air Pollution
Hazel S. Kaufman and Peter B. Kaufman
19

Questions for Consideration

1. What are the best ways to control air pollution from industrial smokestacks?
2. What is the Environmental Protection Agency doing to reduce air pollution?
3. What modifications have been made in internal combustion engines that could result in reduced air pollution by automobiles?
4. What alternatives do we have to the internal combustion engine to reduce air pollution from cars? Are they practical and feasible?
5. What can you do to reduce air pollution?

Political Control of Air Pollution

Political control of air pollution is not a new concept. In 1273, King Edward I of England promoted the passage of a law outlawing the use of certain types of coal. The law, however, was never enforced. In the late fourteenth century, King Richard III put a high tax on coal. Charles II, in 1661, commissioned John Evelyn to make a report on London's air pollution. His *Fumifugium* described with great accuracy the effects of London's smoke on respiratory diseases, the reduction of sunshine, and the corrosion of materials. It has been reprinted recently and is recommended to students of air pollution, as it proposes controls of air pollution that are still applicable in the twentieth century. Evelyn's proposals were ignored. London's smoke problem continued, and during the nineteenth century, it became notorious. The Public Health Act of 1875 did contain a smoke abatement section, but because of pressure from industrialists, it was never enforced. At the turn of the century, a number of smoke abatement societies were formed. One was led by H. A. Des Voeux, who coined the word *smog* for the fog combined with smoke that occurred so often in London. This type of smog became known as *classical smog* in contrast to that caused by automobiles, which is referred to as *photochemical smog*. Despite some legislation, Britain's smog worsened. Not until after London's disasterous smog episode of December 1952 was effective legislation passed. The Clean Air Act of 1956 gave officials the authority to curtail industrial emissions and to restrict domestic heating to smokeless fuels. Coal smoke was dramatically reduced, and London today is a far cleaner city, with 40 percent more sunshine.

Some American cities were also gaining notoriety for their smokiness at the turn of the last century. In 1867, St. Louis passed an ordinance requiring smokestacks to be at least 20 feet higher than adjoining buildings. Probably the first smoke ordinance in the United States was passed by the City Council of Chicago in 1874. Many of these early laws specified violations in terms of grayness of smoke as determined by a Ringelmann chart. This chart has four numbered shades corresponding to 20, 40, 60, or 80 percent blackness. It is to be held at a certain distance in line with the smoke being observed and the percentage blackness is determined. These charts are quite unscientific, especially since they do not indicate the constituents of the smoke.

Pittsburgh was one of America's smokiest cities because of the extensive use of soft coal for home and industrial use. The Mellon Institute made an extensive study of smoke damage, and in 1913 published a report documenting the reduction of sunshine caused by the smoke particles, their irritating effect on the respiratory tract, and the depressing effect of the constant blackness of the city. However, because of the opposition of industrial and railroad corporations, no effective legislation was passed until after World War II. Today, with the use of smokeless fuels and pollution control devices, Pittsburgh's appearance has also changed dramatically.

California was a pioneer in state air pollution control. In 1947, it passed the Air Pollution Control District Act, providing for countywide control districts. The Los Angeles and San Francisco districts have been particularly active. Many state and local governments continue to take an active part in air pollution control. In 1966, the City Council of New York passed a law requiring more extensive use of low-sulfur fuels and the upgrading of incinerators. They also passed a law banning all smoking in public places after November 1, 1974. Such places include supermarkets, elevators, classrooms, hospitals, and recreational, religious, and social gathering places involving 50 or more persons.

No federal air pollution legislation was passed in this country until 1955, and this only provided for research and for technical assistance to state and local governments. The first important national legislation was the Clean Air Act of 1963, which granted permanent authority to federal air pollution control activities and authorized $95 million for expenses during the following 3 years. Amendments followed in 1965, 1967, and 1969, with the most important additions made in 1970. The latter provided for the establishment of national air quality standards and their achievement by July 1975 through implementation plans drawn up by air quality control regions and states. It provided for 90 percent reductions of automobile hydrocarbon and carbon monoxide emissions from 1970 levels by the 1975 model year. Nitrogen oxide emissions were to be reduced 90 percent from 1971 levels by the 1976 model year. It also provided for studies of aircraft emissions and noise pollution.

In July 1970, President Nixon set up the Environmental Protection Agency (EPA), and in December of that year, William D. Ruckelhaus became its first administrator. In April 1971, EPA announced the standards for major air pollutants. They are given in Table 19-1. Two levels of achievement were set by EPA, designated as primary and secondary standards. Primary standards are levels of air quality with an adequate margin of safety to protect the public from adverse health

TABLE 19-1 / National Air Quality Standards

	Primary	Secondary
Suspended particulates		
Yearly geometric mean	75 $\mu g/m^3$	60 $\mu g/m^3$
24-hour maximum	260 $\mu g/m^3$	150 $\mu g/m^3$
Sulfur dioxide		
Annual mean	0.030 ppm	0.022 ppm
24-hour maximum	0.135 ppm	0.097 ppm
3-hour maximum	—	—
Carbon monoxide		
8-hour average	9 ppm	9 ppm
1-hour maximum	35 ppm	35 ppm
Nitrogen oxides		
Annual mean	0.05 ppm	0.05 ppm
Photochemical oxidants		
1-hour maximum	0.08 ppm	0.08 ppm
Hydrocarbons		
3-hour maximum	0.24 ppm	0.24 ppm

effects. Secondary standards, which are stricter, provide protection from other, less serious consequences and include aesthetic factors.

With the energy crisis brought strikingly to the nation's attention with the gasoline shortage in 1973, the environmental movement was dealt a big blow. The emphasis now was on producing more energy regardless of its impact on the environment, and the manufacturers took advantage of the situation. They now had an excuse to delay taking air pollution control measures. The Energy Supply and Environmental Coordination Act, which passed in 1974, gave EPA discretionary powers to extend compliance deadlines to power plants switching from oil to coal. It gave automobile manufacturers until 1978 to meet the federal exhaust emissions standards. This act allows huge coal-burning power plants to be built in the Rocky Mountains, thus causing the deterioration of air quality in clean areas. The Sierra Club challenged the EPA regulations in court. Others feared that nondeterioration regulations would limit growth and hold back energy production.

Even though air quality standards have been relaxed, the American people still consider a healthy environment a top national priority, and hopefully, they will do something about obtaining it. Industry can no longer use the environment as a handy dumping place for its wastes. We do have laws to protect the environment and courts to enforce them.

Air Pollution Control Techniques for Smokestacks

Building tall smokestacks was one of the first methods used in an attempt to control industrial smoke pollution. They are quite effective in the immediate vicinity of the industrial plant, provided that there are no temperature inversions. People living downwind from them do not fare so well. The use of cleaner fuels was also an early control method for both home and industrial smoke control. Pollution has been abated at times by a change in the manufacturing process or by the replacement of worn-out equipment. Sources of air pollution have been removed to less populated areas. It has been suggested that power plants be located near the source of fuel; however, there are difficult problems in transporting electricity over long distances.

Perhaps the most common method of controlling industrial air pollution today is the addition of devices in smokestacks to remove pollutants before they escape into the air. The type used varies with the pollutant. Generally, particles greater than 50 micrometers in diameter may be removed satisfactorily by cyclone separators or wet scrubbers, whereas particles smaller than 1 micrometer in diameter are handled effectively by electrostatic precipitators and fabric filters.

In the cyclone separator (Figure 19-1) the incoming polluted air forms a vortex from which the particles are removed by centrifugal force and fall to the bottom, where they are collected. It is effective for large particles but not for small ones. Efficiencies on a total weight basis vary from 50 to 95 percent.

In the wet scrubber, the exhaust gases pass through a fine spray of water which very effectively removes particles. It also removes some gases. Efficiencies vary from 75 to 99 percent. The wastewater, however, often produces a water pollution problem.

More effective in removing small solid particles are fabric filters. Exhaust gases are forced through bags of very fine clothlike mesh which are capable of capturing tiny particles. Removal efficiencies may be as high as 99 percent. These are used widely, especially in cement, carbon black, clay, and pharmaceutical plants.

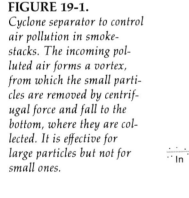

FIGURE 19-1.

Cyclone separator to control air pollution in smoke-stacks. The incoming polluted air forms a vortex, from which the small particles are removed by centrifugal force and fall to the bottom, where they are collected. It is effective for large particles but not for small ones.

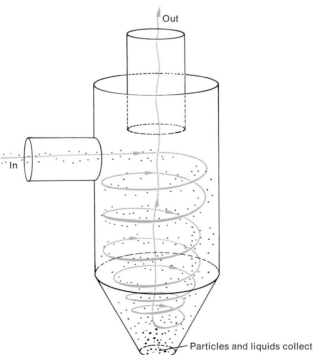

Out

In

Particles and liquids collect

The removal of gaseous pollutants is a more difficult problem. They may be absorbed or converted into a less offensive substance. The wet scrubber used to remove particles is also effective in absorbing gaseous pollutants, such as ammonia, hydrogen sulfide, hydrogen chloride, chlorine, and hydrogen fluoride.

Activated carbon is used to remove polluting gases, since it attracts thin layers of gases and vapors to its surface, leaving the carbon unchanged. The carbon, after being saturated with the collected pollutant, can be regenerated quickly with a steam bath. It is used in removing hydrocarbons. If the pollutants are of any value, they can be collected and marketed. In some cases, sales of the collected pollutants can pay for the antipollution process.

Perhaps the most efficient device for removing particles is the electrostatic precipitator (Figure 19-2). The incoming waste gases pass over high-voltage electrodes, giving particles an electrical charge. Then they pass between oppositely charged collector plates, to which they are attracted. Vibration of the plates causes the collected matter to fall down into large bins. Collection efficiencies by weight are from 80 to 99.5 percent. Power plants burning fossil fuels often use electrostatic precipitators and can achieve a removal of more than 99 percent of all fly ash. The huge amount of fly ash produced in this country presents a waste problem, but ways are being found to make use of it. The initial cost of the electrostatic precipitator is high, but their power requirements and maintenance costs are small.

On Independence Day of 1845, Henry Thoreau went to live at Walden Pond to try to find the meaning of his life away from the mechanized culture that American technology had produced. In the book, *Walden,* he wrote of that experience. The railroad serves as one of the chief symbols of natural violation of a machine-oriented society, "We do not ride on

Air Pollution Control Techniques for Automobiles

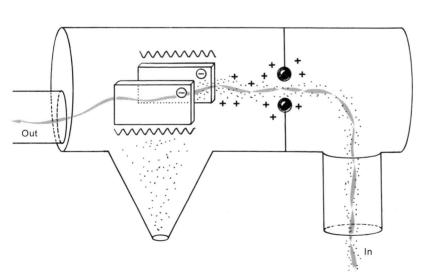

FIGURE 19-2.
Electrostatic precipitator to control air pollution in smokestacks. The incoming waste gases pass over high-voltage electrodes, giving particles an electrical charge. Then they pass between oppositely charged collector plates, to which they are attracted. Vibration of the plates causes the collected matter to fall down into large bins.

Out

In

the railroad," he wrote, "it rides upon us." However, Thoreau was no simple-minded primitivist who believed in the possibility of turning the clock back to a preindustrial age. "We have constructed a Fate," he acknowledged, "that never turns aside. Let that be the name of your engine." The question for Thoreau, as for so many since, is how to find a humanly satisfying mediating point between technological progress and the environmental cost it entails. When Thoreau felt he had found it, he returned to society.

Today, the dilemma is basically the same, although the focus has switched from the railroad to the now-popular automobile. Everyone knows that the automobile, as well as many of our other modes of travel, have an adverse effect on the environment as well as upon ourselves, but up to now, relatively little has been done about it (see Chapter 9). In this section, we shall discuss possible ways by which these problems could be alleviated. These will include both better management of the systems and technologies that we now possess and a look at the new developments in transportation technology that are now under consideration as possible alternatives to current systems.

Although the auto has been around for quite some time, emissions were not decreased until 1968, when the government forced action to this end. Fuel consumption has seen no overall major improvements. One might suspect that the auto industry, with its tremendous resources and technology, might have been able to do a lot better than it has up to now. Indeed, there is quite a bit of evidence that the auto industry has intentionally tried to maintain the status quo as best it could, even if it meant harm to the environment. Alan C. Nixon, President of the American Chemical Society, explained in a letter to *Science* (Crossland, 1974) how he had come up with a simple, logical process that would greatly reduce emissions in the standard internal combustion engine (ICE). By the simple process of injecting air into the engine at high pressure and 21.3°C (70°F), he found that he could burn up all the hydrocarbons (HCs), convert carbon monoxide (CO) to harmless carbon dioxide, and reduce nitrous oxides (NO_xs). He found nothing in the journals on this subject, but a patent search revealed five patents taken out by General Motors Corporation (GM) in the early 1960s that included most of his system. General Motors had chosen to do nothing about these patents except to claim the rights to them.

There is much additional evidence that the auto industry has been purposefully resistant to change. One trick that has been used is the public display of prototype alternative engines, but these demonstrations appear to be attempts to show the superiority of the present ICE, as these prototypes are usually poorly constructed (Esposito, 1970). In the face of new engine designs that are obviously superior in emissions control, fuel economy, or general efficiency, or all of these, the auto industry remains steadfast behind its engine. The cost of retooling the factories to the new engine type is one factor; however, it has been estimated that this cost is only equal to the cost of retooling for the usual annual style changes for 2 or 3 years (Esposito, 1970). Another factor is probably the loss in the "aftermarket"—the sale of replacement parts—which constitutes a large part of the industry's profits. This loss would occur because most proposed engine types are so much *simpler* than the ICE, so less could go wrong. Finally, since most of the proposed engines would use much less, or even no, gas or oil (such as an electric car), the effects on the oil industry would be profound, as private autos account for almost one-third of all petroleum use in the United States (Edel, 1973). Over the years, a natural alliance has developed between the auto and oil industries, as their interests are so similar, and the auto industry is naturally unwilling to injure its "sister" industry (Esposito, 1970).

However, beginning with the Clean Air Act of 1963, the federal government had begun to finally regulate emissions, starting with the 1968 model year (Horowitz and Kuhrtz, 1974; U.S. Department of Transportation, 1974). Its 1970 amendments gave the Environmental Protection Agency, EPA, the power to set emission standards, finally forcing the auto industry to do something about pollution (Crossland, 1974). The standard set by the EPA, compared to precontrol (pre-1968) cars can be seen in Table 19-2. Because of pressure from the automobile companies, emissions control requirements have been postponed many times. According to the Clean Air Amendments of 1977, the 1977 HC standard of 1.5 gpm (grams per mile) will be extended through 1979. The 1977 CO standard of 15 gpm will be extended through 1979, requiring 7 gpm in 1980 and 3.4 gpm in 1981 with waivers possible for 2 years. The 1977 NO_x standard of 2 gpm will be extended through 1980 with a final setting of 1 gpm by 1981.

One of the basic problems in trying to control emissions from the standard piston engine is this: When trying to vary the fuel/air ratio to minimize emission of CO and HC, such a fuel/air ratio tends to favor an increase in the emission of NO_xs. More exactly, a greater proportion of air in the mixture (a "leaner" air/fuel mixture) favors a more complete combustion of CO and HCs (along with an increase in temperature). This presence of additional O_2, however, favors, the production of more NO_xs. When using a richer fuel/air ratio to minimize NO_xs, much more of the other two gases is emitted. This naturally causes technological problems. When more powerful engines were developed, as in the period from 1954 to 1959, when average HP doubled, richer air/fuel

TABLE 19-2 / National Primary and Secondary Ambient Air Quality Standards

Pollutant	Air Quality Standards[a] (ppm)	Averaging Time
Hydrocarbons	0.24	3-hr average concentration
Carbon monoxide	9	8-hr average concentration
	35	1-hr average concentration
Nitrogen dioxide	0.05	Annual average concentration
Photochemical oxidant	0.08	1-hr average concentration

[a] Primary and secondary standards for these four pollutants are identical. Standards are not to be exceeded more than once per year.

Source: Horowitz and Kuhrtz (1974), p. 4.

ratios were used for more power (Crossland, 1974). However, since 1970, leaner air/fuel mixtures have been instituted. Some advances in emission control have been made recently; for example, positive crankcase ventilation valves (PCV valves) have been standard on most cars since 1968. These work by routing HCs, that would normally escape into the air from the crankcase back into the engine; these vapors would otherwise account for one-fourth of the HC emissions (Esposito, 1970).

Another positive development has been the EGR (exhaust gas recirculation) device, which many cars have been equipped with, starting in 1973. Part of the exhaust gas is recycled through the engine, reducing combustion temperature and the formation of NO_xs. With these and other changes, the 1973 standards could be met (Crossland, 1974). However, the 1975 and 1976 standards were stiffer, and they need further improvements for emission control. What are generally considered "tack-on" devices were developed to try to meet these standards. These devices are not very efficient, so the standards could not be met in time and had to be suspended. Basically, these devices were attached to the tailpipe and acted on fumes that had already left the engine. They used a catalyst (popularly platinum) that converted the three major gases to harmless products; that is, they completed the combustion of CO and HCs to CO_2 and H_2O, and reduced NO_xs. However, the catalyst depends on optimum conditions to function efficiently (e.g., the catalyst will not work properly when too hot or cold). And since lead is still commonly used in gasoline, it causes another problem. It rapidly "poisons" the catalyst so that it becomes ineffective. Even if a car is within the 1976 standards originally, it is often beyond them before 10,000 miles have been traveled. In addition, the converters are pollution sources themselves: metal-containing particles can form when gases pass over them, and H_2SO_4 mist was found to form in connection with the converters intended for use on 1975 models. Both of these pollutants can be harmful to respiration.

These catalytic devices obviously are an inefficient means of reducing pollution, and, in addition, they appear to cause an increase in fuel consumption (Crossland, 1974; Organization for Economic Cooperation and Development, 1974). Even ignoring their deficiencies, they at best only delay the inevitable: If clean air is to be achieved, and private ownership of automobiles is to continue, an alternative engine type must come into use. Probably, the only way for this to happen is either that the federal government will have to have standards for emission, which are impossible to meet with the conventional ICE, or that a competitor with an obviously superior engine will have to appear on the market, either dominating the market itself, or forcing the industry to develop a similar engine in order to compete. Quite a bit of progress has already been made in this direction. Present alternative engines can be broadly classified as either internal or external combustion engines. To demonstrate how inefficient the standard piston engine really is, it will be useful to examine some of these new developments, many of which have not been around long enough to be developed to their highest potential.

Basically, there are four new types of ICE in existence: the *rotary, stratified charge, diesel,* and *gas turbine* engines. The rotary engines developed so far (e.g., the Wankel) have little advantage over the conventional engine in either fuel economy or emissions. The stratified charge engine seems promising, in part because it is simply a modification of the basic ICE, and therefore would not involve a major factory redesign to produce it. It solves the problem of finding the proper fuel/air mixture to reduce emissions of all three major gases by varying the ratio throughout the engine for the most efficient combustion. Honda Corporation of Japan has come up with a version of this engine, but Detroit has so far planned none. The diesel engine, which has been around for awhile, primarily for use in trucks, shows some definite advantages in both fuel economy and reduced emissions, but it is noisier and less powerful than a conventional engine of equal weight. A standard Mercedes 220D (diesel) auto with a few modifications met the CO and HC standards for 1976 in an EPA study but not the NO_x standards. Gas turbine ICEs, so far only developed for larger vehicles, again are superior in reducing emissions, except for NO_x emissions.

The Rankine and Stirling engines are the two major *external combustion engines* that have been developed up to now. The Rankine engine, a type of steam engine, seems promising. An external flame boils fluid inside a closed container—it is alternatively boiled and cooled (condensed). The increased pressure from the vapor drives the system. Although so far its fuel economy is as bad or worse than the present ICE, it is supposed to surpass it with a few modifications. In addition, it can burn a lower grade of fuel (Esposito, 1970). Its advantages lie in the realm of emissions control—at its present state, it is capable of emissions well *below* the original 1976 standards for *all three* major gases (Crossland, 1974). In addition, it emits no lead, and since it needs no clutch, the asbestos problem could be eliminated (Esposito, 1970). The Stirling engine, first patented in 1816 (Crossland, 1974), is developed to the point where it can meet all but the NO_x emission standards. This engine also has good fuel economy. However, there are certain technological problems that have to be worked out before this engine can be mass-produced. Ford Motor Company was working on a car with this engine type, but has discontinued research on the Stirling engine.

Another alternative, an electric car, has been developed in the USSR. According to Tass, the Soviet News Agency (May 20, 1975), Soviet scientists have developed an electric car that does not need bulky storage batteries. A prototype station wagon, run on alternating current, is expected to "operate normally" at speeds up to 80.5 kilometers per hour (50 mph) carrying a payload of half a ton. Tass says, "The singularity of the design lies in the machine's storage batteries being charged from common 380-volt industrial alternating current source using a tiristor system (a type of convertor), a device substantially reducing the weight and avoiding the need for bulky charging units. The car can be charged at any of the points along the road where simple plug and socket devices will be mounted. When the vehicle is on the move, the direct current from its storage batteries will be converted again into alternating current and fed to the asynchronous alternating current engine."

We have seen that the technology is available for improving the automobile. However, even improved models will probably still contribute to pollution and drain the world's dwindling fuel supply, and will certainly still further urban congestion and damage caused by road building and maintenance, including salting in the winter. Since it is still questionable whether we shall see any of these improvements in the near future, most solutions now being considered either treat autos as they now exist, or they try to offer alternative systems to the automobile. These solutions can be divided into broad categories: (1) noncapital solutions that involve only minor changes in the present system; (2) restructuring of present systems, often involving capital and technological advances, designed to shift emphasis to less destructive modes of travel (generally away from the auto); and (3) completely new systems involving technological advances and, often, quite a bit of capital. In general, most of these solutions somehow are designed to make some mode of travel more attractive than the automobile, or they simply minimize auto use (much of the first category).

Americans, especially, seem to greatly value the feelings of power and independence that the automobile affords. Privacy, constant availability, protection from adverse weather, and the pride of private ownership are among the benefits that every car owner realizes. This strong orientation toward cars has to be considered in any solution presented to alleviate existing conditions. Any other transportation system, to successfully compete with the car, must offer obvious advantages over it. This is hard to do. Many solutions that have been offered combine both increasing the attractiveness of an alternative system with making the auto less attractive (e.g., changes in laws that would make the car more expensive to drive). Any plausible solution that includes luring the driver away from his or her car has to take this into account.

The first category of solutions accepts the car and simply tries to alleviate problems caused by it. These are mainly short-range solutions that could quickly reduce problems somewhat without major reworking of the system. One of the simplest methods, which would only serve to reduce urban congestion, would be to stagger work hours (U.S. Department of Transportation, 1974). Carpools are a slightly better solution, as they could help to reduce both congestion and pollution. The average occupancy per car in the United States is 2 persons, and only 1.4 per work trip (Horowitz and Kuhrtz, 1974). It is obvious that occupancy could easily be increased, as most cars carry at least 4 persons. It has been estimated that doubling auto occupancy would cut fuel consumption 11 percent (U.S. Department of Transportation, 1974). However, the present low occupancy figures indicate that an incentive would be needed to increase occupancy. Incentives such as preferential treatment on streets and freeways, restricted or more expensive parking, or preferential parking for carpools have all been shown to be effective in increasing occupancy (Horowitz and Kuhrtz, 1974). One other fairly simple solution, which would mainly have an effect on fuel consumption, would be to reduce autos in weight, which is the most important variable in fuel consumption: A 1997-kilogram (4400-pound) car consumes twice as much fuel as a 998-kilogram (2200-pound) car (Crossland, 1974). Differential taxes on cylinder size, and higher fuel prices, have led to the dominance of the smaller car in Europe (Horowitz and Kuhrtz, 1974).

In Brazil, one new solution gaining wide favor is the use of ethanol produced from sugarcane and manioc, two widely grown crops (Hammond, 1977). Alcohol, despite its low caloric heat content compared to gasoline, is a competitive fuel when burned in a properly designed internal combustion engine. Pure alcohol delivers 18 percent more power per liter than gasoline and is consumed at a rate of 15 to 20 percent faster, thus making both fuels approximately the same from an energy standpoint. However, ethanol has a clear-cut advantage: It is more completely combusted and thus gives off as much as 50 percent less CO and NO_xs; moreover, it gives off essentially no hydrocarbons and no lead, since tetraethyllead is not required to increase the octane. Given the present high cost of gasoline and the probability of its supply becoming greatly limited with consequent higher costs, the use of ethanol in autos in Brazil as a substitute fuel has tremendous merit.

Finally, we should mention the use of hydrogen as a fuel, especially since gasoline is becoming increasingly scarce and more expensive. Robert E. Billings, president of Billings Energy Corporation of Provo, Utah, lists several reasons why hydrogen may become the ultimate and ideal automobile fuel:

1. "Hydrogen is the simplest and most abundant element in the universe." A memo from the American Chemical Society states, "Hydrogen can be produced in an almost endless supply from the oceans; every 100 pounds of water yields 11 pounds of hydrogen."
2. Hydrogen can be produced at a cost equivalent to less than 50 cents per gallon of gasoline by the coal gasification process. A California company, Hydrogen Fuels, Inc., claims it has developed a way to mass-produce hydrogen with sunlight, at an estimated cost of 6 cents per gallon equivalent.
3. "Recently, the city of Provo implemented the world's first hydrogen transit bus. For 25 cents, passengers traveled in a mass-transit vehicle, equivalent in every respect to conventional hydrocarbon-fueled versions, except for one extremely important difference. The hydrogen bus generates no air pollution. The by-product is water vapor, while engine operational efficiencies are increased as much as 50 percent" (Coffin, 1977).

The second category of solutions—restructuring, often with capital and technological changes—is much broader. One suggestion that has been made is to transfer autos to public service, where rates would vary depending on the time of day (degree of congestion) and area. Fewer cars would be needed, as they would be returned when not in use, so congestion and area needed for parking could be greatly reduced (Huttman, 1974). However, this will surely not prove a popular solution, as most people tend to be against restrictive measures of this type.

Major emphasis has gone recently toward mass transit, which would have to be made attractive enough to outcompete autos in urban areas. Having fewer vehicles, with more occupants per vehicle, would obviously reduce congestion, need for parking space, fuel consumption, and possibly pollution. Replacing the auto with electrical mass transit would reduce emissions of the three controlled gases. However, mainly because of increased auto ownership, transit patronage has dropped to one-third of what it was 25 years ago (UMTA, 1975). In 1970, in urban

FIGURE 19-3.
Subway car, older and dirtier than some. New York City, August 1973. (Courtesy of the Environmental Protection Agency— Documerica; photograph by Dan McCoy.)

areas only 9.7 percent of all trips was by means of public transportation and about 4.5 percent of all trips was in taxis and school buses (U.S. Department of Transportation, 1974). With this decline in passengers, money was lost, services declined, and a cycle was begun. Fare increases only increased passenger loss. Since most of the poorest people in the cities do not own autos, they are compelled to use mainly shoddy bus lines and subways (U.S. Department of Transportation, 1974; UMTA, 1975). (See Figures 19-3 and 19-4.)

The federal government is now doing something about these problems. Starting in 1961, some federal assistance was available, and, in November 1974, the National Mass Transportation Assistance Act of 1974 was signed, establishing a fund of $11.8 billion over the next 6 years for mass transit. These funds help improve the quality of service, both by directly improving the present systems and through improved technology (Figures 19-5 to 19-10). It has been demonstrated that riders are more responsive to better service than reduced fares (U.S. Dept. of Transportation, 1974). For example, cutting fares in half increases transit use by less than 10 percent (Edel, 1973).

Included in the federal government assistance program is a new development in mass transit (Wells, 1975) known as the U.S. Standard Light Rail Vehicle (SLRV). This streetcar was designed, tested, and funded with the help of the federal government, and San Francisco and Boston put them into operation within a year. The rails are set at curbside, as copied from many cities of Europe and Mexico; hence, loading and discharging of passengers does not interfere with traffic and is much safer. In addition, parking along the curb is eliminated. This system eliminates the noxious fumes emitted by present-day buses, and each SLRV vehicle has an average life of 35 years, unlike the service life

FIGURE 19-4.
Subway car at rush hour. This is one of the newer, cleaner cars. New York City, August 1973. (Courtesy of the Environmental Protection Agency—Documerica, photograph by Dan McCoy.)

of 12 years for a bus. One of the most important features of this system is that the SLRV runs on clean energy that is more efficient to produce. Streetcars did not before, and will not now, guzzle gas.

Even with vastly improved transit systems, however, other incentives will surely be needed to lure drivers away from their autos. Possibilities are special lanes reserved for buses, "minibus" transport between the home and bus station, and restricted parking areas. A more extreme measure might be to ban the auto from the downtown area altogether.

If a much larger portion of freight were carried by waterborne vehicles, fuel consumption as well as pollution could be greatly reduced. A gallon of gas (or fuel equivalent) will carry a ton of freight 16 highway miles (25.6 kilometers), 178 rail miles (285 kilometers), and 306 water miles (490 kilometers) (Lewis, 1975). However, systems such as the St. Lawrence Seaway have so far been encumbered by regulations that prevent them from being able to compete fairly with other modes of freight transport (U.S. Department of Transportation, 1974; Adams, 1975).

Probably the most innocuous form of vehicular travel is bicycle riding (Figure 19–11). It is admittedly a fair-weather sport (usually), but most people enjoy it as a relaxing and inexpensive method of travel. However, the main problem up to now has been that, to make a trip of any real length, one usually had to risk life and limb on the highway. This occurred because any bike trails that existed were usually scattered and disconnected. Now, however, bicycles have become popular enough so that the government is starting to allot money for bicycle trail development (15.3 million bikes were sold in 1973; Horowitz and Kuhrtz, 1974). A federal decision made in 1973 will allot $120 million for bikeways over the 3 years following (Opre, 1974). One method of building trails would

FIGURE 19-5.
Air trains. (Courtesy of the U.S. Urban Mass Transit Authority.)

FIGURE 19-6.
Urban mass transit system, Morgantown, West Virginia. (Courtesy of the U.S. Urban Mass Transit Authority.)

involve utilization of abandoned railroad rights-of-way; this is being considered in Michigan and now being carried out in Wisconsin. Possibly the most impressive project so far is a trans-America bicycle trail, 3500 miles in length, that was planned by the Bikecentennial organization in 1976. The idea is to have this trail serve as the main artery for a system that will eventually pass through all 48 contiguous states (Figure 19-12). The route will be made more useful by following the more interesting areas of the region, and by having readily accessible accommodations along the route (Figure 19-13).

The third category of possible solutions consists of entirely new developments in transportation. One new development is the Witkar, which is powered by a 2000-watt, 24-volt electric motor (Senger, 1974). It occupies one-third of the road space of the conventional auto and weighs only 389.3 kilograms (858 pounds). This car needs about 5.5 minutes of charging time per mile-and-a-half ride and can cruise at about 20 mph. They are being used on a trial basis in Amsterdam, the Netherlands, on a rental basis. The cars are fairly cheap to build (about $2850 now, and this

FIGURE 19-9.

Express bus from Maryland suburb to Washington, D.C., near Washington D.C., 7:25 A.M., May 1973. (Courtesy of the Environmental Protection Agency—Documerica; photograph by Yoichi Okamoto.)

should drop with increased volume of production and sales) and very cheap to run: One test resulted in an energy cost of only $4.28 for 450.5 kilometers (280 miles). This computerized system has already had enough success to be touted as a possible alternative to the auto, especially in presently congested urban regions.

Another recent technological advance has been the concept of Personal Rapid Transit (PRT). A PRT system would be designed, hopefully, to successfully compete with the automobile (Anderson, 1974). It, like the Witkar, would not be dependent on petroleum fuels. The cars would not seat more than four, and often would carry only one person. The guideways that it would run on would be branched enough to minimize the need for transport to the system, and the system would provide on-demand service at all hours. Studies indicate that the system would be much cheaper to build than a conventional rail transit system. It would also be a computer-run system. There would be a great energy savings over the auto, and running the system, if it could divert enough passengers from their autos, would save a great amount of money. All in all, it seems like an extremely promising solution to our present urban problems. One version of the PRT system is now under full-scale testing in Morgantown, West Virginia (UMTA, 1975). (See Figures 19-6 to 19-8.)

Air Pollution Control by Individual Action

Since we are all polluters, we must not only demand action from our government but must take action ourselves. Antipollution tactics must become a way of life. Unbelievable as it may seem, in the United States, air pollution from personal sources is greater than that from industry and utilities combined. In any major American city, pollutants emitted by automobiles exceed those from all other sources. The workers at a

FIGURE 19-10.
New transbusses for urban mass transit. (Courtesy of the U.S. Urban Mass Transit Authority.)

FIGURE 19-11.
Bike racks in front of the Time–Life Building, Avenue of the Americas and 50th Street, New York City, August 1973. (Courtesy of the Environmental Protection Agency—Documerica; photograph by Dan McCoy.)

factory probably produce more pollution driving to and from work than the factory produces all day. We all pollute the air with our furnaces, fireplaces, outdoor barbeques, incinerators, and open burning of refuse and leaves. Snowmobiles and motor boats add more pollution. All living things cause pollution; even the caveman with his first fire polluted the air. Some pollution is inevitable. We cannot eliminate all things that pollute, but we can do much and still have a pleasant life-style and a more healthful one. We can walk, ride a bicycle, or use mass transit, and only when necessary use our automobiles fueled with unleaded gasoline

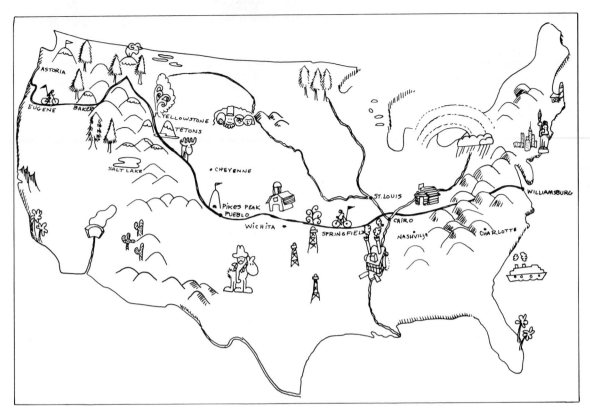

FIGURE 19-12.
Map depicting the routing of the trans-America bicycle trail. (Courtesy of Bikecentennial '76, P.O. Box 1034, Missoula, Montana 59801.)

and kept in good repair. We can use only smokeless fuels in our homes and recycle much of our waste instead of burning it. Cross-country skiing, canoeing, and sailing are healthier and more environmentally sound ways to enjoy the snows of winter and the rivers and lakes in summer than the snowmobile and motor boat.

Summary

Air pollution is one of our most serious enviromental problems. The sources vary from smokestacks and automobiles to noise and Freon-containing aerosols. In this chapter, we have described briefly the history of air pollution control, starting in the thirteenth century in England when a law was passed outlawing certain types of coal. In the United States, the biggest stride forward was achieved with the formation of the Environmental Protection Agency in 1970.

Now, alternatives must be considered, tested, then put into practice. Some of these alternatives mentioned in this chapter include the following:

1. Using cyclone separators, wet scrubbers, or electrostatic precipitators in smoke stacks.
2. Achieving more complete combustion of fuels in the auto's internal combustion engine.

3. Developing other types of engines for autos that cause less pollution of the air.
4. Increasing our use of electric cars, mass transit, bicycles, and, of course, walking.
5. Banning the use of aerosols that contain highly reactive chlorofluoromethanes that break down the ozone layer and increase the incidence of skin cancer.
6. Setting up strict standards and limits for noise pollution at airports, in industries, and in other places where noise seriously impairs our hearing.

7. Planting trees near buildings and around parking lots to absorb dust and other air pollutants.
8. Influencing legislators to pass legislation resulting in stricter air pollution laws to improve the quality of our air.
9. Enforcing the existing air pollution control laws more effectively.

Selected References

ADAMS, B. 1975. The shameful state of transport. *Reader's Digest,* February, 61–66.

ANDERSON, J. E. 1974. PRT. *Environment* **16**(3), 6–11.

BENNETT, S. 1973. *A Trail Rider's Guide to the Environment.* The American Motorcycle Association, Weterville, Ohio.

COFFIN, T., ed. 1977. The future of the automobile. *The Washington Spectator and Between the Lines* **3**(11), 3.

National Wildlife Federation. 1974. Highway salt and the environment. *Conservation News* **39**(6).

CROSSLAND, J. 1974. Cars, fuel, and pollution. *Environment* **16**(2), 15–27.

EDEL, M. 1973. Autos, energy, and pollution. *Environment* **15**(8), 10–17.

ESPOSITO, J. C. 1970. *Vanishing Air.* Grossman Publishers, Inc., New York.

FRI, R. 1973. *Statement to the Public on Clean Air and the Automobile.* Environmental Protection Agency, Washington, D.C.

HAMMOND, A. L. 1977. Alcohol: A Brazilian answer to the energy crisis. *Science* **195**(4278), 564–565.

HEATH, R. 1974. A look at snowmobile damage. *The ORV Monitor* (Environmental Defense Fund, Berkeley, Calif.), June, p. 7; also, "ORV's and the California desert," p. 3.

HOROWITZ, J., and KUHRTZ, S. 1974. *Transportation Controls to Reduce Automobile Use and Improve Air Quality in Cities.* Environmental Protection Agency, Washington, D.C.

HUTTMAN, J. P. 1974. Public utility for autos. *Environment* **16**(3), 42, 43.

LEWIS, R. 1975. Item from the *Ann Arbor News* entitled, "Low-Energy St. Lawrence System Losing to Other Transport," Mar. 6.

OPRE, T. 1974. State planning . . . trails. *Detroit Free Press,* Jan. 17, p. 3Da.

Organization for Economic Cooperation and Development. 1974. Improved transportation systems. *Environment* **16**(2), 28–33.

SENGER, W. M. 1974. Silent cars in Amsterdam. *Environment* **16**(8), 14–17.

THOREAU, H. D. 1910. *Walden.* Houghton Mifflin Company, Boston.

U.S. Department of Transportation. *1974 National Transportation Report Summary.* Washington, D.C.

U.S. Department of Transportation. 1975. Background information on the UMTA. UMTA, Washington, D.C. April.

Wells Newsletter **34**(11), June 1, 1975.

20 Recycling at Work
Peter B. Kaufman

Questions for Consideration

1. What is a methane generator, how does it work, and how can it contribute to meeting our energy needs as an alternative source of energy?
2. How does a compost pile work to produce humus? How do the decomposition reactions in compost piles differ from those in methane generators?
3. What is the process involved in making recycled paper? How does the quality of recycled paper compare with that of paper made initially from virgin pulpwood?
4. What are the ways by which newspaper, glass, metal, and plastic recycling can be made economically practical?
5. In recycling glass, metal, plastics, and paper, how can these materials be separated from each other as an industrial process?
6. How are energy, carbon, and nitrogen recycled in nature?
7. What are some things you can do as an individual to help recycle paper, plastic, glass, rubber, and metal?

The face and character of our country are determined by what we do with America and its resources.

THOMAS JEFFERSON

For decades, the face of America has been blighted by the waste products of our industrialized society. The throw-away ethic has dominated our life-style all too long. The incredible waste of resources and energy is now catching up with us. Faced with vanishing resources and an energy crisis, we are finally beginning to develop a new ethic: one of recycling all kinds of products of our industrial society. Such efforts at recycling have been going on for a long time in European and in Asian countries. In this chapter, we shall look briefly at some of the means now being used to recycle our plant debris, paper, glass, plastic, rubber, garbage, sewage, and other products. We shall make this a very practical account because one of the purposes of this book is to get our readers to do something about our environmental crisis, especially to practice recycling.

One of the best places to start recycling is in the home garden with a compost pile (Rodale, 1960). Compost, by definition, represents a mixture of plant material, soil, eggshells, straw, manure, and other materials that can be broken down (decomposed) by microorganisms (bacteria and fungi). Its main value for use in the home garden, after decomposition has converted it into rich humus, is to improve soil texture, increase soil water-holding capacity, serve as a mulch, prevent or retard weed growth, and protect plants from excess winter injury due to low temperatures or desiccation. It also may serve as a source of nutrients, especially if ashes and manure have been added to the compost pile. However, the fertilizer value of most compost humus is low; a typical analysis might show 0.5 percent nitrogen, 0.4 percent phosphorus, and 0.2 percent potassium.

What kinds of things (substrates) can one add to a compost pile to make good humus? The list is impressive: eggshells, citrus rinds, coffee grounds, kitchen vegetable garbage, tops of cut perennials, leaves, grass clippings, weedy annuals, ashes, straw, manure, sod, crushed granitic rock, soil, inorganic fertilizer, seaweed, overripe fruit, water, and, of course, earthworms. Such a mixture of garden and home by-products will contain many types of fungi and bacteria which act to decompose the compost substrates. However, to make sure that decomposition occurs satisfactorily, many compost makers add "compost starter," which is an aqueous suspension of bacteria and/or fungi that are adept at breaking down vegetable matter. The decomposition action by these organisms works best under aerobic conditions in warmer weather where there is a favorable supply of moisture. Like secondary treatment of sewage, the microbial action breaks down the vegetable and other matter to humus plus CO_2 and, in the process, liberates much heat, generating temperatures typically between 50 and 80°C.

How is a compost pile constructed? There are many ways, but basically, one makes alternating layers of vegetable matter and soil. To stimulate microbial action, one should periodically add some fertilizer and water to the compost pile, especially during periods when the air temperature is warm. The soil is primarily added to provide a source of microorganisms and to hasten the microbial decomposition action by these microorganisms. It also retains moisture needed by these organisms to grow.

In recycling compost, it is best to use compost that has decomposed completely to humus. In northern temperate climates, this may take a year or more, depending on the nature of the substrate(s) in the compost pile and the rapidity of microbial action. Oak leaves, for example, take much longer to break down to humus than grass clippings. One rule of thumb for these northern temperate areas is to use compost in the spring that was started a year or two before.

We have several suggestions to make concerning compost piles. One is to bury at least half of the substrate underground. This acts to insulate the pile against low air temperatures of spring and fall, to make it less unsightly, and to catch water more effectively. Regarding the latter, it is best to make the upper surface of the compost pile concave so as to effectively trap more water. If you want to hasten the breakdown of

FIGURE 20-1.
Biogas plant (40 cubic feet per day), based on home organic waste matter. Gas is used for lighting and cooking. (Courtesy of Ram Bux Singh.)

vegetable matter, a shredder is an excellent device to subdivide this matter, especially if it is bulky, as with large perennial shoots. It is imperative, of course, to leave woody twigs and branches *out* of the compost pile; they break down too slowly in comparison with the other material we have cited above. Finally, if you want to be innovative with your composting efforts, grow celery or other vegetables directly in a well-matured compost pile. It works beautifully. A second compost pile

FIGURE 20-2.
Biogas plant capable of producing 200 cubic feet of gas per day. (Courtesy of Ram Bux Singh.)

can then be established for purposes of modifying your garden soil or for mulching.

Recycling and Energy Production with Methane Generators

Our second approach to recycling involves the use of animal wastes (manure) and plant debris (including algae) to produce methane (CH_4) and organic fertilizer for urban, farm, and garden use. Methane generators have been in widespread use in Germany, France, India, and Japan, and are now coming into their own in the United States (Singh, 1973, 1974; Fry, 1974). They can be used in both northern temperate and in tropical regions. Several of the big advantages to their use is that they utilize readily available substrates; they help to solve local energy needs especially when used in combination with wind generators and solar panels; they provide useful fertilizer by-products; and they help to recycle immense quantities of manure and garbage that are spewed out on the landscape from large feedlot operations and municipal garbage open landfills.

Now, let us look at how methane generators work. In contrast to composting, an aerobic operation, methane generators involve breakdown of organic matter in the absence of oxygen (anaerobic conditions). The latter, therefore, involves a well-known process called fermentation. The main products of this fermentation are CH_4 and CO_2, with trace amounts of hydrogen sulfide and nitrogen gas also given off. The conversion occurs at or around normal atmospheric pressure and at temperatures ranging between 15 and 50°C. The heat of combustion of the substrate(s) ranges between 500 and 700 British thermal units per

cubic foot (Btu/ft^3) of gas produced. The heating value of methane is about 1000 Btu/ft^3.

In general, between 4.5 and 6.5 ft^3 (at standard temperature and pressure) of methane can be obtained from the conversion of 453 grams (1 pound) of dry organic materials; or 9 to 13×10^3 ft^3/ton of organic materials. The latter corresponds to a heating value of 9 to 13×10^6 Btu/ton and represents 60 to 80 percent of the heating value of the original organic materials.

Figures 20-1 to 20-5 represent several methane generators now in use in India, as provided for illustration here by Ram Bux Singh, one of the great pioneers in the field. In India, these generators are called biogas plants. Singh has designed units of all sizes for use in home gardens and on large farming operations and in villages and cities. In cold climates, these plants are partially buried underground and well insulated, as shown in Figure 20-6. All of them can be operated throughout the year, although in cooler areas, the rate of production of methane drops during winter months.

Figure 20-7 illustrates a system for the production of methane from activated sludge produced from secondary treatment of sewage. Such

FIGURE 20-4.
Biogas plant capable of producing 500 cubic feet of gas per day. (Courtesy of Ram Bux Singh.)

methane generators are coming into increasing use in sewage treatment plants. The methane produced from them can be used to meet at least part of the energy needs for running sewage treatment plants. In some areas, methane generators, used in conjunction with wind generators and solar panels, *could* provide for all energy input needed to run a modern sewage treatment plant, including in such plants processes for tertiary treatment of sewage in addition to the usual primary and secondary treatments.

Preliminary estimates of methane costs, using as a model system fermentation performed by algae on a sewage pond rather than a generator per se, are between $1.50 and $2.00 per 10^6 Btu's. Table 20-1 gives a comparative summary of the costs of biogas installations in India and the United States.

In addition to the mitigation of sewage disposal difficulties, advantages of methane generators include the low energy requirements of the

FIGURE 20-5.
Batch feed digesters. (Courtesy of Ram Bux Singh.)

FIGURE 20-6.
*Design for a biogas plant
used for cold climatic areas.
(From Singh, 1973, 1974.)*

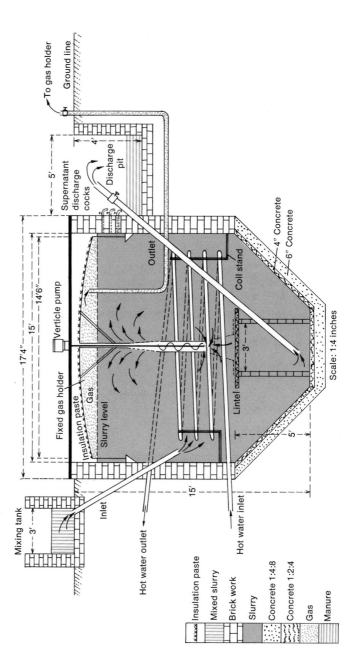

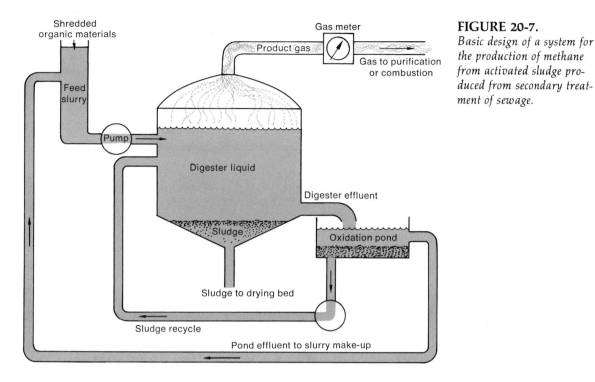

FIGURE 20-7.
Basic design of a system for the production of methane from activated sludge produced from secondary treatment of sewage.

system (i.e., atmospheric pressure and temperatures in the range of 20 to 50°C), the production of a single hydrocarbon gas which is easily purified, and almost complete conversion of the substrate(s) to CH_4. Two disadvantages of methane generator operations include the relatively large size of most installations and the relatively slow rate of reaction.

Many municipal sewage treatment systems are increasingly becoming overloaded and archaic. Some are still wasting huge amounts of energy through failure to recycle properly the activated sludge that is derived

Recycling Sewage

TABLE 20-1 / Installation Cost of Different Sizes of Biogas Plants—Estimations Based on 1973 Levels of Prices

Size of the Plant (ft^3 gas/day)	Approximate Cost of the Installation in India	Approximate Cost of the Installation in the United States	Remarks
100	$ 140	$ 400	Family size
250	350	900	Joint family size
500	600	1800	Farm size
1200	1500	4000	Industrial size
2000	2250	5500	Industrial size

Source: Ram Bux Singh, Gobar Gas Research Station, Ajitmal, Etawah (U.P.), India, 1973.

from secondary treatment of sewage. Some sewage treatment plants actually burn this sludge and dump the ash in sanitary landfills. Any modern sewage treatment plant today not only recycles its activated sludge to make methane to help meet its energy needs and fertilizer from the fermented sludge, but also utilizes one of the several methods for tertiary treatment of sewage to produce potable (drinkable) water. Some of these tertiary treatment methods, illustrated in Figures 20-8a and b, have been put into practical use, as in the South Lake Tahoe, California, sewage treatment plant, one of the most modern in the world. In this plant, the basic system includes chemical addition (lime), nitrogen removal (by an ammonia stripping tower), mixed media filtration, and activated carbon adsorption. The carbon adsorption unit is composed of eight carbon columns, each containing about 20 tons of granular activated carbon. The final effluent from this system is colorless, odorless, and free of suspended solids. Over 99 percent of the biochemical oxygen demand (BOD) and 96 percent of the chemical oxygen demand (COD) are removed. The initial investment in such a system is expensive, but the final product, recycled water that is almost clean, is worth it to human health, to the goal of cleaning up our rivers, and for utilization in our drinking water systems.

Recycling Industrial Products

PAPER

In the United States, we recycle about one-fifth of our paper. In contrast, Japan recycles about one-half of its paper. The primary deterrents to paper recycling in the United States are economic. Right now, harvesting trees to make pulp for paper manufacture costs less than the recycling process. Removal of ink is now feasible but expensive, and recycled paper is generally not useful in making white bond or slick paper for magazines. There has been such a glut of recycled paper on the market recently that the net monetary return for recycling centers has made paper recycling relatively unprofitable. However, this will change as pulp becomes more scarce, labor costs increase, and recycled paper comes into greater use.

One of the most telling arguments for recycling paper is that it results in a reduction of solid waste and a reduction of the demands placed on the nation's forests and woodlots. The rate at which timber is being cut in our national and state forests for lumber and pulp is indeed alarming. Let us examine the environmental impact for 1000 tons of low-grade paper here in order to compare virgin pulp with 100 percent recycled paper (Table 20-2). The data clearly show that much less energy is expended, less pollution is caused, and no trees are used in the process of making low-grade paper through recycling.

What is involved in the process of recycling paper? Hodges, in his book, *Environmental Pollution*, illustrates well the process for making cardboard from recycled paper. This process basically involves use of a hydropulper to clean and recondition the paper. First, the paper stock is diluted with water. Then, the paper fibers (mostly cellulose) are reconditioned by screening, steaming, and heating. The paper stock is then

TABLE 20-2 / Environmental Impact for 1000 Tons of Low-Grade Paper

Environmental Effect	Virgin Pulp as Source	100% Recycled Paper as Source
Trees	17,000	0
Processed water used	24 million gallons	10 million gallons
Energy consumption	17 billion Btu	5 billion Btu
Air pollution	42 tons	11 tons
Water pollution		
BOD	15 tons	9 tons
Suspended solids	8 tons	6 tons

Source: *Newspaper Recycling—An Economic Analysis,* based on a study done by the Monmouth Ecology Center, Asbury Park, N.J., 1973.

thickened while it forms on cylinders and water drains on a screen mesh. The waste paper is then formed into paperboard that can be cut and molded into cartons or rolls of cardboard.

Aside from the recycling of paper in industry, what can we do as individuals to recycle our own paper? The possibilities are enormous. Here are just a few suggestions worthy of consideration:

1. Use newspaper as a mulch in the garden to control weeds and to prevent loss of soil moisture. It can be held in place with compost, stones, soil, or cut tops of dead herbs.
2. Use large shopping bags to get well-rotted manure for the home garden. They are easy to carry, but must be emptied within several hours after filling them with manure.
3. Use mimeograph or ditto paper for copies for your letters or for taking notes or composing script.
4. Recycle your paper at local ecology centers.
5. Use shredded newspaper for stuffing in boxes.
6. Use newspaper for papier-mâché—in molding animals and other figures.
7. Use newspaper instead of plastic sheets as a protectant when painting a room or painting furniture.
8. Use magazines and old telephone directories when pressing plants to obtain dried plants or flowers.

PLASTICS

Plastic wastes constitute about 2 percent of the total solid wastes produced in the United States, approximating 3 million tons in 1970. Of this, packaging comprised about two-thirds, industrial wastes one-sixth, and such items as housewares, construction, toys, appliances, furniture, transportation, footwear, and records the remaining one-sixth. What do we do about recycling all this plastic?

Unsanitary dumping is the most widely used method of waste disposal. In the United States, there are about 12,000 land disposal sites. Some

FIGURE 20-8a.

Physical–chemical sewage treatment components (flow chart at bottom) and how they operate. (Courtesy of the Environmental Protection Agency.)

PRETREATMENT—Preliminary screening of floating debris and settling of sand, grit, and other large particles. This component is also used in conventional-type treatment plants.

FILTRATION—The chemically clarified wastewater is passed through beds consisting of sand or crushed anthracite coal, or a combination of the two materials. This step will result in virtually complete removal of the remaining suspended solids not removed in the preceding step.

CLARIFICATION—This component is normally a basin or series of basins where chemicals such as alum or lime are added to coagulate the waste particles into large masses that rapidly settle out of the wastewater stream. Virtually all settleable solids are removed in this step.

ADSORPTION—In this step, the wastewater is passed through columns of activated carbon granules to remove dissolved organic material. When the capacity of the carbon to remove further material is exhausted, the carbon is heated to burn off the collected material. This enables the plant to reuse the carbon again and again.

DISINFECTION—The addition of disinfection is for destruction of harmful bacteria. Disinfection is also performed at conventional treatment plants.

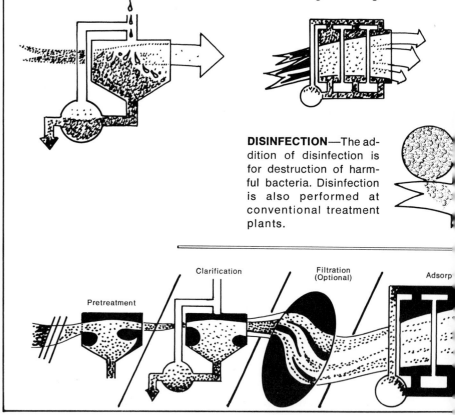

Pretreatment

Clarification

Filtration (Optional)

Adsorp

plastics do degrade; there is even one that degrades in water. However, nondegradable plastics, for the most part, are put in land disposal sites. They make a relatively stable material for the landfill because (1) they do not break down and pollute underground water systems with decay-

FLEXIBILITY

There is a great deal of flexibility in using a physical-chemical treatment system. Filtration may be used either before or after the carbon adsorption component or both before and after depending upon the required quality of the treated wastewater. Many other variations are possible to meet the specific water quality needs of the municipality. In some instances, the carbon columns may be deleted with chemical clarification and filtration providing the necessary treatment. The details as to what treatment system would be appropriate to meet your city's needs should be discussed with your consulting engineering firm.

WHY PHYSICAL-CHEMICAL TREATMENT?

It has become evident in recent years that biological treatment is not the only solution to all municipal waste treatment problems. Biological plants are sensitive to the vagaries of weather and unusually strong wastewaters dumped into the collection system. Either may completely disrupt the treatment process and result in inadequately treated wastewater passing through the plant into the receiving waters. Many biological plants cannot provide the high degree of solids removal and biochemical oxygen demand (commonly referred to as BOD) removal required to meet Water Quality Standards.

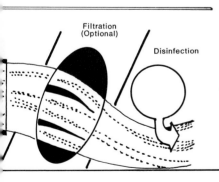

Filtration (Optional)

Disinfection

HOW MUCH DOES IT COST?

The chart below shows the approximate national average costs, including plant amortization,* operation and maintenance for a typical physical-chemical treatment plant. The graph is a general guide and is subject to considerable variation due to geographical locations, labor rates, and site conditions.†

*Based on 24-year economic life at 6 percent.

†Two-stage lime treatment, filtration, carbon, and disinfection. Also included are recalcinating lime and sludge incineration.

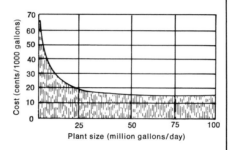

WHERE ARE PHYSICAL-CHEMICAL PLANTS BEING PLANNED?

This is a representative listing giving an indication of the rapidly spreading acceptance of physical-chemical treatment. The number of municipalities planning this type of treatment system is increasing rapidly.

Location	Size (million gallons/day)
Niagara Falls, N.Y.	60
Garland, Tex.	30
Fitchburg, Mass.	15
Rocky River, Ohio	10
Cleveland, Ohio	50
Cortland, N.Y.	10
Owosso, Mich.	6
Painesville, Ohio	5

ing matter or chemical residues (if not burned), and (2) they do not settle appreciably over the years.

According to the plastics industry, the burning of polyethylene and polystyrene are not harmful when efficiently burned in an incinerator.

ACTIVATED CARBON INSTALLATIONS

Locations where activated carbon has been used successfully to provide tertiary treatment include: South Lake Tahoe, California; Nassau County, New York; Pomona, California, and Colorado Springs, Colorado. Municipalities that are planning to utilize the activated carbon process independently of a biological treatment process include: Rocky River, Ohio; Owosso, Michigan; Garland, Texas; and Niagara Falls, New York. Data produced at these places clearly indicate the ability of activated carbon to produce effluents with very low levels of organics. A more detailed discussion of several of these installations follows.

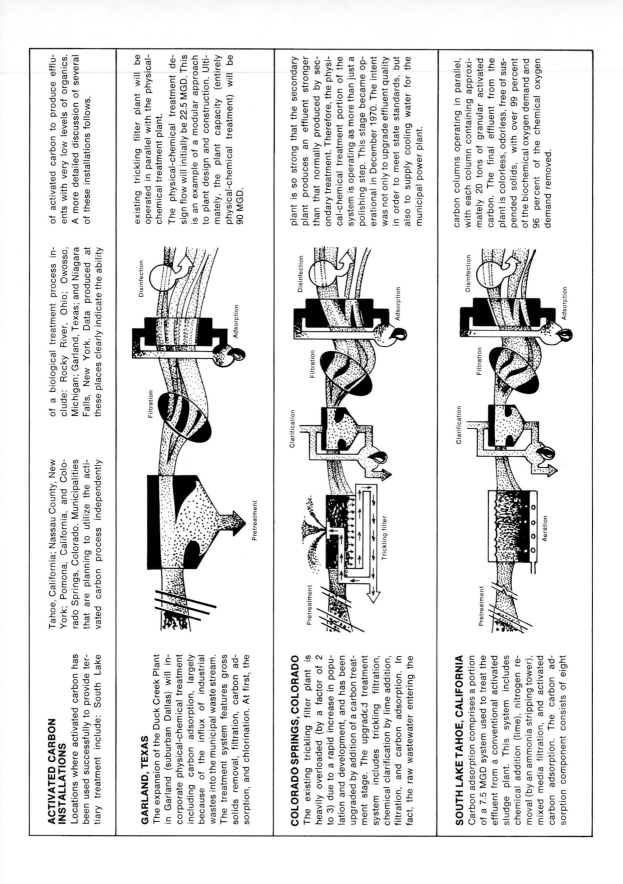

GARLAND, TEXAS

The expansion of the Duck Creek Plant in Garland (suburban Dallas) will incorporate physical-chemical treatment including carbon adsorption, largely because of the influx of industrial wastes into the municipal waste stream. The treatment system features gross solids removal, filtration, carbon adsorption, and chlorination. At first, the existing trickling filter plant will be operated in parallel with the physical-chemical treatment plant.

The physical-chemical treatment design flow will initially be 22.5 MGD. This is an example of a modular approach to plant design and construction. Ultimately, the plant capacity (entirely physical-chemical treatment) will be 90 MGD.

COLORADO SPRINGS, COLORADO

The existing trickling filter plant is heavily overloaded (by a factor of 2 to 3) due to a rapid increase in population and development, and has been upgraded by addition of a carbon treatment stage. The upgraded treatment system includes trickling filtration, chemical clarification by lime addition, filtration, and carbon adsorption. In fact, the raw wastewater entering the plant is so strong that the secondary plant produces an effluent stronger than that normally produced by secondary treatment. Therefore, the physical-chemical treatment portion of the system is operating as more than just a polishing step. This stage became operational in December 1970. The intent was not only to upgrade effluent quality in order to meet state standards, but also to supply cooling water for the municipal power plant.

SOUTH LAKE TAHOE, CALIFORNIA

Carbon adsorption comprises a portion of a 7.5 MGD system used to treat the effluent from a conventional activated sludge plant. This system includes chemical addition (lime), nitrogen removal (by an ammonia stripping tower), mixed media filtration, and activated carbon adsorption. The carbon adsorption component consists of eight carbon columns operating in parallel, with each column containing approximately 20 tons of granular activated carbon. The final effluent from the plant is colorless, odorless, free of suspended solids, with over 99 percent of the biochemical oxygen demand and 96 percent of the chemical oxygen demand removed.

The concept of recycling municipal source plastics, unsorted from other wastes, in a system such as pyrolysis or combustion for energy recovery appears very attractive from a short-range point of view. However, as sorting and property upgrading technology develop and other materials in the pool of municipal wastes find greater return in reuse than in combustion, similar recycling for waste plastics could develop (*Recycling Plastics*, 1973).

One of the most ingenious ways for recycling plastic bottles is in the construction of concrete bridges. Incineration of plastic bottles has been the subject of much controversy among air pollution people because of hydrochloric acid fumes in the smokestack gases. This is not so in Elgin, Illinois. Here, a new bridge will be made of lightweight concrete, using 25,000 scrap plastic bottles. The bridge will be a concrete arch, 30.4 meters (100 feet) long and 2.1 meters (7 feet) wide, used to handle foot traffic and city maintenance vehicles. The project represents the first reported use of plastic in this way. A total of 1815 kilograms (4000 pounds) of 0.31×0.63 centimeter ($\frac{1}{8}$ by $\frac{1}{4}$ inch) plastic chips will be used in the plastic–concrete bridge. Scrap plastic will make up 30 percent of the normal concrete mixture. This will save 817 kilograms (1800 pounds) per truckload of concrete. Furthermore, the *weight* of the bridge will be 9 percent lighter than if regular concrete were used. Plastic concrete weighs 10 percent less than the same volume of conventional concrete and is almost *equal in strength*. Regular concrete with zero slump is harsh and difficult to work, whereas with plastic, this problem is solved because plastic acts as an "internal lubricant."

Individuals can also recycle plastic. Plastic bottles or jugs can be cut, perforated at the bottom, and used as flower pots. I have also seen them used to grow aquatic plants, such as rice, in greenhouses (not perforated, of course). Plastic bottles are also excellent devices for transporting water for use in the home garden, and when cut off at the bottom, for protecting plants from frost damage in the garden. Plastics can also be used as filler in making dikes, berms, retaining walls, or dams in flood-control projects. Indeed, a little ingenuity on your part is all that is needed to recycle your plastics.

WATER

One aspect of recycling not often emphasized, but now of increasing importance in saving energy and money, is water recycling. School site and building planning and construction now incorporate definitive water recycling systems. The way in which this is done is to take rainwater from the school building roof and channel it into an outdoor pond (Figure 20-9). The pond serves three primary functions: It provides a primary landscape feature on the school site; it serves as an important resource for field studies in biology classes on ecology of aquatic plants and animals; and the water in the pond is recycled into the school building, after filtering, for use in showers and toilets. Only drinking water and water for cooking is obtained from municipal water supplies. The volume of water thus saved is immense, especially where the

FIGURE 20-8b.
(Opposite) Several physical–chemical sewage treatment systems in operation at three different locations in the United States. (Courtesy of the Environmental Protection Agency.)

FIGURE 20-9.

Thurston Elementary School Pond and Marsh at Ann Arbor, Michigan. Here, rooftop and grassland water from an adjacent subdivision is brought into a bog-swale area of the site, creating the aquatic habitat shown. [*From* Outdoor Classrooms on School Sites *(U.S. Department of Agriculture Soil Conservation Service PA-975), Government Printing Office, Washington, D.C. January 1972, p. 22; courtesy of Karl W. Grube.*]

daytime consumption of water is very large, in schools with several thousand students.

Another example of water recycling is seen in a project established by Michigan State University for the city of Lansing, Michigan. Here, they have set up three large ponds. Into the first, effluent activated sewage sludge is delivered. The water from this pond is used to fertilize field crops. The water from this first pond then drains into a second pond. This second pond is used to grow plants and fish that can be harvested for livestock feed, after processing. Water in the second pond gets greatly purified through sediments settling, natural percolation of water through soil, and plants absorbing nutrients from the water. This water then flows into a third pond, which is fit to be used for recreation, such as boating and swimming. What a magnificent way to recycle "wastewater," without having to treat it further, letting natural processes gradually improve its quality.

RECYCLING GLASS, METAL, AND RUBBER

Various devices have been designed to separate ferrous from nonferrous metals, using magnets: metal from glass, using various combinations of pulverization, screening, slurrying, and centrifugal force; and different colors of glass by optical means. One such separation flow system, incorporating some of these principles, is illustrated in Figure 20-10. The optical system for separating glass works on the following principles: "A continuous stream of individual particles are dropped through an optical box. In the optical box there are three photocell assemblies set at 120° intervals, and suitable illumination sources. Opposite each photocell head is a background with variable shades of color. Each particle passes through the viewing area, and if there is a change in its reflectivity with respect to the background standard, either lighter or darker as desired, a blast of compressed air is triggered to deflect the off-color particle from the main stream." Two separators can be operated in series. The first can separate the flint (clear glass) from the colored glass, and the second can separate the colored glass into amber and green.

Of course, the most economical means of achieving this separation of glass from metal, and of different colors of glass, is in individual homes, restaurants, and other places where glass and metal containers are used. This is being done in many communities now where recycling centers exist. There exists a crying need to get legislation passed that will ban nonrecycled bottles and cans, as has been done in Oregon and Michigan. Industrial and labor people say this will eliminate jobs; the answer to this is that it will create new jobs in recycling centers and in companies that reuse glass and metal from such centers. Anyone can see what an unsightly mess exists along our roadsides and highways from throw-away bottles and cans.

Broken glass can either be recycled into the manufacture of more glass bottles or, alternatively, it can be used as a recycled glass slurry seal. This consists of crushed glass. It is used in place of sand to coat heavily

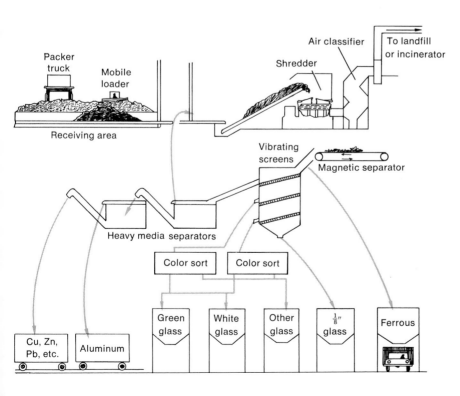

FIGURE 20-10.
System for separating a mixture of trash composed of ferrous and nonferrous metals and glass. This recycling system was planned by the National Center for Resource Recovery. [From Down to Earth, 1(4), *1973. Published by the Jos. Schlitz Brewing Company; reproduced by permission.]*

traveled streets (Anonymous, 1972). In February 1972, a 440-foot length of heavily traveled street in New Orleans, Louisiana, received a coating of glass slurry seal, using crushed glass from bottles instead of sand, as aggregate. The emulsion used was a "quick-set" anionic type. The ground glass was blended with a lightweight aggregate. Seventy percent of the volume was represented by glass. Such a road surface thus saves sand (and decreases large-scale sand mining operations) and is equally effective as a siliceous surface (both bottles and sand are made up of silicates).

Aluminum cans, widely used for beverages in the United States (Figure 20-11), are easily recycled. Aluminum Co. of America has devised very efficient ways of recycling aluminum cans (Figure 20-12). Thus, picking up such cans along roadsides, and highways, and taking them to recycling centers for reuse, makes eminently good sense. These centers exist in many municipalities, at Alcoa, Coors Brewing Co. (Golden, Colorado), and other brewing companies. If each of us helped in this effort, we could begin to cope with the unsightly littering of roadsides with aluminum beverage cans and get paid for it at the same time. Environmental groups, Scout, and other groups, as well as individuals, could all help to recycle aluminum cans (and bottles) to clean up a national environmental disgrace.

In 1973, at least 200 million tires were discarded in the United States (Whitaker, 1974). How can all these tires be recycled? At Trenton, New Jersey, Hydrocarbon Research, Inc., is grinding up and hydrogenating used tire casings for payloads of oil and carbon black. They were also found to be suitable for reuse in the manufacture of synthetic rubber. It was calculated that payout on the plant would, with a "destructive distillation" process, take only 4 years. Firestone Corporation is able to

FIGURE 20-11.
The geography of aluminum beverage cans in the United States. (Courtesy of the Aluminum Co. of America; published in Alcoa pamphlet, "Aluminum Can Recycling: Desirable and Profitable.")

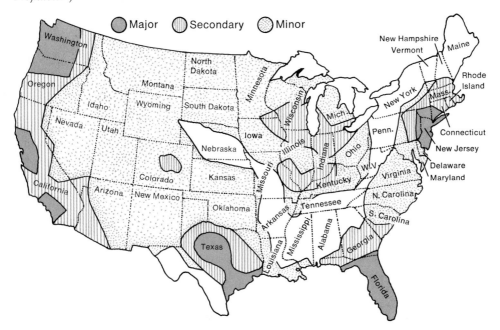

recover every last ounce of used tire and to upgrade the value of the recovered materials. Goodyear Corporation burns old tires in a pyrolysis plant (3000 casings per day) to provide steam for the manufacture of new tires. One other use of tire casings, proposed by the Environmental Protection Agency, is to use them to create artificial reefs in marine habitats. A rubber reef is built by binding old tires into 12 unit modules, drilling holes in them to permit air to escape, and dropping them into shallow tidal water. They soon become encrusted with barnacles and other organisms that attract larger species of fish, which in turn provides fishing. Off the coast of Florida's San Marcos Island, a rubber tire reef 1.6 kilometers (1 mile) long has attracted 33 species of warmwater fish! Finally, used tires can be shredded, ground, minced, and perhaps

FIGURE 20-12.
The environmental "closed loop," "Yes We Can" aluminum can reclamation and recycling. (Courtesy of the Aluminum Co. of America.)

cooked, and then made into whatever is most feasible economically. Some shredded tires are mixed with rubber-base stock, colored, patterned, and used for playground surfacing.

Summary

Recycling has been practiced by nature since life began on our planet. Witness the carbon cycle, the nitrogen cycle, the food chain, and the flow of energy through the ecosystem. People can learn from nature about the business of recycling. This will help us to live *with* nature rather than destroy it. We have tried to provide some practical examples of ways in which recycling can be practiced. The sooner we get into this life-style of recycling our products and natural resources, the faster we shall reach the goal of living within the bounds of what nature has provided for us. Recycling requires ingenuity and can be very fulfilling. We hope you have gotten the "recycling bug" from reading this chapter.

Selected References

Anonymous. 1972. Recycled glass slurry seal. *The American City.* July, p. 30.

Fry, J. L. 1974. *Practical Building of Methane Power Plants for Rural Energy Independence.* Standard Printing, Stanford, Calif.

Hodges, L. 1973. *Environmental Pollution.* Holt, Rinehart and Winston, New York.

Newspaper Recycling—An Economic Analysis. 1973. Based on a study done by the Monmouth Ecology Center, Asbury Park, N.J.

Recycling Plastics. 1973. The Society of the Plastics Industry, Inc., New York.

Rodale, J. W. 1960. *The Complete Book of Composting.* Rodale Press, Inc., Emmaus, Pa.

Singh, R. B. 1973. *Bio-Gas Plant Designs with Specifications.* Gobar Gas Research Station, Ajitmal Etawah (U.P.), India.

Singh, R. B. 1974. *Bio-Gas Plant Generating Methane from Organic Wastes.* Gobar Gas Research Station, Ajitmal Etawah (U.P.), India.

Whitaker, H. R. 1974. New uses for old tires. *Sci. Digest,* October, pp. 67–68.

Tapping Our Energy Resources: Alternative Energy Sources
Dale M. Grimes

21

Questions for Consideration

1. What is the cost of output of electrical energy versus cost of solar cells in the southwestern United States if the interest rate on the initial investment is 10 percent? Repeat for the northeastern states.
2. What is the potential energy in a stream of water as a function of its head and quantity of flow? Assuming 90 percent conversion efficiency, contrast the energy output per kilogram of flowing water for heads of 200, 50, and 3 meters.
3. What are the ecological effects of the following?
   a. A plume of cool nutrient-laden water near an oceanic thermal gradient plant.
   b. Covering 60 percent of a desert region with lattice-mounted solar cells, thereby shielding the desert floor.
   c. Forcing high-pressure water down a dry well, then extracting heat to run a geothermal plant.
   d. Damming an estuary to obtain energy from tidal surface height variations.
4. What is mariculture, and why is it of major significance in energy conversion?

Water Energy Conversion Systems

Until the nineteenth century, available power from flowing streams was limited to geographic sites along riverbanks, where direct mechanical linkage from a turning waterwheel could be used to drive factory belts or feed mills. Then, discovery of efficient means of producing, distributing, and utilizing electrical power removed the riverbank-siting restriction and inexpensive power could be transferred anywhere within range of an electrical power grid. One large hydroelectric plant could service a large consuming area. And with the added flexibility of location came a sudden and great increase in demand for what the world had not known before: inexpensive, readily available power.

Only a few years later, wherever people lived and worked, whether the land was flat, hilly, or mountainous, they were damming rivers and using hydroelectric generators. Probably the most striking engineering achievement of the first half of the twentieth century was the unprecedented increase in hydroelectric power. In a very real sense it is the availability of inexpensive and plentiful power that distinguishes present world culture from all previous cultures. And initially, much of that power was hydropower.

Turbogenerators can be very efficient machines. Modern ones in the Niagara Falls generating complex convert as much as 93 percent of the water's energy into electricity. Hydroplants commonly have efficiencies in excess of 90 percent. Hydroelectric generators come in all sizes and utilize a dramatic range in heads, or change in water elevation. Grand Coulee Dam on the Columbia River in the state of Washington has a present generating capacity of 2.1 gigawatts, while others range down to a few kilowatts. Lac Fully in Switzerland has an operating head of 1650 meters, while others range down to only a few meters.

In this chapter, all units used are those recommended by the International System of Units (SI), as approved and updated in 1971, and as recommended by both the American and British governments. A complete listing will be found in *NBS Special Publication 330, 1972* ed., published by the U.S. Department of Commerce.

Temperatures are listed in kelvins (K), or in degrees Celsius. Other symbols are c, centi, 10^{-2}; μ, micro, 10^{-6}; n, nano, 10^{-9}; k, kilo, 10^3; M, mega, 10^6; G, giga, 10^9; terra, 10^{12}; W, watt; A, ampere; J, joule; m, meter; s, second; and h, hour.

Sixteen percent of the electric power consumed within the United States in 1971 was hydropower. In mountainous regions, such as the western United States, Canada, and the Scandinavian countries, the percentage is much higher. In 1972, hydroelectric power accounted for 99.75 percent of all electrical power produced in Norway and 75 percent in Canada. Montana, a relatively dry state, obtained 68 percent of its power from hydroplants.

Hydropower is a steady and predictable source of power. In contrast with burning fossil fuels, it adds no aerosols to the atmosphere, which block ultraviolet radiation, and later fall as acidic rains. It contributes nothing to atmospheric dust loads that alter local climates. It requires no supportive strip mining. It creates no radioactive waste to be "permanently" stored. There is no possibility of nuclear poisoning, nuclear explosions, or of nuclear blackmail. Hydro dams often supply irrigation water to local farming areas, and in many places act as a means of flood control.

Still, there are problems. The principal one is a lack of large, appropriate, conveniently located sites left in politically desirable and stable areas of the world to provide power for all of man's desires — such sites have all been used. Prime remaining areas are separated from population centers by mountains, oceans, forests, or other natural barriers. Dams also interfere with the natural river-basin biosphere. Nor can dams last indefinitely. Silt and sediment slowly form behind them, filling the artificial lake and interfering with desired operation.

Artificial lakes sometimes lose much of their water permanently by undesired means from the river system. Evaporation losses are often important and, in hot arid regions, severe. Hydraulic pressure on previously dry riverbanks sometimes creates water seepage into underground water systems, with a permanent loss from useful applications. In other areas, seepage caused by increased hydraulic pressure has lubricated long-standing but unstable dirt or rock formations, causing mudslides and earthquakes — quakes dangerous to the area and to the dam itself.

Every few years the Federal Power Commission issues a report on the status of hydroelectric power in the United States. The 1972 report indicated that about 30 percent of the potential sites had been developed. Their ultimate figure, 178.6 GW, included all potential sites not located in national parks, national wilderness areas, or national scenic areas. It did not, however, take into account economic factors, including distance from generating to potential consumption sites; environmental issues; the wisdom of damming limited water supplies in hot, arid regions; and so on. As of the date of writing, the United States was consuming electrical power at about twice the rate that could be generated were all such potential hydro sites developed.

Hydropower Plants

The power that can be developed from a flowing stream increases directly with head, which is the usable difference in elevation between the water on the input and output sides, and the quantity of water flow. In most instances a workable head requires a dam. For a given power output, the higher the head, the less the water that must be handled, the smaller can be the pipelines and the turbine, and the smaller can be the amount of water stored during the wet season, or the season of melting snow, to permit continuous operation during the dry season. But, although high heads are desirable, they are not essential, and since most of the world's population does not work and live near mountainous regions, most of the world's hydroelectricity comes from low- and medium-head units.

In low-head units, with heads of around 10 m, turbines often look similar to a ship's propeller and are mounted on a horizontal shaft. Such propellers are not particularly subject to damage by water-carried grit. Turbo-generator units are often placed inside a hollowed cave in the riverbank just below the dam. The units are normally coupled with nearby plants, and together they provide for both normal and peak daily loading. The power grid must be designed to incorporate hydro unit output when stream flow is at its lowest seasonal ebb. Such units have a small power output and commonly are scattered throughout a region and are remotely operated and controlled from a central control center.

Medium-size units, with heads of 10 to 50 m are normally located in a dam site similar to but higher than the low-head units. In this case, however, pressure is so great that water must be sent through steel pipes and turbines that are concrete-mounted for stability. Water turbines are quite similar to steam turbines in many respects. High-pressure water enters the turbine all around its rim, so for this reason turbines are often mounted on a vertical shaft. Vanes are adjustable to control output speed, which is typically quite high. Units of 100-kW are common. Efficiencies vary from 70 percent at quarter load to more than 90 percent at full load. Since high-speed turbine blades hold rather close tolerances, they are subject to damage by water-carried grit.

Units with heads in excess of 150 m commonly use a contained waterwheel. The wheel is mounted with a horizontal axis and with cups on its outer edge. High-speed water is ejected through a nozzle at the end of the water supply pipe directly onto a cup supplying torque to the wheel. Efficiencies are in excess of 90 percent, and that efficiency is maintained even under lower loading. They are less vulnerable to water-carried grit. Were the load to suddenly change, and a sudden change in water flow through the pipe to result, the contained momentum of the water would be likely to result in plant damage. Therefore, many high-speed units are fitted with a surge tank to either absorb or supply water as needed to satisfy the requirements of a sudden load change.

Pumped Hydroelectric Storage

An important consideration for any utility company that supplies electrical power to the public is the uneven rate of power consumption. Instantaneous changes occur when larger individual loads are added to or subtracted from the power grid, and even more important are the predictable cyclic changes, which vary daily, weekly, and seasonally. Ratios of daily minimum to maximum electrical power consumption in major metropolitan areas are typically about 0.6. Maximum usage occurs in the late afternoon, with a secondary peak in the morning, while minimum usage occurs in the early hours of the morning. On a weekly basis, consumption is largest in midweek through Friday, with lower values on the weekend. A northern hemisphere seasonal maximum power consumption occurs around New Year's Day, and a second maximum occurs during the peak air-conditioning month of July.

Consumption variations require the utility company to be capable of supplying the maximum demanded power. Yet average sales are much less. Instantaneous load variations require rapid and sometimes abrupt switching on and off of electrical generators. Steam turbines cannot be switched on and off rapidly. Their operation requires a load fairly fixed, or at least slowly varying, and ranging within narrow limits.

Hydropower, on the other hand, is relatively easy to turn off and on and has less severe consequences. If a flowing stream of dammed water can be held for future use, fuel is not wasted. For these reasons hydro and fossil or hydro and nuclear plants are often joined together to service a given power grid. In recent years the use of hydropower as ballast for utility loading has gone a step further; during low load periods, the hydrogenerators are reversed and run as motors to force water back up to the higher elevation from whence it came, so that it can later be extracted as needed. In other places, totally new artificial lakes have been created and filled for this purpose. As of January 1972, the United States had a total capacity of 3.9 GW in pumped storage systems, and an additional 6 GW under construction.[1] In the United States, pumped storage capacity is being increased at a faster rate than hydroelectric generation. The sudden increase is due to an awareness of the potential role of controlled hydropower in helping to meet cyclic load variations, with the high reliability of hydroelectric generators, and the relatively large number of potential water storage sites available around the country—far more than

[1] Notable pumped storage systems are TVA's Racoon Mountain project and the Ludington, Michigan, unit with reversible capacities at 382.5 MW and 1.65 GW, respectively.

there are hydro generation sites. Pumped storage promises to play an increasingly important role in power grids during the coming decades.

Tidal Energy

The principal tidal force is the gravitational attraction of the moon, about 2.25 times stronger than that originating with the sun.

The moon travels around a spot on the earth each 24.8 hours, causing, as it goes, high and low tides twice each day. Periods of rising water levels are called flood tides. Dropping water levels are called ebb tides. The moon's orbit is elliptical and tidal extremes vary with lunar distance. Maximum tidal changes occur when lunar and solar positions are best aligned near the time of the equinoxes in September and March. Smallest tidal changes occur near the solstices in June and December.

Tidal height on specific coasts depends upon tidal forces and the land and sea interface. As the tidal bulge sweeps through the ocean toward the land, tidal effects are minimal if the approach to land is through deep water. If, on the other hand, a basin of just the right depth lies between land and sea, water momentum builds up within the basin and tidal variation is amplified. Restricted bays and estuaries can have particularly high tides.

Historical records of Western European devices to harness tidal energy go back at least to the eleventh century. Tidal mills were common along the northern coasts of Europe and North America in the early twentieth century. The advent of inexpensive, reliable electrical power in the early twentieth century rendered nearly all tidal units incapable of economic competition, and they were soon discarded. In the mid-1970s the only large operational tidal electrical power plant was a 320-MW unit located at the mouth of the Rance River in northern France. There, maximum tidal variation is 13.5 m. Twenty-four generating units are built into a retaining wall across the estuary. The units have adjustable blades that permit electrical power generation by water flowing in either direction. The average annual output is 1970 TJ.

Tidal power plants are perhaps more closely related to pumped storage than to hydro generators, since most must effectively and efficiently handle water flowing in two directions.

Since power availability is determined by the position and phase of the moon rather than desires of human beings, some means of energy storage is essential for a useful output.

Tidal generators must be built to withstand saltwater corrosion. The Rance unit is constructed of high-quality concrete, stainless steel, and an aluminum bronze. Construction took from 1959 to 1967, and all systems have proved to be capable of resisting their environments. Generally, in cold regions, ice must be kept out of the turbine blades, while in warm waters, fouling by marine life presents problems. Builders must consider these timing and environmental factors when choosing a construction site.

Although there is a large number of likely locations for tidal generators around the shorelines of the world, the total available power

Tapping Our Energy Resources: Alternative Energy Sources   423

generating capability is small. Therefore, although tidal power could become an important source in scattered localities, tidal power is not destined to play an important role in total world energy economy.

Sea Thermal Gradient Systems

The oceans form continuing current flows: flows that travel in all directions, north–south, east–west, and up–down. Surface currents are particularly large where they pass through restricted channels. The northbound Gulf Stream between Miami and Bimini reaches speeds of 8 m/s, although it has an average speed of about 2 m/s, and carries more than 30 million m³ of water per second—1000 times the flow of the Mississippi and more than 50 times the total flow of all freshwater rivers in the world.

One direct means of harnessing energy from oceanic circulatory patterns would be to anchor propellers in the stream and use them to turn electric generators, thereby utilizing the kinetic energy of the stream in a way similar to hydropower. But the problems are overwhelming. The stream does not stay in a fixed narrow channel but, instead, meanders back and forth over several kilometers. Average stream velocity is small, and a pressure head is nearly nonexistent. Anchoring a screw would be difficult. In common with other sea installations, undue interference with shipping must be avoided; all parts are subject to marine corrosion and biological fouling. And energy densities in the stream reach a maximum of about 2 kW/m², so a very large energy-conversion system would be necessary to obtain important amounts of energy. We conclude that this is not a fruitful direction at our present technological level.

Turning to another means of converting energy from the oceans, we find that there is as much as 22°C temperature difference between waters on the surface of a tropical ocean and those at a depth of 500 m. This difference gives rise to the possibility of obtaining useful energy from an ocean thermal energy-conversion system. Estimates of the total power that might be extracted from the tropical oceans without damaging local ecologies run very high—into the thousands of gigawatts. Indeed, utilization of the seas as a giant receptor of solar energy from which to supply power to the world seems hopeful.

In the late 1920s, a French scientist successfully demonstrated a heat engine that worked on a small temperature difference between two tubs of water during a lecture at the University of Michigan. His later attempt to build a demonstration 40-kW plant near Cuba in 1930 was abandoned after his cold-water pipe was lost during a storm. In 1966, J. H. Anderson, Sr., and J. H. Anderson, Jr., were issued a patent for a system that heated and cooled a secondary working fluid by seawater at appropriate stages in the engine cycle. Working fluids suggested include ammonia, propane, and isobutane. The principal drawback is their need for large heat exchange between the fluids. Once that is accomplished, a rather compact heat engine can be operated. Expected temperature differences limit the thermodynamic efficiency to about 3 percent.

In 1975, E. J. Beck applied for a patent on a device that eliminates the need for a heat engine. He suggested passing warm seawater through a restriction that produces bubbles and thereby turns incoming warm seawater into foam. The volume of the exiting foam is nearly all vapor although its mass remains nearly all as liquid. The expanding volume is used to force the foam up a vertical column (Figure 21-1). Maximum column height obtainable when working between 25°C and 5°C is about 296 m. Recommended height for maximum power extraction is 197 m. Above the top of the column, the vapor is released and condenses on the cold-water surface, while the liquid is channeled to a separate vertical column where it descends, rotating a hydraulic turbine as it returns to sea level.

Beck suggests that seaborne operation might proceed with a smaller vertical height and use some of the pressure generated by the foam to assist in rotating the turbine. He suggested Guam as a likely land site for a trial unit.

Generally speaking, the temperature difference decreases rapidly with increasing latitude greater than 10°. Also, higher latitudes are subject to

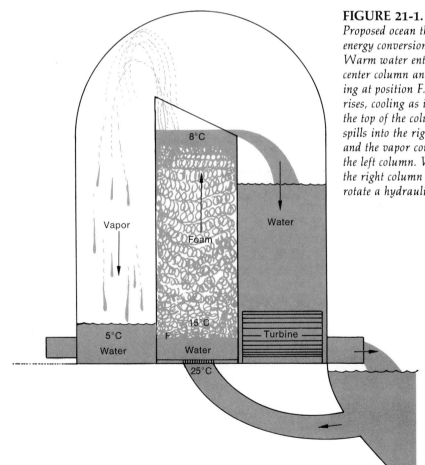

FIGURE 21-1.
Proposed ocean thermal energy conversion system. Warm water enters the center column and is foaming at position F. The foam rises, cooling as it goes, to the top of the column. Water spills into the right column and the vapor condenses in the left column. Water in the right column is used to rotate a hydraulic turbine.

occasional severe storms, variously named hurricanes, typhoons, or cyclones. Design requirements to operate on smaller temperature differences and to withstand such storms make operation at latitudes higher than 10° less promising. Other system problems are avoidance of shipping lanes, marine corrosion, biological fouling, and assuring that waters once used depart the generating area.

Mariculture and Energy Conversion

Ocean thermal energy conversion is particularly convenient when useful sites can be found near tropical shores. Such locations involve no important problems in anchoring the sea plant. Such power plants could, as a by-product, form the stable base for a mariculture industry as seafood farms. Mariculture would use the cool nutrient-laden water effluent from a power plant to support a contained chain of marine life.

In general, photosynthesis in the tropical oceans is limited by the near depletion of necessary ingredients from oceanic surfaces. These nutrients have been removed by earlier biological processes and, therefore, in spite of the abundance of sunshine, nothing grows or lives there, and vast oceanic areas are biologically barren. Deeper waters, on the other hand, are commonly blessed with the necessary minerals for life, which have settled from above, but lack the necessary sunshine to sustain a life chain.

Wherever cold, nutrient-laden waters enter a natural upwelling, the combination of nutrients and abundant sunshine supports major concentrations of marine life. Higher life forms feed on lower ones in a food chain based upon the photosynthesis of simple plants (see Chapter 2). For example, just one major upwelling off Peru supplies 20 percent of the world's fish harvest. Equally as important for a controlled environment is that predators and parasites of shellfish, man-made pollutants, organisms producing diseases in man, fouling organisms—all systems that have played havoc with commercial seafood businesses—are not present on a significant scale in deep waters.

According to Othmer and Roels, an experimental biological station on the north shore of St. Croix has been successful in the use of cool deep water in mariculture. Experiments in tanks and isolated pools have shown that best results are to be expected by passing the power plant effluent through a series of four holding ponds as shown in Figure 21-2. The first pond is to occupy half the total area, and is to be used to grow diatoms. After seeding, the waters develop about 1 million diatoms per cubic centimeter. The diatom-laden water is next fed into the shellfish pond. Shellfish, including clams and scallops, have been found under these controlled conditions to grow at rates well in excess of those in their native locations. A panel of seafood experts found them to be superior in taste and quality to those harvested from their native sites. Shellfish filter the diatoms from seawater and produce meat at conversion efficiencies ranging upward from 60 percent.

After periodic harvesting, the culls are to be passed along for food in the next step, the crustacean pond. Native Virgin Island spring lobster and the New England lobster have both been raised with marked success. Each has shown high growth rates under these controlled

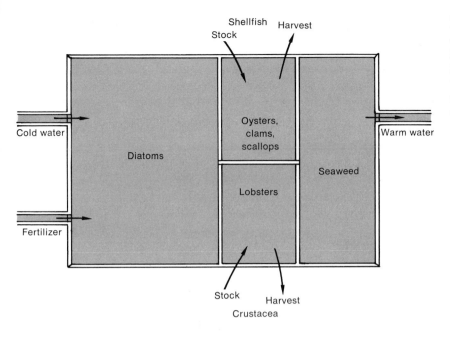

FIGURE 21-2.
Schematic diagram showing proportional areas of sea ponds to be utilized for various functions in the symbiotic production of seafood.

conditions. Were the effluent from the crustacean pond to be returned directly to the sea following lobster harvest, there would be danger that hostile microorganisms would develop and contaminate incoming deep waters. Therefore, a final stage is added, which is devoted to the growth of useful seaweed, based upon the effluent of the meat-producing ponds. The output would be a marketable product and clean, warm water. Since shallow pond water is warmer than oceanic surface water, the final effluent would probably supply warm water to the power plant.

The most promising species of marine life are still being chosen. By using these species, and by adding artificial nutrients as needed while controlling the depth of the holding ponds, it is expected that such mariculture will produce average proceeds of $12 per square meter ($50,000 per acre) of holding pond.

Ocean-Born Industry

In many ways the oceans are ideal sites for energy-conversion plants. They are out of sight of human beings and occupy no valuable real estate. Seaborne units are, by their nature, movable, and there is a great deal of oceanic area available for use. As of 1975, the watery 70 percent of the earth's surface is essentially uninhabited by things useful to man. These advantages will surely become of even greater importance as population pressures continue to mount. There are also disadvantages. Construction of usable sea sites is expensive. They are subject to motion under heavy sea conditions, and there is no agreement on how any nation can acquire the right to permanently occupy a portion of the open sea. Nonetheless, sea occupancy pressures are large and are growing.

The use of oceans for industrial sites has been pioneered by the oil companies and their offshore drilling rigs. In recent years, the U.S. government has supported a development program at the University of

Hawaii's Oceanic Institute for study of a possible stable floating city. The objective is a stable oceanic base on which could be placed airports, industrial or waste-processing sites, resort complexes, energy-conversion complexes—in short, whatever might be useful additions to land-short urban areas. The University of Hawaii program envisages a 300-m-diameter floating city consisting of 10 connected, wedge-shaped sections, each supported on three flotation units. A total deck area of 18,000 m² is planned in four stories. Steel-reinforced concrete is the most attractive construction material.

Such platforms could support a series of sea thermal conversion power stations or solar cell varieties (to be mentioned shortly). Such sites could be permanently positioned or slowly moved with the seasons in order to remain in the best geographic location. Synthetic fuel, fresh water, and harvested seafood could be exported from these remote locations.

Electrolyzing Water

All can agree that the most convenient fuel now in common use at fixed power installations is natural gas. Its combustion results in exhaust gases of water vapor and carbon dioxide. There are no appreciable sulfides or other health-impairing components and essentially no particulate matter. Natural gas is obtained from ground wells and is piped throughout much of the Western world directly to consumers through networks of underground pipelines.

Since gas-fired electrical generating stations operate at high temperatures, they are efficient. They are clean enough and safe enough to permit their siting near population centers.

The only problem with natural gas is a simple and elementary one: We are running out of it. Production in the United States peaked in the early 1970s. New and permanent sources of the gas are needed.

Were gaseous hydrogen to replace methane in the pipelines of the world, the only effect on the consumer would be that nozzle sizes in existing gas burners would need to be altered. Changeover from methane to hydrogen would be relatively easy and inexpensive. Indeed, manufactured coal gas, once widely used both in the United States and other countries, contains about 50 percent hydrogen. Its use was routine and without special problems. Final burning of hydrogen supplies only water vapor as an exhaust product if combustion is carried out with oxygen.

Although chemical procedures for extracting hydrogen from steam do exist, a likely means to obtain it from water is by direct electrical hydrolysis.

Distilled water contains some positive and negative ions. When a voltage is impressed across the water, the positive hydrogen ions are converted to gaseous hydrogen, which bubbles up through the water. The negative ions supply gaseous oxygen, which, in turn, bubbles to the surface. The two gases are kept separate. By this means, energy is transformed from its high but unstorable electric form to an easily storable chemical form. This separation device is called a *hydrolyzer.*

Hydrolyzer units have been designed to match solar collectors atop individual houses while operating at efficiency of 60 percent. In another plan a household might receive all its energy as gaseous hydrogen, which would supply directly all heating needs, while other hydrogen would pass into the 60 percent efficient hydrolyzer operating in reverse, where it combines hydrogen with oxygen to produce the household electricity.

Still another alternative is to retain the present fuel distribution system by first combining hydrogen with carbon to produce methane. The necessary carbon might be extracted from the air as carbon dioxide, gathered from ground plants on land sites, from seaweed as part of a mariculture program on ocean sites, or from carbonate deposits.

Solar-Fired Furnaces

One use of solar energy is in a solar furnace. By this means, furnaces are fired in the total absence of exhaust gases, which are found in furnaces fired by fossil-fuel combustion. Solar furnaces are able to produce temperatures much higher than those produced by electrical heating. There are many solar-fired furnaces in operation around the world today. The principal problem with such furnaces is not technical, but is their nondependable nature: They are only functional while the sun shines.

The world's largest functional solar furnace is located in the western portion of the Pyrenees mountains of southern France. An array of wall-sized mirrors are mounted each on a rotatable pedestal that is electrically aimed so that the sun's rays reflect onto a large, fixed mirror. The fixed mirror is on the side of a nine-story building and is 40 m high and 54 m wide. It, in turn, reflects light to a focus within the furnace chamber.

About 70 percent of the visible radiation that strikes the first mirrors arrives at the focal plane. In full sunlight the total usable power is about 1 MW. A maximum average temperature of about 4100 K, 68 percent of the sun's surface temperature, occurs across the central 6 cm of the focused beam; there, special restrictive working conditions must be in force, since no materials are solid at that temperature. Conventional furnace techniques can be used with lower temperatures.

Solar Steam Plants

As the crisis in worldwide energy production caused by petroleum shortages came closer during the late 1960s and early 1970s, several groups of people proposed development work leading toward large-scale conversion of solar to electrical power. One particularly interesting proposal came from the University of Houston in Texas. The proposal calls for the construction of a system in many ways similar to the solar furnace, but which permits electric power to be generated by means of a steam turbine.

The Houston proposal is simple in concept. It calls for the construction of a high tower within a flat land area that receives large average amounts of sunshine. The tower is to be surrounded by mirrors and topped by an absorbing boiler. Each mirror is to be mounted on an electrically driven pedestal to rotate with the motion of the earth so that its reflection is always directed toward the top of the tower. Atop the tower would be a transparent window, circular about the tower axis and sweeping upward in a way designed to allow maximum light entry from the mirrors. At the center would be a cylindrical boiler whose exterior is blackened with tungsten. The boiler would contain a working fluid, probably water. Hot steam would pass through insulated pipes to ground level, where it would rotate a turbine, which, in turn, would drive an electrical generator. Cooled water would be pumped back to the boiler for reheating.

About 87 percent of the light incident upon the window is expected to be transmitted through to the boiler and, of that, 95 percent to be absorbed by the tungsten-coated boiler surface. The boiler, therefore, is expected to absorb about 30 percent of the incoming solar power to strike the collector area, about 760 MW. If an overall turbogenerator conversion efficiency of 33 percent is assumed, the electrical power output is to be 250 MW.

Another method of producing electricity from sunlight and a steam turbine has been proposed by the Meinels, a husband-and-wife team of optic specialists. They proposed a series of collectors for acquiring the solar energy, but without a central tower. The Meinels' design calls for a large thermal storage system in which heat from the collectors is stored for use as needed.

Their proposal includes large-scale use of "solar farms" to capture sunlight. Consider an area of land in the southwestern American desert, and let us use the Meinels' value of 30 percent conversion efficiency. The average solar power input is about 260 W/m². By using the proposed land area of 120 km on each side, the average solar power input is 3.7 TW. If it is assumed that 60 percent of the land will be covered with solar collectors operating at 30 percent efficiency, the average power output will be 670 GW. In this case, desert land unsuitable for conventional farming would be particularly attractive.

As a further example, consider the Four Corners' power plant. It was placed in the sparsely populated country near the juncture of Colorado, Utah, Arizona, and New Mexico, near coal fields, and where viable pollutants from it would not land in highly populated areas. Generated electricity is shipped long distances to points of consumption. This power plant has the questionable distinction of being the only man-made polluter large enough to be observed by astronauts' unaided eyes. Coal to fire the plant is obtained by strip mining. The Meinels report that there is more land currently under lease for strip mining around the plant than would be required for an equivalent operation by solar energy in perpetuity and, with solar collectors, the land could also be used for sheep and cattle grazing. The Meinels also point out that the 1200 km² of land currently leased by the U.S. government for strip mining of coal could be used in conjunction with their approach to produce most of the electrical power expected to be consumed in this country in the year 2000.

The Meinels propose to locate their plants close enough to an oceanic water supply so the extraction of the necessary waste heat could be profitably used to desalinate water. In contrast with the Houston proposal, the Meinels system would be easily adaptable to seaborne operation. Large oceanic areas lie relatively unused and unproductive. Systems of the Meinels design might be located on permanently seaborne rafts and the output energy used to synthesize fuel, to be shipped back for consumption.

Of the various possible continuing energy sources, the most promising from the combined viewpoints of cost, widespread applicability, and absence of pollutants is direct solar-to-electric energy conversion by means of silicon solar cells.

Silicon Solar Cells

Silicon solar cells were first developed in 1954. The cells are electronic devices that, when illuminated, change incoming light energy directly into outgoing electrical energy. Silicon solar cells have had to face severe competition that until now has been overwhelming. For until now, there have been available many reliable, inexpensive ways of producing electrical power.

Early cell applications were relegated to powering remote telephone stations, and cell development was continued, in the main, only within the Bell Telephone Laboratories.

In the meantime, the U.S. government, along with others, was putting billions of dollars into the energy source which popular wisdom said would one day replace our fossil fuels with electrical power dirt cheap: nuclear power from uranium. Thirty years after the start of large-scale and sustained nuclear development programs, the United States today obtains nearly as much energy from burning wood as from controlled uranium fission!

Solar cells found early widespread applications in light-sensing and monitoring devices, but their use as power converters awaited the space program. In space, solar cells had important advantages they did not have on earth; the sun never clouded over and seldom set. Power was continuously available from a lightweight, easily deployable package, and purchase cost was not an overriding criterion. After some preliminary tests, all space vehicles were powered by them on long missions. The step from solar-cell-powered spacecraft to proposed solar cell power on earth was taken in 1970.

A principal problem with solar cells has been their cost. In 1972, solar cell production was still small, only about 400 m²/year, and nearly all of it was used in space vehicles. On a terrestrial site, with an output efficiency of 15 percent, maximum power output is about 150 W/m². Units available for commercial application, in early 1973, cost about $6000 per square meter. There is, however, every reason to expect costs to tumble rapidly were production to increase. Using an ultimate cost estimate based upon the known corresponding cost reduction of silicon transistors, the MITRE Corporation, under contract with the National Science Foundation, estimated an ultimate cost of about $15 per square meter.

Silicon forms an efficient base for solar cells, and because it is the second most available element in the earth's crust, it is widely and readily available. Its properties are probably better understood than those of any other metal, thanks to previous extensive study efforts by the electronics industry. Active areas of research and development are how to decrease the cost of purifying silicon, how to decrease waste, and how to decrease the cost of processing. Tyco Laboratories are working on the development of a less costly means of producing appropriate single crystals. In 1974, Tyco succeeded in making high-quality single-crystal silicon ribbons that were 2.5 cm wide and 240 μm thick in lengths up to 2 m; the length was limited by the height of the ceiling of the room in which the work was done. Use of ribbons avoids the previous necessity for sawing and polishing the product, thereby avoiding capital and labor costs and the wastage of silicon sawdust. The Tyco program's ultimate objective is the simultaneous growth of several ribbons 5 cm wide, 125 μm thick, and up to 50 m long.

The development of the Tyco Laboratories process eliminates the previously very costly way of making silicon solar cells used until 1975.

Silicon solar cells are sensitive to wavelengths from about 400 nm to about 1100 nm, a range containing about 65 percent of incoming solar energy. Maximum response occurs at about 800 nm, a wavelength at which the human eye does not even function. It is this that permits construction of silicon light-sensing devices, such as television cameras operating in light that is to human eyes total darkness.

The most efficient energy-conversion level is reached with cool, well-ventilated silicon cells. Generally speaking, silicon solar cells are more efficient during winter than summer use.

A good operating energy-conversion rate is about 15 percent of incoming sunlight; very good cells may run at a conversion rate up to about 18 percent as of the date of writing.

Although, like other nonfocal devices, panels of solar cells suffer somewhat from surface reflection, forward scattering of light does not affect output. Therefore, they are not particularly susceptible to the influence of surface dust and minor degradation. They have no moving parts; no turbines are required, nor are rotating electrical generators. They need no vacuum or specially designed blanketing surface films. They emit no noxious gases or other waste products. The cells would not be susceptible to damage by humidity; there is nothing to wear out: A high-quality cell has an indefinite lifetime in an earth environment. Typical output values per solar cell are a voltage of 0.6 V and, under full illumination on earth, an output current density of about 270 A/m^2.

Southwestern and northeastern United States receive solar input power on horizontal surfaces averaged over daily and seasonal variations of 260 W/m^2 and 150 W/m^2, respectively. These values are the extremes of areawide averages in the contiguous United States. If cells are 15 percent efficient in energy conversion, the annual electrical energy output from each square meter of covered surface will be 1.2 GJ and 0.7 GJ, respectively, at the two locations. Pedestal-mounted devices are preferable, since they can be electronically controlled and driven to remain always normal to incoming solar rays.

Using pedestal-mounted devices, 64 km^2 of southwestern desert area at latitude 32°, a region 8 km by 8 km, could provide an annual electrical energy output equal to that from the Grand Coulee Dam, the nation's largest hydro source. The United States and the rest of the world have numerous surface areas 8 km on each side that could be set aside for use as solar energy farms and be covered with solar cells. It is estimated that during 1975 the United States consumed energy for all purposes at the average rate of about 2.8 TW, which is equivalent to an annual energy consumption of 88.4 million TJ. This much energy could be obtained from an area 215 by 215 km in the southwestern desert.

One means of obtaining energy without incurring further environmental damage would be to use exposed areas of existing man-made structures or other large areas covered by concrete or asphalt. As an example of what might be done, consider a house fully electrically equipped and air conditioned, though not electrically heated, in the Washington, D.C., area. A typical suburban house has about 160 m² of horizontal roof area and consumes 4300 MJ (1200 kWh) of energy each month during the air-conditioning months of May through September and 2500 MJ (700 kWh) each month from October through April. Total electrical energy consumed is 39 GJ, a value that could be produced from 45 m² of horizontal roof space. If the house were covered with 140 m² of cells, it would produce over three times as much electrical energy as it needs during the year. A slanted roof designed to catch more of the sun's light would increase electric production still more. Since the solar cells only deliver steady amounts of energy when exposed to light, energy storage systems would be installed to carry on during cloudy and dark periods.

There seems to be no reliable estimate of the total land area under roof in the United States, but estimates range from 35,000 to 70,000 km². Were the lower figure to be covered with 15 percent efficient solar cells and exposed to an average input of 250 W/m², it would produce an average power output in excess of 1 TW, roughly one-third of the anticipated total 1975 power consumption for all purposes, and with no balance-of-payments problem. Extending this idea, southeastern Michigan, including Detroit, consumes electrical power at the average annual rate of about 5 GW. It has 800 km of divided expressways about 0.2 km wide. Let us assume that the expressways are covered as is a covered bridge, in an esthetically pleasing manner, and the roofs covered with solar cells. Average annual power output from the expressways alone would be about 5.2 GW—more power than is used even in that highly industrialized area with its adverse solar climate.

About 80 km east of Miami, Florida, on beyond the deep-water cut of the Gulf Stream, begins an area of shallow seas unique among the oceans of the world. The banks extend in a southeasterly direction along an arc from Cape Kennedy, Florida, toward Puerto Rico for a distance of about 1700 km. Most of the area lies under the political jurisdiction of the Bahamas.

Dry land in that area permanently occupies only about 12,000 m², and most of that lies barren and unused. The southernmost islands do not even attract tourists, as their better-equipped northern neighbors do. The islands are, however, abundantly blessed with sunshine.

The island of Great Inagua in the Bahamas exports salt obtained by evaporation of ponds of seawater from some of its 1500-km² land area.

There are no careful records published of annual solar input, but an estimated average normal component is 350 W/m². On that basis, were the island of Great Inagua to be 60 percent covered by solar cells, its annual energy output would be 1,400,000 TJ—the energy equivalent of 16 Grand Coulee dams. Then, too, collectors need not be on land, but could be arrayed on the abundant shallow seas—seas that can support only limited surf. By using them, the Bahamas and their neighboring islands and banks could become one of the world's major energy-exporting areas.

Still another promising site for solar cell arrays is the southwestern corner of the Commonwealth of Puerto Rico. The area is semiarid, relatively sparsely populated, and blessed with abundant sunshine through a low air-mass sky at only 18° latitude. A moderate-size solar plant located there could easily supply all the energy needs of the island, and more for export. Were Puerto Rico's expressways to be covered, they, too, could easily supply the island's needs.

Moving further afield, large low-latitude areas are used little by human cultures because they are too arid and hot. The Sahara region of Africa, most of Arabia, and much of Australia are examples. Although there are important exceptions, commonly the populace of these regions is very poor, since there is little to do there that is of value to others. Were solar farms to become commonplace, such areas could prosper as permanent energy suppliers to a world hungry for their product.

If mankind is to continue to be blessed with large amounts of expendable and inexpensive power, the most flexible, abundant, and least ecologically and environmentally damaging means for its provision now appears to be via conversion of solar power to electrical power through solar cells grouped and arrayed to best fit the contingencies of particular sites and situations.

Solar Satellites

The first proposal for a solar satellite came in 1968. The proposal was for a satellite power station with a 10 GW electrical output. A large solar collecting panel was to cover an area of 97 km², with two solar cell panels each 4 km by 4 km and inexpensive reflectors covering the remaining 65 km². The cells would receive a continuous incoming solar power density of about 4.1 kW/m². The array was to be located in a fixed equator orbital position above the surface of the earth. A cell conversion efficiency of 11.5 percent was assumed. Electrical power was next to be turned into microwave power and radiated to the earth as a microwave beam. A conversion efficiency between direct and microwave power of 86 percent was anticipated. Collecting-beam diameter on the earth was chosen to be 7.5 km, and it was to receive 12.7 GW of microwave power. Power density in the receiving area would average 288 W/m², a value about equal to the average solar power input in the southwestern United States and well below the 1 kW/m² maximum recommended microwave power exposure level for living creatures. Such a beam intensity would leave life within it undamaged. Power from one satellite was to be enough to service a city the size of New York, or, say, the state of Georgia.

An advantage of a satellite power system is the ease of switching the microwave beam between locations of major power consumption, say from New York to Los Angeles. Cities could be added or dropped from a beam in phase with daily power-consumption cycles.

Perhaps someday such satellites will be constructed and become sufficiently numerous to ring the earth in the manner of Saturn's rings. But not in this century—for the cost both in dollars and in necessary energy expenditure would far outweigh the expected return.

Wind Power

Windmills became commonplace throughout the Great Plains of the United States shortly after their early settlement and remain so today. The principal function has been to pump water. Prior to the construction of areawide electrical power grids, propeller-type windmills were often used in rural areas to power small electrical generators which charged batteries that, in turn, supplied electrical power to the house and farm. By the early 1950s such units were commonly available for purchase in sizes up to 10 kW.

Experimental work on large-scale power conversion from wind to electrical power began in France in the 1920s with the construction of a 20-m-diameter two-bladed propeller. Work next developed in Russia, Mongolia, and throughout western Europe. In 1941 the most successful model to date was built. It was a 53-m-diameter two-bladed propeller, which was mounted on top of a mountain known as Granpa's Knob, near Rutland, Vermont. It ran successfully for 4 years before one of the blades failed and the experiment ended. But development work continued. The U.S. government sponsored a low level of research at New York University and Stanford University. Reasons for the ultimate Vermont breakdown were pinpointed, and the Department of the Interior planned to build a larger prototype machine, a plan that fell victim to government neglect during the Korean War. And there, in the main, the matter rested. It rested during nearly two decades of increasing power consumption until a former Naval captain turned professor of civil engineering at the University of Massachusetts, William E. Heronemus, entered the picture.

Heronemus was associated with the Navy's nuclear propulsion program and knew firsthand some of the problems with nuclear power: problems that were not yet accepted by the technical community. He started looking for alternative means of providing power for New England. He proposed a major wind-power system to be placed on offshore banks and in floating positions off the New England coast. The continental shelf extends for hundreds of kilometers east and southeast of Cape Cod to form Nantucket Shoals and George's Bank. These banks are particularly attractive sites for wind generators, sites totally out of sight of land dwellers in regions where winds are large, dependable, and unhindered by local topography. Heronemus has since extended his original plan to cover many other regions of the United States, including Long Island, the upper Pacific coast, eastern Wisconsin, and so on.

In 1972 Heronemus made a specific proposal to supply much of New England's energy needs by wind power before the year 1990, with important contributions beginning as early as 4 years after go-ahead. New England consumed an average electrical power of 8.4 GW during 1968. The proposal calls for an annual average wind-generated power of 38.2 GW. It is to be produced by 83 clusters of offshore wind stations, each of which contains 165 towers, each tower supporting three 60-m-diameter wind blades and driving a 2-MW generator, from which is to come an average of 900 kW. The wind stations are to be located on the continental shelf. Each of the 83 clusters is to be 5.5 km in diameter and is to have its own electrolyzer station for supplying synthetic fuel, in this case, probably hydrogen. The hydrogen is to be either stored in a deep-sea storage facility or piped ashore for consumption as demand warrants. Individual towers would be either floated and anchored or built directly on the continental shelf.

A second mill-tower configuration found to be more suitable for the lower-speed winds found over the continent has 18-m-diameter propellers and top axes that are 215 m above ground level. Each tower has 50 windmill–generator combinations; such units rotate to remain always into the wind. Scattered throughout likely sites from Montana and North Dakota, southward through Oklahoma and Texas, rough calculations show that they could produce all electrical power consumed in those regions with excess for sale to other parts of the country.

The Wisconsin Senate received a detailed systems analysis of a possible wind-generator network for eastern Wisconsin in October 1973, under contract with Heronemus. He recommended a system supplying an annual average output of 7.4 GW by extracting only about 0.25 percent of the available wind energy over the affected area. Some of the generating stations would be floated offshore in Lakes Michigan and Superior, while most would be tower-mounted, straddling highways. Cities located in the tradewind zone, such as San Juan and Honolulu, seem destined to draw much of their power from windmills located on nearby mountains.

Wind Trains

Best wind machine designs for the topography and needs of the Great Plains are quite different from those of New England for which Heronemus first generated his proposal. On the Great Plains, high winds are a commonplace and almost continuous problem. Plant growth is hindered and soil erosion accelerated or created by wind erosion. High towers holding windmills aloft would be visible for long distances and provide little surface relief from the wind.

Wind trains consist of an oval train track about 8 km long and 1.5 km wide loaded with flat cars. Airfoils are mounted vertically on each of the flat cars. The airfoils are to be automatically trimmed by computer control from an external wind monitor to power the train.

The University of Montana has run computer simulations of possible power outputs from such a system, using records of wind data from Great Falls, Montana. They found the overall efficiencies for the tracked vehicle–airfoil concept of power, depending on the system used, to vary from 54.8 to 62.8 percent. The tracked vehicle–airfoil systems simulated seem to show a significant increase in overall efficiency compared with the windmill.

Such units would be scattered over the Great Plains and, as would be the case with the Heronemus plan, produce more electrical power than the area consumes. Side benefits would be reduction of mean downwind windspeed and commensurate reduction of wind-induced plant damage and soil erosion. Visual pollution would be a little more severe than that caused by trains, and such systems would not interfere with low-flying aircraft. Land interior to the tract could be farmed. Inertia of the train will keep power output up during momentary wind lulls. Tying the outputs from many such systems scattered over a large geographic area would produce a network-wide continuing power output throughout most wind conditions. Maximum seasonal output would coincide with the January peak in power consumption.

Geothermal Availability

The potential availability of geothermal energy is vast. However, at the present time, techniques have been developed only for energy utilization in areas that provide natural steam. From them, it appears that a worldwide total of only a few tens of gigawatts of continuous power can be drawn. This power could be important locally but not particularly important on a global scale. However, techniques are being developed to utilize lower-temperature water reservoirs, both for heating of buildings and for power generation using dry rocks directly. If these procedures are perfected, the geothermal energy made available could equal hydroelectric power in magnitude and in worldwide importance.

The first use of hot geothermal waters for electrical power generation dates within the twentieth century. In the Tuscany area of Italy, an experimental steam turbine unit was built by the Larderello Company and successfully tested in 1904. Subsequent development has continued, and the area is now serviced by 160 live steam wells supplying about 400 MW of electrical power.

A similar geothermal steam area exists in Sonoma County, California, in an area called "The Geysers." First attempts to obtain electrical power from the Geysers began in 1956 and culminated in commercial operation of a 12-MW unit beginning in 1960. A power output of 500 MW was reached in 1974, and it was expected to reach 900 MW during 1977. Power generated there now is the least expensive power in the Pacific Gas and Electric Company's power grid. A sustained output capacity of 2 to 3 GW is believed to be possible from the steam wells. Elsewhere, lower-temperature steam beds that contain hot water, as well as steam, are also being used. Such a bed near Wairakei, New Zealand, now generates about 200 MW, 8 percent of that nation's electrical power, and its output is limited only by local demand for power. A plant near Cerro Prieto, Mexico, 30 km south of Mexicali, generates about 150 MW of electrical power, enough for that portion of Baja California.

Exploration for possible geothermal sources has been slowed by the seeming rarity of surface hot water or steam outcroppings. Recently, it has become apparent from exploratory drillings that many more regions of subterranean hot waters exist than are obvious on the surface. Many areas of the western United States are now known to have significant geothermal potential; an immense single bed was found under part of Oregon, and another underlying part of Montana.

Directly usable geothermal areas are divided into four types: *dry steam, wet steam, hot water,* and *hot rock.* Dry steam indicates that only live steam is ejected from the well. It can be piped directly to a steam turbine and requires little additional investment or attention. Rather conventional steam turbines can be used. The Larderello and Sonoma sites are dry steam sites. Wet steam sources are those where both steam and hot water are ejected from the well. Such sites appear to be about 20 times as abundant as dry ones and will ultimately be more important. In the Wairakei plant, the steam is used and the relatively pure water is simply discarded. This process is inefficient, since heat is wasted. Also, the water from most wet wells contains too many dissolved or particulate materials to permit its direct discharge into the environment. In many cases excess hot water could be passed through a flash boiler to produce drinkable water. In a plant near El Tatio, Chile, the plan is to pass the effluent from the flash boiler into solar evaporation points for mineral separation.

A hot, underground sea has been discovered in the Salton Sea trough underlying southern California's Imperial Valley and running south to Mexico's Cerro Prieto. Water pumped from it reaches the surface at about 300°C, with a salinity about seven times that of seawater. It is heated by magma upwelling between the Pacific and North American plates. Although there is no accurate measure of the energy that might be extracted from the basin, educated guesses run from an amount equal to half of the known world oil reserves down to a few hundred gigawatts for a period of 20 years.

Turning to other potential sites, a single hot water reserve under the Texas Gulf Coast was found which covers several hundred square kilometers. It seems certain that similar large thermal reserves underlie the west coast of all the Americas, under Japan, Indonesia, Turkey, the eastern Caribbean rim, under Hawaii, and in short, wherever plates meet or have met in the past 100 million years. Power developments are now being actively pursued in Ethiopia, the Tatum area of Taiwan, two different sites in the Philippines, in central Java in Indonesia, at Moyuta and Zunil in Guatemala, and on the island of Guadeloupe on the eastern Caribbean rim.

On a lower scale of usage, hot water sources show promises for use in heating of buildings. This is now done on a restricted scale in Idaho and is widespread in Iceland. Investigative studies are underway in France and in the USSR.

Hot Rock Systems

Hot rock geothermal systems separate the generation of useful power from the natural occurrence of groundwater. Although drilling to hot rocks could occur anywhere on earth, in practice only those regions where shallow formations of hot, solid rock closely underlie the surface

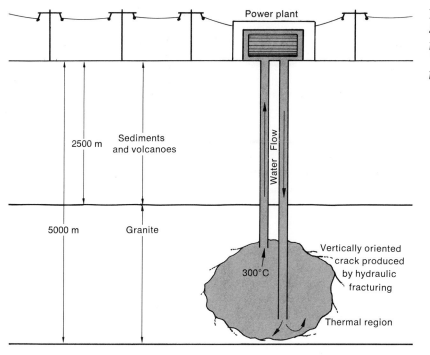

Power plant

2500 m — Sediments and volcanoes

Water Flow

5000 m — Granite

300°C

Vertically oriented crack produced by hydraulic fracturing

Thermal region

are practical for use. Shallow wells are required to keep costs low enough, and solid rock is required to avoid rock lubrication and slippage, which generates local earthquake activity. The system to be used consists of two wells (Figure 21-3). The first is dug into a region deeper and warmer than the expected working temperature to be used. The second terminates just above the region from whence heat is to be extracted. Cool water is then forced down the first well under pressure. The high-pressure water cracks the base rock, and seepage out the first well and up the second one begins. The effluent is kept under pressure while it is passed through a heat exchanger, where its useful heat is extracted before it is pumped back down the first well in a continuing process. As heat is extracted from rocks lying between the two lower well openings, they cool and crack still further. This permits the water to seep through a larger volume of rock and thereby extract heat from an increasing volume of rock. Such cracking must occur for sufficient heat to be extracted to justify the expense of the operation.

Maintenance of high pressure on both sides of the exchanger will keep many dissolved chemicals in solution and thereby keep exchanger damage at a minimum. The output side of the exchanger will use a working fluid; the particular fluid chosen will depend upon the anticipated temperature of operation, but isobutane and ammonia seem to be likely prospects. There are problems in keeping the circulatory system open, avoiding damage to exposed metals by the hot, saline water and proper choice of well size, depth, and location to obtain an acceptably large energy output. The procedure shows much promise, but it needs extensive testing before it can be applied on a widespread and systematic basis.

Summary

Perhaps late in the twenty-first century our civilization will extract its power from continuous natural power flows that surround us. There will be no resulting harmful effluent gases or particulate matter, and our fossil fuels can be saved for use as a partial supply of the necessary feedstock for the chemical industry. Perhaps next century, but not in this one. Although many of the world's governments and certain other organizations are supporting research and development programs with that objective, in all cases only routine procedures are being used, and in no case is sufficient emergency effort being expended so that time to meet this objective will be dramatically decreased.

Nonetheless, in principle, this happy state is possible, and there appears to be no insurmountable barrier, either physical or economic, to its accomplishment. The important natural power flows and the energy conversion means are

1. Hydroelectric.
2. Solar, direct thermal.
3. Solar, solar cell.
4. Wind.
5. Geothermal.

Each of these systems is briefly discussed.

At this time, by far the largest utilized of these sources is hydroelectric. However, effective harnessing of that resource seems to be nearly complete, and it is woefully inadequate to supply all desired needs. Each of the other systems may become important, but the largest single hope lies with solar cells. It is anticipated that subsequent price drops will permit energy conversion, using solar cells on a major scale sometime early in the twenty-first century.

Selected References

ELDRIDGE, F. R. 1973. *Solar Energy Systems.* MITRE Corporation, Washington, D.C.

Federal Power Commission. 1972. *Hydroelectric Power Resources of the United States.* Government Printing Office, Washington, D.C.

HAMMOND, A. L. 1974. Solar energy: Promising new developments. *Science* **184,** 1359–1360.

HERONEMUS, W. E. 1972. Pollution-free energy from offshore winds. Eighth Annual Conference and Exposition, Maine Technology Society, Washington, D.C.

OTHMER, D. G., and O. A. ROELS. 1973. Power, fresh water, and food from cold, deep sea water. *Science* **182,** 121–125.

POWE, R. E., H. W. TOWNES, and D. O. BLACKKETTER. 1973. Development of a large capacity wind powered electrical generating system: A concept. Research Report, Mechanical Engineering Department, Montana State University, Bozeman.

ROUNSEFELL, G. A. 1974. *Handbook of Marine Science*: Section III, *Mariculture,* Vol. I and II. CRC Press, Inc., Cleveland, Ohio.

Applications for Solar Energy and Wind Power
Richard MacMath

22

Questions for Consideration

1. What are the primary advantages of solar energy over other conventional energy sources that depend on fossil fuels?
2. How can solar power, wind generators, and methane generators be combined to provide part or most of the energy needs of a farm? Solar power and wind generators for a home in Michigan versus Arizona?
3. What other techniques can be employed in house and building construction to conserve energy besides storm windows and insulation?
4. How can solar and wind power be used to meet the energy needs of a greenhouse that you have for growing flowers, vegetables, and other crop plants?
5. How do solar panels work to provide energy for heating water or heating one's home?
6. How do wind generators work to provide electrical energy?

At a time when fossil fuel supplies are rapidly becoming scarce and the ecological consequences of our present system of energy production are threatening life on this planet, there remains relatively untapped an energy source that is renewable (not limited in supply) and whose use results in negligible wastes and pollution. By means of a continuous fusion reaction taking place at a safe distance of 93 million miles at no economic or environmental cost, this source produces a clean and freely distributed form of energy. This source, of course, is the sun. We already know that plants have been harvesting light energy from the sun for millions of years for photosynthesis (see Chapter 3 and Figures 22-3 to 22-5).

At present, most all of our necessities (and luxuries) are totally dependent on the production of energy from fossil fuels—oil, coal, and natural gas. Almost 90 percent of the energy utilized in providing for our basic needs of food, heat, light, and electricity is produced from fossil fuels. We continue to demand more and more energy for the life-style we have grown accustomed to. The fact which, until recently, we have failed to recognize is that there is a *limited* supply of fossil fuels on this planet. We are only now reluctantly realizing that the present system of energy production cannot be maintained indefinitely. According to geological survey expert Hubbert (1971) worldwide consumption of fossil fuels is

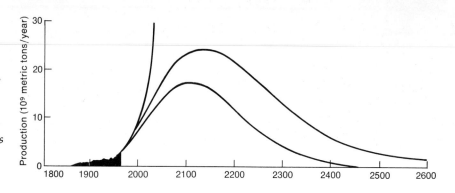

FIGURE 22-1.
Cycle of world coal production plotted on the basis of estimated supplies and rates of production. The top curve reflects Averitt's estimate of 7.6 × 10¹² metric tons as the initial supply of minable coal; the bottom curve reflects an estimate of 4.3 × 10¹² metric tons. The curve that rises to the top of the graph shows the trend if production continues to rise at the present rate of 3.56 percent per year. The amount of coal mined and burned in the century beginning in 1870 is shown by the black area at the left. [From M. K. Hubbert, "Energy Resources of the Earth," Sci. Amer. 225(3), 1971; used with permission of W. H. Freeman and Company, San Francisco.]

now doubling once per decade. Since fossil fuels are essentially fixed in supply, one must question how long such consumption can continue. From estimates of the amount of fossil fuels present in given regions on the basis of geological mapping, the probable length of time that fossil fuels will be available can be calculated. From these studies the cycle of world coal production can be plotted (Figure 22-1). Although known world reserves are large, supplies will be depleted in a few hundred years if present rates of consumption continue.

The cycle of world oil production can also be plotted on the basis of estimates of the amount of oil that will ultimately be produced (Figure 22-2). According to these estimates, 80 percent of all oil produced will be consumed in less than 65 years at present rates of consumption.

Even these figures may be overly optimistic, according to Odum (1973), if the concept of net energy is considered. The true value of energy, says Odum, is the net energy, which is that amount remaining after energy costs of getting and distributing that energy are subtracted. Worldwide inflation is driven in part by the increasing fraction of our fossil fuels that have to be used in extracting and mining, continues Odum, and if the energy reaching society for its general work is less because so much of it has to go immediately into transporting and refining processes, then the real work to society per unit of money

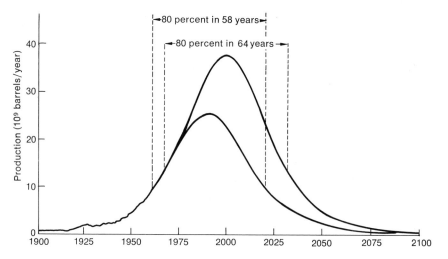

FIGURE 22-2.
Cycle of world oil production plotted on the basis of two estimates of the amount of oil that will ultimately be produced. The top curve reflects Ryman's estimate of 2100 × 10⁹ barrels and the bottom curve represents an estimate of 1350 × 10⁹ barrels. [From M. K. Hubbert, "Energy Resources of the Earth," Sci. Amer. 225 (3), 1971; used with permission of W. H. Freeman and Company, San Francisco.]

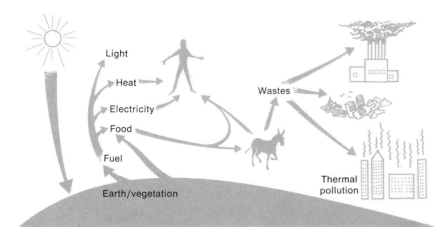

FIGURE 22-3.
Energy flow diagram — present systems. Inherent in our present system of energy production is the problem of wastes. Technological solution to these problems drain our resources and often increase the costs of goods, further spurring inflation. (Courtesy of Richard Mac-Math, Sunstructures, Inc., Ann Arbor, Michigan.)

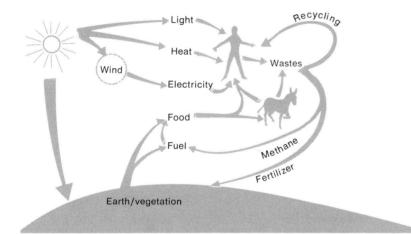

FIGURE 22-4.
Energy flow diagram — alternative system. This system of energy production could depend primarily on nonpolluting, renewable, and freely distributed sources of energy generated by the sun to meet our energy needs. (Courtesy of Richard MacMath, Sunstructures, Inc., Ann Arbor, Michigan.)

circulated is less. Many calculations of energy reserves that are supposed to offer years of supply are as gross energy rather than net energy, and thus may be of much shorter duration than often stated.

The realization that we depend upon fossil fuels for 90 percent of our energy needs and that there is a limited supply of these fuels rapidly being depleted leads us to seek an alternative system of energy production. Inherent in our present system of energy production (Figure 22-3) is the problem of wastes. Technological solutions to these problems drain our resources and often increase the costs of goods, further spurring inflation (e.g., catalytic converters).

Clearly, other energy sources must be tapped to provide for our needs. Solar energy — and its manifestations in wind, water, and plants — remain to be utilized by human beings on a widespread scale. An alternative system of energy production (Figure 22-4) could depend primarily on nonpolluting, renewable, and freely distributed sources of energy generated by the sun to meet our energy needs. This system also regards wastes as a valuable resource and utilizes them for fuel, fertilizer, and solid recycled materials. Such a system of energy utilization is a *positive* approach to our present energy/environmental problems that does exist and can work.

Solar Energy

HISTORY

Considered by many to be an "exotic" and unreliable source of energy, solar energy has actually been put to work by humans for thousands of years. The earliest applications were probably developed to dry fruit and crops, distill water, and heat dwellings. The thick adobe walls of many American Indian shelters, when exposed to sunlight all day, are able to store heat and slowly reradiate it into the living space during the cold nights to keep the space warm until morning. Many mountain and cliff dwellings employed large natural overhangs to block the high summer sun, shading the dwellings and keeping them cool, while exposing them to the low winter sun, warming living spaces during the cold months of the year.

These early applications of solar energy (and more contemporary ones described later) that employ no mechanical equipment or apparatus and utilize easily obtainable, inexpensive, local materials are called *passive systems*. As early as the mid 1800s, other passive systems were developed employing manufactured materials. One such large-scale operation was a solar distillation plant constructed in 1872 in northern Chile. The plant covered 51,000 square feet of land and provided fresh water from salt water for a large number of workers at nearby mines. The plant operated for 40 years, providing up to 6000 gallons of fresh water per day until the mines were abandoned.

Mechanical systems employing solar energy were also being developed at that time. At the 1878 Paris World Exposition, a large solar steam generator was displayed. It consisted of a large parabolic mirror that reflected incoming solar radiation to a focal point which produced temperatures high enough to generate steam. The steam was used to power a pump and other types of mechanical equipment. Similar applications were built in the southwestern United States in the early 1900s, also producing steam to run mechanical equipment.

These examples illustrate that the utilization of solar energy to do work is not new and "exotic." The first oil well in the United States was drilled in 1859 in Pennsylvania. Less than 20 years later, large-scale projects utilizing solar energy were already under way. The use of the sun's energy is not new, but we now have new materials, new knowledge and experience, and a broader understanding of the earth's ecosystem and the worldwide need for a safe, clean, and widely available energy source.

NATURE OF SOLAR ENERGY

Energy from the sun is created by the fusion of hydrogen into helium at temperatures reaching above 1 million degrees Fahrenheit. The earth's atmosphere receives solar energy at the rate of 1.5×10^{18} kilowatt hours

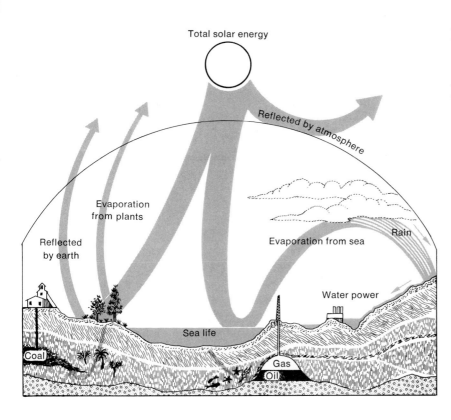

Total solar energy

Reflected by atmosphere

Evaporation
from plants

Rain

Evaporation from sea

Reflected
by earth

Water power

Sea life

Coal

Gas

Oil

FIGURE 22-5.
Diagram depicting how energy, originally derived from the sun, is used or dissipated on earth. About 30 percent of the radiation reaching the atmosphere is immediately reflected back into space. Approximately 50 percent is absorbed by the atmosphere, land, and ocean masses to contribute to the temperature of the environment. About 20 percent is used in the evaporation, convection, and precipitation processes of the hydrologic cycle. A small fraction powers the movement of air (wind) and the circulation of oceans. An even smaller amount is converted into plant energy in the chlorophyll of green leaves; it is this tiny fraction that has produced all the fossil fuels on earth, along with our food and other vegetation. (Courtesy of Richard Mac-Math, Sunstructures, Inc., Ann Arbor, Michigan.)

per year. The total output of all human energy production in the world in 1970 was less than four thousandths of a percent (0.004 percent) of this total. About 30 percent of the radiation reaching the atmosphere is immediately reflected back into space (Figure 22-5). Approximately 50 percent is absorbed by the atmosphere, land, and ocean masses to contribute to the temperature of the environment. About 20 percent is used in the evaporation, convection, and precipitation processes of the hydrologic cycle. A small fraction powers the movement of air (wind) and the circulation of oceans. An even smaller amount is converted into plant energy in the chlorophyll of green leaves. It is this tiny fraction that has produced all the fossil fuels on the earth, along with our food and other vegetation.

Every day, the sun's energy reaching the earth is several thousand times as much as we can possibly use. In 1909, the total electrical energy consumption of the United States could have been supplied by the solar energy incident on 0.14 percent of the U.S. land area, based on the U.S. *average* solar incidence and a 10 percent conversion efficiency. Even the small amount that falls on a house roof is many times the amount of energy coming in from the utility lines. In less than 3 days the solar energy reaching the earth is greater than the estimated total of all known reserves of fossil fuels.

The problem is not that there is too little energy available. The energy reaching us from the sun is actually much more than we need. The problem is how to harness and store this energy for use on demand. There are many different types of "collectors" of solar energy, too numerous to mention here. The reader must consult the references listed at the end of the chapter for a comprehensive survey. This chapter will briefly describe flat plate collectors, focusing collectors, and a few other small-scale applications. Flat plate collectors are usually stationary and can utilize heat from diffuse solar radiation as well as direct, making it possible to collect heat on bright, cloudy days. Focusing collectors must be moved during the day to track the sun remaining normal to the sun's rays for they utilize only direct radiation, but can produce much higher temperatures.

FLAT PLATE COLLECTORS

There are many variations in design and materials of flat plate collectors. All, however, are built on the same principle. Solar radiation is absorbed by a blackened metal surface, usually copper, aluminum, or steel, heating it to temperatures as high as 149°C (300°F) depending on materials used and climatic conditions. This heat is transferred to a fluid, most often water or air, flowing over the metal surface. If water is used, it is usually circulated through tubes welded or molded into the metal plate, or it flows down channels or grooves in the metal surface. In solar air heaters, the air flows over a baffled or corrugated surface. The baffles provide more surface area for the transfer of heat to the moving air than a flat metal surface.

The collector is insulated on the back side (away from the sun) to prevent unwanted heat loss, and is covered on the front side (facing the sun) with one or two panes of glass. The glass traps the absorbed heat, creating a "greenhouse effect," much like a car with its windows rolled up on a hot, sunny day. The glass transmits the short-wave solar radiation but reflects the long-wave radiation (heat) emitted by the black collector surface.

One of the simplest solar collectors presently in operation is the Thomason system, invented by Harry Thomason near Washington, D.C. He uses water as the heat-collecting and storage medium (Figures 22-6 and 22-7). His collector is constructed on his south-facing roof of panels of corrugated aluminum with the channels running vertically down the roof, painted black for efficient heat absorption, and covered by single-pane glass. Water pumped up to the top of the roof absorbs the heat from the collector as it flows down the grooves in the corrugated surface. The water is then collected at the bottom of the roof and then is directed down to a storage tank located in the basement of the house. The solar-heated water first enters a small storage tank, which holds the domestic hot water heater. Here, the solar-heated water "heats" the domestic hot water supply before it passes to a large hot water storage tank. This larger tank is surrounded by coarse fist-sized stones. The heat from the water, which can be stored for up to 4 or 5 days, is transferred to the stones surrounding the tank. When heating is required in the house, air can be circulated around the stones to pick up the stored heat and then this hot air can be distributed by ductwork throughout the house. The cooler water in the large storage tank is pumped back to the top of the roof to begin the cycle over again. As Thomason says, there is

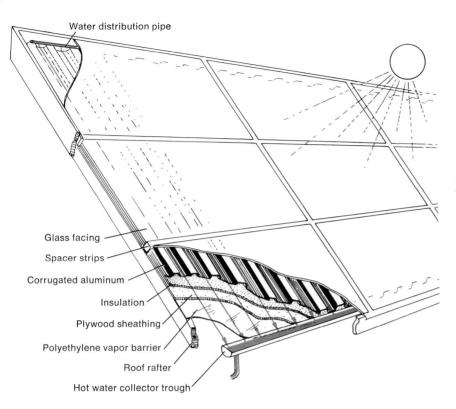

Water distribution pipe

Glass facing
Spacer strips
Corrugated aluminum
Insulation
Plywood sheathing
Polyethylene vapor barrier
Roof rafter
Hot water collector trough

FIGURE 22-6.
Construction details of the solar collector used in the Thomason system, invented by Harry Thomason near Washington, D.C. This system used water as the heat collecting and storage medium. (Courtesy of Richard MacMath, Sun-structures, Inc., Ann Arbor, Michigan.)

nothing less expensive than "sunlight, rain from the heavens, and stones from the fields."

The pitch of the south roof on which the solar collector panels are located influences collector efficiency, and the optimum pitch varies with latitude and application. A good rule of thumb is to have a roof pitch equal to the latitude of the house plus about 10°. In lower Michigan, for example, the pitch would be near 55°. However, a variance of ± 10° will not significantly hamper the solar heating system's overall efficiency. The reason for this pitch is that the collector surface should be at right angles to the winter sun, which is low in the sky (about 30° above the horizon in December in lower Michigan), to receive the maximum solar heat available. However, if the system is to be used primarily for air conditioning, such a steep roof pitch is not necessary because the sun is higher in the sky during the summer months.

FOCUSING COLLECTORS

Most focusing collectors are nearly spherical or cylindrical in shape and usually parabolic or circular in cross section. The focusing may be done with glass lenses on curved mirrors or metal surfaces. Parabolic mirrors reflect the incoming parallel rays of direct solar radiation onto a focal point (or line in the case of cylindrical collectors), achieving very high temperatures. This type of collector can achieve temperatures approaching 3000°C or less at the focal point, depending on the area of the collector.

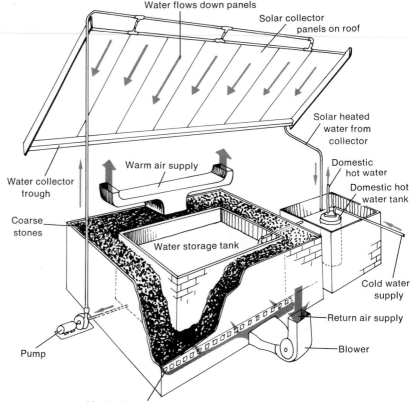

Small focusing collectors [1.2 or 1.5 meters (4 or 5 feet) in diameter] made of polished steel, aluminum, or chrome, or many small mirrors are often used in nonindustrialized countries to boil water and cook foods. Larger collectors are common in many African countries and the Australian continent for industrial purposes. The high temperatures generated usually produce steam to operate steam-powered equipment. Perhaps the largest focusing collector is located in France, where a solar laboratory operates a focusing collector that is 10 stories in height.

OTHER APPLICATIONS

Perhaps the simplest of all systems for converting solar radiation into usable energy has been developed by Steve Baer in New Mexico. Built into the south wall of the Baer house are 207.9-liter (55-gallon) oil drums, each containing about 53 gallons of water. There are 90 drums stacked in racks with the exterior ends painted black and the interior ends painted white. Just outside the stacked drums are double plates of glass. The drum wall can be covered or exposed by means of large, movable insulation panels or doors. These are fabricated from a styrofoam core of insulation and covered with a highly reflective aluminum skin. When the sun rises in the morning, the doors are opened, allowing solar radiation to strike the blackened drums and heat the water inside them. The water, having excellent heat storage capacity, can store the heat for at least a couple of days. When the doors are closed (during cloudy days and at night), the only way for the water to give off its heat is into the house. The closed insulating panels prevent the heat from escaping. It is as simple as that. To prevent too much heat from entering the house, heavy curtains can be drawn on the insides. The process is reversible—the house can be cooled by opening the panels at night and keeping them closed during the day.

One of the most recent developments using the sun's energy has been the production of solar cells. Originally developed for use in the space program (on Skylab, for instance), there are now experiments being carried out for using solar cells to generate electrical power in a typical residence. These proposals range from roof-mounted solar cells for single home use to "solar farms," which would be central power plants that generate electricity with solar cells and distribute it to a large number of homes, businesses, and industries. At present, the generation of electricity by solar cells is prohibitively expensive, and extensive research is needed to decrease this cost. Wind power is a less-expensive means of generating electricity on a residential scale.

HISTORY

Wind Power

As with solar energy, wind power is also considered by many to be an energy source that is "too valuable" and therefore unreliable. It was put to work centuries ago for pumping, irrigation, and other types of mechanical work. Large Dutch windmills and small, multivaned wind pumps found on farms throughout the midwestern United States are a familiar site to everyone. It was not until this century that extensive experiments with wind generators began—converting wind energy to electrical energy by transferring the rotational motion of the blades to a generator or alternator. In 1950, there were an estimated 50,000 wind generators in use in the Midwest alone, but rural electrification programs made them (temporarily) obsolete. At present, there are a number of commercially manufactured wind generators of various sizes (200 to 6000 watts) available for residential use.

On a larger scale, research on supplying electric power from the wind for towns and large urban areas has been going on for at least 50 years in Denmark, Germany, and England. The most massive experimental wind generator to date was built in the United States in the early 1940s in Vermont. The wind generator was 33.5 meters (110 feet) high, and the blade diameter was 53.3 meters (175 feet). The project ended in 1945 when one of the two 8-ton blades sheared off.

More recently, there has been a number of proposals for developing a series of offshore generating plants to provide electrical power for the entire New England area, and a series of plants in the Great Plains to provide electricity in the midwestern United States.

PRINCIPLES

Basically, a wind generator converts wind energy to electrical power by transferring the rotational motion of the blades to an alternator or generator which produces electricty. At present, a three-bladed propeller design (Figure 22-8) is believed to be the most efficient and therefore most often used, although two-bladed models are not uncommon. The entire blade and hub assembly, called the rotor, rotates in a direction perpendicular to the wind. The speed of this rotation varies with the blade design (sails, airfoils, etc.), with the fastest speed obtained by a propeller–airfoil design. The rotor is usually connected to the alternator by means of a gear or belt-drive assembly. Gears are used to increase the relative rotational velocity of the alternator. For example, if the gear ratio is 1:5, the coil of the alternator rotates five times for every rotation of the rotor. A few wind generators are direct drive, meaning that there is no gearing between the rotor and the generator. These designs, however, use specially made low-speed generators which are both very heavy and expensive to build. Most alternators and generators are designed to work only at high rpms, and therefore make the gearing necessary. The alternator generates current by rotating a metallic coil (usually copper) through an electromagnetic field. The induced current is then distributed

FIGURE 22-8.
Construction details of a wind generator. This system employs a three-bladed propeller, which is believed to be the most efficient and, therefore, most often used design. The wind generator converts wind energy to electrical power by transferring the rotational motion of the blades to an alternator or generator which produces electricity. (Courtesy of Richard MacMath, Sunstructures, Inc., Ann Arbor, Michigan.)

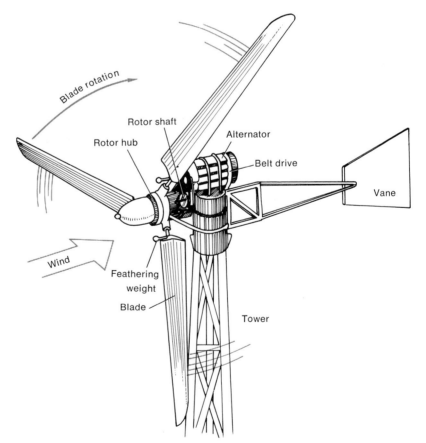

to a control panel, which measures the amount of power and current generated (Figure 22-9). The electrical power is stored in a battery storage bank where 5 days of electrical energy may be stored for use during calm spells. Batteries store electric power in the form of direct current. The power can be used directly in this form or may be converted to alternating current by means of an inverter. Most appliances have both direct- and alternating-current ratings, but a few require just alternating current. A gas-fired alternator is usually included in the system to provide electric power for unusually long periods of calm weather.

It is clear that there are abundant sources of energy all about us in the sun and its manifestations in the wind, water, and oceans. Each of these are free, renewable, nonpollutant, decentralized sources of energy which can be utilized and integrated into a whole, self-sufficient system for survival, dependent only to a small degree on fossil-fuel sources.

Although energy from the sun and wind is plentiful, there are still problems to be overcome in utilizing and storing energy from these sources in some areas of the world. To make solar energy and wind power immediately practical and feasible requires a policy of energy conservation to minimize the energy demands of a building or dwelling. Instead of designing all glass buildings and relying on fossil-fuel-powered equipment to control the interior environment, we must begin

Summary: Energy Conservation

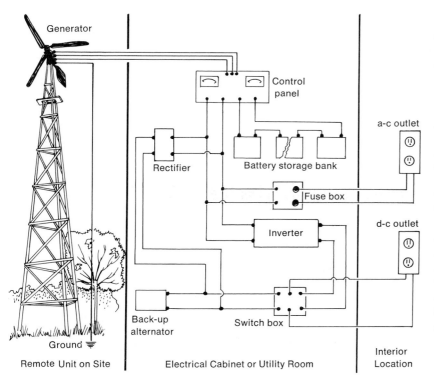

FIGURE 22-9.
Schematic diagram for a wind power–battery storage system. The electrical power is stored in a battery storage bank, where 5 days of electrical energy may be stored for use during calm spells. The batteries store electric power in the form of direct current; the power can be used directly in this form or may be converted to alternating current by means of an inverter. Most appliances have both direct- and alternating-current ratings, but a few require just alternating current. (Courtesy of Richard MacMath, Sunstructures, Inc., Ann Arbor, Michigan.)

to design with "energy-conscious" techniques to reduce our need and consumption of energy. The use of proper materials and design techniques responding to the sun and wind can significantly reduce the energy needs to heat and cool a residence (Figure 22-10). For example, a dwelling in the form of a trapezoid with the long side facing south (toward the sun in the northern hemisphere) and the short side exposed to the cold north, can contribute significantly to the heating of the living space during the winter. Glass and window areas on the long south side will enable sunlight to enter and heat the space during the cold months of the year. Window overhangs of the proper length can block the high summer sun and shade these windows during the warm months of the year. If many of the living spaces are below the frost line, they will remain cool in the summer and warm in the winter; for below frost line, the earth's temperature remains at a constant 55 to 60°C. A sod and earth roof can reduce the heating load of a dwelling during the winter and cooling load during the summer. The sod provides an excellent friction surface for snow to collect during the winter. Snow, an excellent insulating material, can pile up to 4 or 5 inches on the roof, adding insulation to the dwelling during the coldest time of the year. During the summer, dew collects on the sod overnight and begins to evaporate in the morning and early afternoon. Whenever water evaporates, it absorbs heat, and the evaporating dew picks up heat from the living spaces below, keeping them cool during the hottest months of the year.

There are many other "low-energy techniques" to significantly reduce the energy demands required to heat and cool a building. Techniques such as these, which respond to the forces of nature and local climatic conditions rather than trying to conquer, and which use local, natural building materials, are perhaps the most important steps toward an effective energy-conservation approach which makes solar heating systems and wind-powered electric systems feasible and practical. There are no new technological developments necessary to begin designing and

FIGURE 22-10.

Combined wind generator and solar collector system on a residence, utilizing sun and wind to significantly reduce the energy needs to heat and cool a residence. (Courtesy of Richard Mac-Math, Sunstructures, Inc., Ann Arbor, Michigan.)

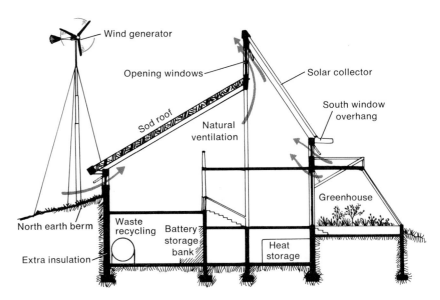

building with these techniques. The only new development would be the integration of all these techniques and alternative energy sources into one living unit. It can be done; all that is needed now is the commitment to do it.

BRANLEY, F. M. 1957. *Solar Energy.* Harper & Row, Publishers, New York.

DANIELS, F. 1964. *Direct Use of the Sun's Energy.* Yale University Press, New Haven, Conn.

Energy Primer. 1974. Portola Institute, Menlo Park, Calif.

GOLDING, E. W. 1955. *The Generation of Electricity by Wind Power.* E. F. Spon Ltd., London.

HALACY, D. S. 1973. *The Coming Age of Solar Energy.* Harper & Row, Publishers, New York.

HUBBERT, M. K. 1971. Energy resources of the earth. *Sci. Amer.* **225**(3), 60–70.

ODUM, H. T. 1973. Energy, ecology and economics. *Ambio* **2**(6), 220–227.

PUTNAM, P. C. 1948. *Power from the Wind.* Van Nostrand Reinhold Company, New York.

Selected References

23

Environmental Education: A View from the Community Nature Center

Charles Nelson, Charles Barnes, and Susan Schick

Questions for Consideration

1. What is environmental education?
2. What is a community nature center?
3. What role does the nature center have in the community?
4. How does a nature center respond to a community's environmental educational needs?

Environmental education has some interesting historical aspects. The term *environmental education* is fairly new, being defined in the very first issue of the *Journal of Environmental Education* (Stapp, 1969) as "communication aimed at producing a citizenry that is knowledgeable concerning our biophysical environment and its associated problems, aware of how to solve these problems, and motivated to work toward their solutions." The concepts of environmental education are not very new, although before the late 1960s, the ideas were dispersed between many disciplines and called many different things. The environmental crisis of the 1960s did spawn public awareness and the sudden readiness of many people to accept and, indeed, to demand a process of education dealing with the environment and their relationship to it (Debell, 1970). To most people, the problems of environmental deterioration seemed so new that it appeared a logical solution should be brought forward. The deterioration of our environment is not new, although it has now reached a critical point, where our daily lives and life styles have been affected. The result of this environmental awakening was a new and eager audience ready to act. All the philosophies and definitions of the new environmental education can be found in the old schools of conservation education, natural history, nature study, and outdoor education. Environmental education now encompasses all these fields, but adds more stress to the objectives of generating favorable attitudes and behaviors toward the environment.

In 1973, a statewide environmental education plan titled, "Michigan's Environmental Future," was developed by a governmental task force appointed by Governor William Milliken. For the first time, a detailed plan defined the parameters of what environmental education would mean in Michigan. The scope of the plan was, indeed, broad and all-encompassing and envisioned environmental education as a new structure of government-appointed councils, boards, and regions that would help implement changes in agriculture, business, industries, citizen organizations, elementary and secondary education, government, higher education, mass communications, professional and trade associations, and religious and youth organizations. These changes would affect the very core of our culture and create within us a more correct image of ourselves and how we relate to our environment. There is a danger here in believing we have identified a discipline that can deal with all aspects of our environment and our relationship to it. That task is too large, the field too scattered, the fronts too numerous. We must realize that environmental education has been diffused over a wide spectrum of many disciplines.

Examples of the frustrations, frequent failures, and difficult successes of single agencies trying to wrestle with the entire spectrum of environmental education are well documented in a report from the Educational Services Department of the National Audubon Society entitled, *A Cooperative Effort Between Audubon Centers and Schools to Define Possible Roles of Nature Centers and Schools in the Integration of Environmental Education into the Total Curriculum* (National Audubon Society, 1974).

The editor of the *Journal of Environmental Education* (Schoenfeld, 1975) noted with some alarm that "environment" was defined at a national conference to now include not only the degradation of physical environment—noise, air, land, and water pollution—but also "starvation; malnutrition; health care; the oppression of women, minorities, and third world peoples; housing; transportation; discrimination; nuclear power; and occupational safety." The editor also states that "We must recognize that we can't do everything at once; painful choices have to be made."

Community Nature Centers

As we have seen, environmental education is a process that involves many agencies working in many different disciplines. Nature centers can be a key part of the environmental education process. It must be stressed, however, that no two nature centers will operate in the same manner. Each center must respond to its social and physical surroundings, the method and extent of its funding, and the philosophies of its governing board.

The first step in developing a successful nature center is a careful analysis of the community and its environmental education needs. If there are no elementary, secondary, or adult programs of environmental education available, the primary purpose of a nature center would be to develop a high-quality environmental educational program for people of all ages. The second purpose would be to work for the protection and designation of an ecologically intact area in which to teach environmental education.

At this early stage in developing policy, it should be realized that there are many areas of environmental education in which it must be decided whether or not to become involved. The field of environmental action is one such area. Environmental action plays an important role in the correction of immediate environmental situations that have become intolerable. In an analysis of the community and its needs, it may be found that environmental action groups are already in existence. The need for an additional action organization may be limited. On the other hand, a large urban area may find sufficient support from environmental action-oriented groups to begin a nature center. It is most important to obtain the maximum support possible, to ensure a solid base from which to operate. An educational program that satisfies basic curiosities about the natural world, cultivates an interest in environmental causes, and, in time, creates a positive attitude in the community will benefit everyone concerned.

Once the goals and purposes have been defined, the methods of achieving them have to be decided. Knowing that the whole spectrum of environmental education is much too broad a front, we have to narrow our views and aim at what can be realistically accomplished within the community.

If a nature center is to have a useful and successful program, it must recognize its community's needs, realize its best resource potential, have a basic understanding of the entire environmental education field, and make a wise choice as to which aspect to focus on. A nature center must also go one step further; it must have staff whose training complements the nature center program. For example, a nature center that exists for the sole purpose of educating young people may want a staff highly trained in educational techniques, methodology, and nature crafts. On the other hand, a nature center taking on the additional responsibility of higher education and interested in an adult program will need a staff that is better trained in the biological fields. The success of any interpretive program will depend, to a great extent, on the background training of the naturalists who will be working directly with the public.

The roles of *interpretive naturalist* and *environmental education specialist* are often confused or appear ambiguous to students entering the field of environmental education. There is a distinct difference. The naturalist interprets specific natural elements and relates them to the whole environment. A naturalist usually works directly with people who are learning about specific aspects of their natural world. An environmental education specialist, however, is a *coordinator* who develops environmental education programs for entire school systems or other educational organizations (U.S. Fish and Wildlife Service, 1975). The specialist does not usually deal directly with the out-of-doors or with the direct teaching of students. Duties of the specialist might include setting up programs between teachers and foresters, soil conservationists, or naturalists. An environmental education specialist is an administrator, whereas a naturalist is a teacher, often used as a resource person. A naturalist may take on the additional responsibility of being an environmental education specialist, but in doing so, he or she will assume two distinct duties.

Any meaningful success of a nature center depends on the ability of the interpretive naturalists. In choosing naturalists, most nature centers that concentrate on the natural history aspects of environmental education prefer to employ wildlife biology, botany, or ecology majors. It is important to look for people with a deep and intense love for the out-of-doors. Prospective employees who canoe, backpack, and have wilderness background training are particularly valuable because of their potential abilities for teaching the educational programs in these areas. As far as formal education is concerned, those students who have had courses in ornithology, dendrology, plant taxonomy, entomology, and wildlife ecology will be better qualified for a position as naturalist than those students with classes only in environmental teaching techniques and resource management.

People with special emphasis in any one of the biological fields are also useful in an adult education program. Starting with a sincere enthusiasm for the natural world, then adding some intensive training and liberal amounts of enthusiasm, it is possible to develop naturalists who are eager to share their knowledge with anyone who will take time to listen.

A good deal of energy should be expended on creating and maintaining "esprit de corps" among the staff of naturalists. Given a well-informed and enthusiastic team of naturalists, the task of environmental education becomes a matter of developing those programs and teaching techniques that would be most effective for the wide range of age groups visiting a nature center.

The Natural History Approach

There is currently much controversy over what techniques and methods should be used while teaching in the out-of-doors. One technique of teaching involves the strict identification and categorization of organisms and natural phenomena (Smith, 1974). On the other end of the spectrum, the sensitivity or "acclimatization" approach deals primarily with concentration on sensory involvement with the natural environment, striving to include inputs to all five senses (Van Matre, 1972). The "natural history approach" includes a blend of both techniques.

ELEMENTARY EDUCATION

Many elementary schools do not have well-developed environmental education programs of their own. There is, however, a growing interest from schools and some support from informed communities to begin such a program. It is difficult, however, for a community nature center to develop an effective environmental education program for the many school districts it often has to serve. A better alternative would be to convince the various school systems that they need an environmental education specialist. Nature centers should, if all else fails, work directly with the schools, but this function would be better accomplished from within, with the nature center helping primarily in a supportive, advisory capacity.

The elementary programming at most nature centers is quite extensive. The lasting attitudes that will be incorporated into future decision making concerning the environment are formed with this age group. Most students on their first visit to a nature center are given a general introductory tour, the main objective being to instill enthusiasm in the students for the out-of-doors while teaching about the natural communities and related natural history facts.

After the initial visit to a center, other types of tours should be available to the students. A pond study tour involves the actual sampling of pond life with nets, jars, hands, and anything else that will hold water. The excitement of the sheer numbers and varieties of life found in ponds keeps children continually interested. Naturalists merely serve to direct the activities and identify "critters."

Many naturalists and environmental educators downplay the role of naming plants and animals; however, names provide a "handle" by which to remember and store information in the mind. Naming objects is a natural process that should not be suppressed. This is most evident in pond study sessions. The first question is: "What is it?" *This curiosity should not go unanswered.* To be sure, more should be taught than names, but naming is an important function in learning about the natural world.

Seasonal tours pointing out features most evident in the out-of-doors for each season can be important in a nature center's yearly programming. Winter offers animal tracks and signs, animal adaptations, food sources, and shelter requirements. Spring brings wildflowers, ducks, geese, and the rebirth of life-giving processes.

The study of trees and shrubs is another specialized tour. Students are taught the ecology of different species of trees, together with basic old-field succession. Identification also plays an important part in this type of tour.

For all these tours, a nature center should provide the classroom with a "pre-tour packet," which consists of approximately 10 to 20 slides and a read-along script. This is the initial link from the classroom to the out-of-doors. Often, this familiarization opens the door to a productive session at the nature center. Fears, questions, and apprehensions are minimized and the seed of curiosity planted.

Besides these suggested tours, a nature center should attempt to fulfill any request illustrating particular topics that a class may be studying. In all ways, then, the nature center becomes a versatile educational tool to the school community.

SPECIALIZED EDUCATION

Work with inner-city high school and grade school children can provide a real test of the methodology and philosophy behind this natural history approach to environmental education. But how do you instill desirable attitudes about the natural world when these students' ideas of the out-of-doors range somewhere between complete indifference to absolute terror?

When working with inner-city students, the type of attitude the naturalist displays toward the out-of-doors is very important. An understanding, enthusiastic approach to the natural world is extremely contagious and becomes the basis for communicating the ideas and feelings these students should have for the out-of-doors. A real sense of discovery is vital. The urban environment is all-inclusive: Everything relates directly to the city. The out-of-doors or natural world is a totally new world, and if enthusiastically presented, can have a far-reaching impact upon the student's attitudes toward life.

SECONDARY EDUCATION

At the secondary level, programming becomes more involved, to satisfy the older student's interests. The enthusiasm of the interpretive naturalist is still important, but it is applied to a more sophisticated program in the "natural history approach" to environmental education. As knowledge of the out-of-doors increases, the interests of the students become more diverse, and the need for an educational experience aimed at these interests becomes apparent. For this reason, a program of wilderness experiences can put students in a close, personal relationship with the natural world. In the wilderness, a student begins to absorb basic biological lessons, and, perhaps equally important, a strong emotional feeling for the wonders of nature. The solitude of the wilderness, the ever-changing mood of a lake or river, and the predawn stillness are all part of forming wilderness experiences and attitudes. Short wilderness trips are truly "excursions into the textbook of the out-of-doors." Here, all the ecological principles can be readily taught. In these wilderness trips, canoeing and backpacking are the modes of transportation. The awareness of the straining weight on their back or the resistence offered by the paddle are just two of the elements that will begin to influence them in their approach to the natural world and help to mold their perspective of nature and their place within it. Thus a program of wilderness experiences should attempt not only to teach the students, but also to stimulate the formation of important attitudes.

ADULT EDUCATION

In general, adult courses introduce interesting aspects of our environment to the public. Classes in nature photography, wildflower identification, birds, and ecology are all "gateways to the natural world." Classes designed to help teachers understand the environment and acquaint them with methods for teaching in the out-of-doors can also be important.

A wilderness expedition can be the ultimate in a teaching/learning experience. Here, we are immersed in the direct experience of the natural environment. Where wilderness survival is taught, so, too, should emphasis be placed on *survival of the wilderness.*

A far-reaching part of any nature center's program should include talks and slide presentations offered to organized groups in the community. These programs should be light and entertaining, with the environmental message being one that invites and entices the audience to sample a bit of the natural world and to discover why there is so much interest in saving it. Adults often must be startled into awareness of the world around them. Evening woodcock watches, owl whistles, and frog listens are all aimed at awakening interest and concern for the natural world. Here, more than ever, naturalists must be informed about their subject, and able to show a genuine enthusiasm and to offer insights that relate the subject matter to human lives.

Summary

It is now realized that the major goal of environmental education should be changing the attitudes our culture has developed toward the environment. *The realization that human beings are an integral part of the living systems of nature and their continued existence depends on their ability to operate within the rules and regulations imposed by our ecosystem should be the main thrust of the environmental education process.*

How, then, do you go about changing attitudes? The logical progression is first through imparting knowledge about the natural world. With this newly acquired information, a growing interest develops in the functions and interrelationships of the natural world. This increasing interest launches the corresponding change of attitudes that we are striving to accomplish. Here, then, is where the natural history approach of environmental education comes to the forefront in providing the essential foundation and impetus to bring about these attitude changes. Teaching natural history fulfills the role of imparting the essential knowledge that starts the total attitude change that we call environmental education.

Natural history has further attributes that make it an ideal medium for this process. Everyone has an innate curiosity about the out-of-doors which makes the teaching of natural history an acceptable form of education that knows no age, cultural, occupational, racial, or philosophical boundaries, and it really works.

Tremendous community interest exists in nature centers and the environmental education programs they provide. There is continual demand for more educational programs that can fulfill the needs of a community on its way to changing its attitudes toward its natural environment.

Literature Cited

DeBell, G. 1970. *Environmental handbook.* Ballantine Books, Inc., New York.

Governor's Environmental Education Task Force. 1973. *Michigan's Environmental Future, A Master Plan for Environmental Education.* State of Michigan, Executive Office of the Governor, Lansing, Mich.

National Audubon Society. 1974. *A Cooperative Effort Between Audubon Centers and Schools to Define Possible Roles of Nature Centers and Schools in the Integration of Environmental Education into the Total Curriculum.* National Audubon Society, New York.

Schoenfeld, C. 1975. Whither EE? *J. Environ. Educ.* 6(3), 67.

Smith, R. L. 1974. *Ecology and Field Biology.* Harper & Row, Publishers, New York, pp. 692–754.

Stapp, W. B. 1969. The concept of environmental education. *J. Environ. Educ.* 1(1), 30–31.

U.S. Fish and Wildlife Service. 1975. *Notes on Coordinating a Community Program in Environmental Education.* Jenny Publishing Company, Inc., Minneapolis.

VAN MATRE, S. 1972. *Acclimatization—A Sensory and Conceptual Approach to Ecological Involvement.* American Camping Association, Bradford Woods, Ind.

Suggested Readings

ALTMAN, H., and C. TROOST. 1972. *Environmental Education: A Source Book.* John Wiley & Sons, Inc., New York.

CARSON, R. 1965. *A Sense of Wonder.* Harper & Row, Publishers, New York.

LEOPOLD, A. 1966. *A Sand County Almanac.* Oxford University Press, New York.

National Science for Youth Foundation. 1972. *Changing Emphasis in Environmental Education* (Proceedings, Conference X). National Science for Youth Foundation, New Canaan, Conn.

PETZOLDT, P. 1974. *The Wilderness Handbook.* W. W. Norton & Company, Inc., New York.

STAPP, W. B. 1967. *Integrating Conservation and Outdoor Education into the Curriculum (K—12).* Burgess Publishing Company, Minneapolis.

TILDEN, F. 1957. *Interpreting Our Heritage.* University of North Carolina Press, Chapel Hill.

U.S. Fish and Wildlife Service. 1975. *Environmental Education Resources.* Jenny Publishing Company, Inc., Minneapolis.

24

Law of the Land:
Methods for Preserving Ecologically
Significant Land Areas
Richard W. Snyder

Questions for Consideration

1. What are some of the legal devices that a private landowner may use to preserve his or her land in its natural state or as an open space for future generations?
2. What are some of the legal devices available for the preservation of public lands for use as open space or as natural areas?
3. Which of the legal devices available gives one the best tax advantages?
4. Cite some of the famous law cases that established the legal basis for preservation of private or public land as open space or natural areas.
5. Why is land-use planning so essential for agricultural lands, watersheds, lake shorelines, and natural areas?

The quality of an environmental unit, whether it be a nation, state, or 0.4047-hectare (1-acre) parcel, is closely linked with its associated legal system. Through legal measures designed to both encourage and restrain man's actions, lawmakers may chart the course for the consumption and protection of natural resources. Those creating and administering laws, in turn, respond chiefly to interest groups and persons who work or possess particular knowledge in the area affected by the laws. It is important, therefore, for one engaged in the study of ecological and environmental topics to gain an understanding of the legal principles touching upon their specific field of interest.

For the botanist, there is an obvious concern for the preservation of land and waters where plant life grows. As the demand for land and water resources increases and the possible uses for given land or water units proliferate, the importance of legal devices to prevent the overburdening of these resources becomes more apparent. Most often, the botanist and others concerned with the healthy and varied abundance of plant life prefer that land and water resources be preserved in a natural or agricultural state. Unfortunately, pressure on the same areas for conversion to home sites, roadways, industrial and shopping centers, airports, and other uses are often inconsistent with suitable botanical habitats.

Simultaneously, however, the demand for the same land for natural and agricultural purposes continues in an upward spiral. Food production, recreation, scientific research, and preservation of critical ecological sectors are just a few important reasons for maintaining sufficient quantities of open space and natural land. Even the pressure on such areas from these types of uses, which do not significantly alter the character of the land, is becoming intense and more commonly conflicting.

Understandably, as competing interests continue to vie for control of increasingly scarcer natural, agricultural, and other open-space land, legal devices to regulate land use assume a more important role. Legal devices to control land use are not new; to the contrary, principles upon which modern land-use law are based were born among the earliest chapters of Anglo-American legal history. Yet, the application of those legal concepts to the American land-use puzzle has recently blossomed and continues to develop—statute by statute, case by case.

Legal restrictions or incentives assigned to a given unit of land are of two origins, public and private, although the line between the two is not without overlap. Private restrictions, for example, must normally receive some public sanction. So, while the legal concepts to be discussed are not always clearly labeled private or public, for purposes of identifying and analyzing the various legal principles, the classification is helpful. Moreover, in considering alternative solutions to a particular land-use problem, the reader should weigh the probable advantages and disadvantages of public versus private initiated land-use measures. Furthermore, special attention should be given to the comparative difficulty of implementing a specific plan using public and private devices. Finally, consider the course one might pursue, given a basic understanding of the applicable legal structure, to promoting a constructive land-use plan— whether it be for a nation, state, township, or 0.4047-hectare (1-acre) parcel—through the use of both public and private initiated measures.

Privately Initiated Land-Use Planning[1]

Private ownership of property is a cornerstone of the American system. Yet, Americans have adopted a rather short-sighted pattern of exercising this right. Typically, land is purchased and employed to whatever end and to such extent as the owner desires; then, what remains is conveyed by sale or gift, or devised by will or by the laws of the state to another party. Legal measures are available, however, to alter this pattern and to allow private owners of land to participate in the land-use planning process. This section will describe the options available in one jurisdiction, the State of Michigan, for a landowner with a sense of pride in the quality of his property and the desire to preserve it in its natural state or as open space for future generations.

For the purposes of this discussion, we shall assume that Mr. Black owns a piece of wooded property called Greenacre which he wishes to preserve. His ownership of the land consists of a number of rights and obligations regarding Greenacre. First, he has the right to the exclusive use and enjoyment of his land, and if Mr. Black wishes, he may sell all of Greenacre or some portion or interest in it. Of course, Mr. Black's rights in Greenacre are not absolute. The government may affect his use of

[1] Parts of this section are from Dinerstein, Marks, and Snyder, 1973.

Greenacre through zoning, taxes, or other measures when the public interest appears to supplant Mr. Black's property rights.

When Mr. Black makes his decision to preserve Greenacre, he may consider a number of different methods. Perhaps the simplest one is to give all rights and interests in Greenacre (legally called in fee simple) to a conservation organization or a governmental unit. He might give it to a county or township to use as a park, to a school for an outdoor classroom, or to a conservation organization as a natural area. Mr. Black will, of course, have to find some organization that will accept the land he wishes to give. In the case of governmental bodies, authority to receive property for such purposes may be regulated by statute. Michigan townships, for example, may acquire property for parks by "purchase, gift, condemnation, lease . . . ," but taxes may not be levied to acquire property without voter approval [M.S.A. 5.46(14)].

Although a gift of Greenacre to a governmental body in fee simple seems quite easy, there are problems. Mr. Black might have specific desires for the future use of Greenacre. For example, he may want to have it used as a hiking area in its natural state. If he gives Greenacre to Parkless Township, it could decide 10 years later that Greenacre should be used for a baseball diamond, or even that Greenacre should be sold for development. Fortunately, there are a number of legal devices that may be used by Mr. Black to control the future use of Greenacre.

RIGHTS OF ENTRY AND REVERTERS

Mr. Black could give Greenacre to Parkless Township *on the condition that* it be used as a park. Then, if Parkless Township decided some years later to use the land for another purpose, Mr. Black could terminate the ownership by the township and retrieve the land. This is referred to legally as a *right of entry*. Mr. Black places a condition in the deed which, if subsequently breached, allows him to recover the land. If the language of the deed states that Greenacre would be given to Parkless Township as long as it is used as a park, this would create a *possibility of reverter*; then, upon breach, the title would automatically vest in Mr. Black.

After Mr. Black dies, his heirs can enforce the condition against the township, for the right of entry or possibility of reverter passes according to his will or the laws of the state governing the succession of land. The Michigan Marketable Title Act, M.S.A. 26.1271-9, will affect the duration of the right of entry. Under this act the holder of the right of entry would have to file for record notice of his interest once every 40 years unless there is evidence of the interest in the monuments of title of the present owner of the land and the interest has been on record during the 40-year period. If a 40-year period passes without recording, the holder of Greenacre will have full marketable title (M.S.A. 26.1273). The Act is deemed not to effect "any right, title or interest in land owned by the State of Michigan, or any department, commission or political subdivision thereof" (M.S.A. 26.1274).

The principal problem with the right of entry, however, is that someone will have to take some action to protect Greenacre when the township decides to change its use. If, after Mr. Black's death, his heirs move away or simply neglect to observe the usage of Greenacre, Parkless Township will be unrestrained in its utilization of the parcel. The person who holds the right of entry is the only one who may enforce it. Furthermore, the right of entry may be waived by the party holding it, through inaction.

Mr. Black can circumvent this problem by bringing in a third party. He could give his land to a conservation organization, who could, in turn, give the land to Parkless Township subject to the right of entry the organization would hold. A stable organization should exist as long as the township, and can provide a reasonable guarantee that Mr. Black's wishes are followed. This same end may be achieved by Mr. Black's giving the land to Parkless Township on condition that it would be used as a park, and then giving the right of entry to a conservation organization.

There are certain problems with this use of the right of entry. One, of course, is finding a proper organization to hold the right of entry. Another concerns the seriousness of the penalty that results from breach of the condition which Mr. Black has placed in the deed. The courts have long frowned upon restrictions on land that result in forfeiture when a breach occurs. In Michigan, if the condition is nominal and evidences no intention of actual and substantial benefit to the party in whose favor it runs, a forfeiture will not be enforced (M.S.A. 26.46). If Parkless Township changes the use of Greenacre, a court may very well try to avoid finding a breach of the condition that Mr. Black put in the deed.[2] Therefore, if this approach is used, the deed should state very clearly how long the right of entry is to last, what will be considered a breach, and a statement of Mr. Black's purpose in making the restriction.

What if Mr. Black thinks he cannot afford to give away all or part of Greenacre? It might turn out to Mr. Black's surprise that the tax advantages of giving Greenacre for a public use will make a donation of land much less of a burden than he anticipated. Tax implications will be noted in greater detail later. If Mr. Black weighs the advantages and disadvantages of an outright gift and decides that he cannot afford it, there are still several options open to him.

EASEMENTS

One such alternative is to grant an *easement* in the land. An easement is a legal term for an interest in a piece of property less than the entire "fee" or total ownership. For example, Mr. Black might already hold part of Greenacre subject to an easement for a county road. Landowners often grant easements to neighbors who need to have a way to cross a piece of land for access to a road or lake, or to utility companies for the right to traverse the property with wires or pipelines.

There are two types of easements. The easements described above are called "affirmative easements." When Mr. Black grants an easement to

---

[2] See, for example, *Central Land Co. v. City of Grand Rapids*, 302 Mich. 105, when the city was allowed to drill for oil on land given for park use.

Mr. White to cross Greenacre at a certain point, Mr. White has an affirmative right. A road commission that may build a road, or a bather who may cross Greenacre to get to a beach, also has an affirmative easement. The other type of easement is called a "negative easement." Instead of Mr. Black granting Mr. White rights to do certain things on Greenacre, Mr. Black agrees *not* to do certain things on Greenacre. Mr. White's easement is his legal right to force Mr. Black not to do those things.

For example, Mr. Black could grant Mr. White an easement not to cut any trees on Greenacre. If Mr. Black then tried to cut trees on Greenacre, Mr. White could go to court to stop him. "Conservation easement" is a new type of negative easement that is being developed in a number of states. Using this easement, Mr. Black would still have possession of Greenacre, but could not develop it or disturb the vegetation and animals. There are easements similar to conservation easements, such as scenic easements along highways. Under Michigan law, for example, the Michigan State Highway Department may acquire land along state trunkline highways for the restoration, preservation, and enhancement of scenic beauty (M.C.L.A. 252.251-3).

Without specific statutory reference, there may be no authority for conservation easements in many jurisdictions, for historically there was no such thing as a common-law conservation easement. The Michigan Supreme Court recognized that "the House of Lords . . . has stated that the categories of easements expand with the circumstances of mankind," which leaves an opening for novel easements.[3] An easement for nature trails and walkways is much like a traditional easement for a way. An easement for a sanctuary open to the public is much like a charitable trust. These should be easy for a court to sustain. An open-space easement, however, or one that prevents any disturbance to trees or other plants, seriously reduces the alienability of the servient or "burdened" estate. A court must weigh the utility of the easement against the restraint upon alienation (Brenneman, 1967, p. 25). Some states, by statute, have specifically provided for the open-space easement or conservation easement.[4]

There is one more important division among easements. Let us take the example of Mr. Black, who grants a negative easement to Mr. White, stating that no trees may be cut on Greenacre. This results in a burden placed upon Greenacre. No trees can ever be cut. At the same time, if Mr. White is Mr. Black's next-door neighbor, Mr. White's land will benefit from the easement. Mr. White will enjoy the trees on the adjacent property, and also can say to a prospective buyer that he will not have to worry about the trees next-door being cut.

---

[3] *Johnston v. D.G.R. and M.R.R.*, 245 Mich. 65.

[4] The following statutes provide for conservation easements: Maine Public Laws of 1969, Chapter 566, 667-8; Iowa Code Annotated 11D.1-5. Some, in addition, provide for tax relief when the easement exists. West's California Code, Government 51050 et sequa; Connecticut General Statutes Annotated 47-421 et sequa; Burns Annotated Indiana Statutes 56-807; Missouri Annotated Statutes 67-870 et sequa (for open lands); Oregon Revised Statutes 271.710 et sequa; Purden's Pennsylvania Statutes annotated 16 11941 et sequa; Code of Virginia 10-151 et sequa; and Revised Code of Washington 84-34-210 et sequa. In other states the easements are limited to forest or natural districts, Illinois Annotated Statutes 105 466b; Hawaii Statutes 183-32; General Statutes of North Carolina 13A-38-39.

The law says that when this easement benefits Mr. White's land in particular, it is appurtenant to Mr. White's land. Why is this relationship important? Because under the law of most jurisdictions, when the easement is appurtenant, the burdens and benefits run with the land. If we use the example of the easement not to cut the trees, if Mr. Black dies or sells Greenacre, the burden or obligation of the easement passes to future owners of Greenacre. If he dies and the land passes to his son, Mr. Black, Jr., the trees still cannot be cut. At the same time, if Mr. White should die or sell his land, the benefit of the easement remains with the land, and the future owners of Mr. White's land can force whoever owns Greenacre not to cut the trees.

Why is this important? There are two reasons. One is that Mr. Black and Mr. White want permanent restrictions on Greenacre, and this has happened. The other reason is that not all easements are appurtenant. Suppose that Mr. Black grants an easement to Parkless Township that no trees will be cut on Greenacre. The burden of the easement is on Greenacre, just as in the example with Mr. White; but this time Parkless Township holds the benefit of the easement. The benefit is not attached to a particular piece of property as it was in the case of Mr. White's property. The court should find that the benefit is in gross; what this means is that it is not attached to any particular piece of property. Why should we know this? Because in most jurisdictions, if there is an easement in gross and the easement is not of a commercial nature, the burden may be assigned, but normally the benefit may not.

Again, using the Parkless Township easement, if Mr. Black dies or sells Greenacre, future owners may not cut the trees; but if Parkless Township tries to sell or give away the benefit (which is the township's right) to enforce the easement, the easement will no longer be enforceable. In the same way, if the easement in gross were given to a conservation organization instead of Parkless Township and the organization disbanded or was reorganized, the benefit of the easement could not be passed to someone else. The township or organization could, however, be given a small chunk of property as an outright gift, and then the easement could be appurtenant to that small piece of Greenacre. This would get around the problem, provided it is clear that the easement actually benefits the piece of land owned by the township.

The appurtenant-versus-in gross problem also arises in allocating property taxes. Under the example where Mr. Black grants an easement to his neighbor Mr. White, when the tax assessor values Greenacre, he may take the value of the easement and add it to Mr. White's assessment on his property, for Mr. White's property is made more valuable by the easement. In the State of Michigan, where the easement is appurtenant, it is taxed with the land that receives the benefits. In the example of an easement in Greenacre given to Parkless Township, Mr. Black must pay the full tax on Greenacre; this is because where the easement is in gross, it is taxed with the burdened land.[5]

One final note on easements remains. Parkless Township or Mr. White can always relinquish the easement in Greenacre. Thereafter, the future owner of Greenacre would no longer be under an obligation

[5] *Stansell v. American Radiator Co.*, 163 Mich. 528.

to follow Mr. Black's wishes. It might be possible to protect Greenacre against this eventuality by creating what is termed an executory interest in another conservation organization which would vest upon an attempted release.

RESTRICTIVE COVENANTS

Restrictive covenants are promises that will affect the use and enjoyment of Greenacre. As in the case of easements, they may be either affirmative or negative. Affirmative covenants may be much more varied in scope than the affirmative easements that were discussed above. One can make covenants that physically affect the land, as with easements, or there may be covenants that dictate such obligations as who will pay taxes on a piece of property or how the land may be sold. Negative covenants and negative easements, however, are very similar in terms of what they can accomplish.

There are a number of legal devices that come under the heading "restrictive covenants." One is the "common law covenant," which was developed as a method by which property owners could enter into binding agreements respecting their property. Relief for breach of such agreements was in the form of money damages. This limited remedy and other technical difficulties gave rise to the more flexible concept of the equitable servitude that emerged from the courts of equity. Equitable servitudes are essentially the same as the common-law covenant but may be enforced so as to prevent injury before it occurs. This feature makes the equitable servitude a much better device for conservation purposes. As in the case of easements, there are burdens and benefits involved in equitable servitudes. If Mr. Black gives or sells Greenacre to Parkless Township subject to a servitude, the burden will be upon Greenacre, now owned by Parkless Township, and the benefit will be held by Mr. Black.

To be enforced, servitudes must run with the land. This phrase in speaking of servitudes is similar to the phrase appurtenant when speaking of easements — that the benefit goes to a particular piece of land. Unlike easements, servitudes are not enforced when they are in gross. Therefore, if, as described above, Mr. Black gives Greenacre to Parkless Township subject to an equitable servitude requiring the area to be kept in its natural state, and Mr. Black then does not own any land benefited by the provision, the servitude will not be enforced.

Mr. Black could attach the benefit of the servitude to his neighbor's land. Then the servitude would be enforced, but only if Mr. White chose to do so. Another approach would be to have Mr. Black retain a small piece of Greenacre; then the servitude could be enforced by Mr. Black and his successors, assuming, as in the case of easements, that the servitude does actually benefit the retained land.

Unfortunately, equitable servitudes have been somewhat neglected in conservation planning. Often, a property owner like Mr. Black will not be in a position to give away Greenacre. Instead, Mr. Black may sell the land to a developer. If Mr. Black wishes to control the future development of Greenacre, he may use restrictive covenants. Particularly where there is a shortage of available property of a specific character, the owner should still be able to sell land at a reasonable price even with restrictions. In some cases, sensible restrictions could increase the value of the land.

It must be remembered that equitable servitudes are not enforced when they are in gross, that is, not benefiting a particular piece of property. If Mr. Black will be retaining a small part of Greenacre that will be benefited by the servitude, he should have no problem with enforcement, at least in his lifetime. If Mr. Black will be selling all of Greenacre, he may impose mutually restrictive covenants upon the future owners of the lots in Greenacre. Courts in the State of Michigan refer to these mutual restrictions as reciprocal negative easements. To be enforced, it is required that there be (1) a common grantor, (2) a general plan for the land by the grantor, and (3) actual, record, or inquiry notice to the grantee of the burdened lots and their successors. These restrictions can be enforced by any of the lot owners against any other owner who violates one of the restrictions and are normally recorded in the deeds and on the plat of the property.

Mr. Black should be realistic in planning restrictions for his property. He should foresee that a property owner might be reluctant to take his neighbor to court to enforce a restriction against the cutting of trees, where the neighbor is removing them one at a time, or even in groups. There are methods by which a watchdog organization may be provided to enforce the servitudes. If Greenacre is large enough, there could be a property owners' association which could hold title to a piece of common property, a strip of beach area, for example, to which the servitudes could be attached. If the size of the property is too small for a workable property owners' association, it might be possible for Mr. Black to grant an easement to a conservation organization, to which the servitude would be attached.

Equitable servitudes are subject to the doctrine of changed conditions. This means that if the condition of the surrounding land were to change markedly, a court could find that the reason behind the restriction is no longer valid, and that the restriction should not be enforced. Although this risk is always present to one who plans restrictions, the possibility of a later ruling of changed conditions could be considerably reduced, with a clear statement of the purposes for the covenant and the need for natural and open-space land, despite changes in the character of the surrounding area.

LEASEHOLDS

Another device for the preservation of Greenacre is the lease. Although a lease is commonly thought of in terms of month to month or year to year as is common in residential housing, much longer leases may be created under the law. Leases of 99 years or even of 2000 years are possible. A lease for a term to commence at once does not violate the rule against perpetuities, which rule, generally described, imposes a limitation upon the time for which restrictions or reversions may be placed upon real property.[6] A lease may also provide for an option to renew either at the end of a single term or perpetually at the end of each term. Therefore, Mr. Black could grant a 99-year lease of Greenacre to Parkless Township at $1 per year, renewable in perpetuity.

Why would Mr. Black prefer the lease method? One reason is that this is the method with which the courts are fairly comfortable, and which is easy to establish. Also, a leasehold would be useful in aiding Mr. Black in his efforts to control the use of Greenacre. Mr. Black may, by the terms and conditions of the lease, impose restrictions upon the use of the land to maintain and enhance natural characteristics which he considers desirable. In the event the lessee fails to comply with the terms of the lease, Mr. Black could enforce the lease either through an eviction proceeding, by a suit for damages caused by noncompliance, or by an action of an equitable nature to force the performance of terms or abatement of acts inconsistent with the lease. If Mr. Black sold Greenacre, the sale would be subject to the rights of the lessee and Mr. Black could assign his rights in the lease to his purchaser. If the land were donated to a conservation group or governmental body, the lease would provide the appropriate restrictions, and the advantages of the outright gift would apply.

TRUSTS

In general, the trust is a device used for making dispositions of property. When a trust is created, legal ownership of property is transferred from the original owner to a trustee and an equitable interest is created in a beneficiary. A trust can be created *inter vivos*—during one's lifetime—or by will. The components of the trust are the settlor, or donor, who is the person who creates the trust; the res, or the trust property; the trustee, or the person who owns, administers, or controls the trust; and the beneficiary, or the person for whose benefit the trust is created.

The terms of the trust are set forth in the trust instrument or legal document that creates the trust. The settlor can make such provisions with respect to the duties and powers for the trustee and the rights and benefits of the beneficiary as he or she may deem appropriate; and if the provisions do not run counter to any rule or policy of the law, they are valid and will be enforced by the courts. If the exact intent of the instructions of the settlor cannot be ascertained or carried out as stated in the trust instrument, the court, through a doctrine known as "cy pres," will require the trust to be administered to achieve the settlor's general intent (M.S.A. 26.1192).

It should be emphasized that great latitude and flexibility are allowed in the creation of a trust, and almost any type of restriction can be placed

[6] *Toms v. Williams*, 41 Mich. 552.

on the use of the trust property. One important aspect of trust flexibility is that a trust may be made subject to modifications or even revocation by the settlor. This means that not only can the settlor write his own "law," but he can amend it or repeal it as he wishes. The flexible nature of trusts will be particularly appreciated where the settlor has a rather complicated scheme for his land or where, for one reason or another, the use of another legal device does not seem to be feasible.

For example, assume that Mr. Black wishes to impose upon Greenacre a scheme whereby area A is to be used for biological research, area B is to be left totally undisturbed, and area C is to be used for a public park. This will result in the need for different restrictions being placed on the different areas. If an *inter vivos* trust is created, Mr. Black can retain the right to amend the plan any time he wishes or revoke the plan entirely if it does not work out. If the trust functions properly, Mr. Black may give up his control over the trust; the trust would then become irrevocable, and the courts could see that Mr. Black's intent is carried out.

The law classifies trusts into a number of different categories, but primary interest here is for the express trust created for charitable purposes. The case law and statutory law treat charitable trusts rather favorably. They may be created by gift, grant, bequest, or devise, and may acquire or receive both real and personal property. A charitable trust is not precisely defined in the statutory law, but Michigan statute M.S.A. 26.1191 speaks in terms of "religious, educational, charitable or benevolent uses." M.S.A. 26.1200(2)(b) states that a charitable trust means the relationship where a trustee holds property for a charitable purpose.

The distinction between a charitable and a private trust is not critical in the general analysis of trusts, for most of the important legal principles apply to both types. The distinction does become critical in an analysis of taxation and will be discussed later in the chapter. There are differences between the charitable trust and the private trust which should be noted. For example, charitable trusts escape the full brunt of the rule against perpetuities and the rule against accumulations. The doctrine of cy pres, mentioned above, applies only to charitable trusts in some jurisdictions. It is probable that the conservation trust would be construed as a charitable trust, although there has not been a definitive court ruling on this question in many jurisdictions.

The State of Michigan has a charitable trust act, M.S.A. 26.1200(1-37), that provides the principal control over open-land trusts. The act provides control over the administration, operation, and disposition of the assets of all charitable trusts by a system of filing of the trust instruments, inventory, and periodic reports to the state's attorney general by the trustees. The attorney general is authorized to make investigations into the trust administration and to make any additional rules and regulations necessary for the administration of the act. He or she is also authorized to bring suit to secure compliance with the act and to secure proper accounting and administration of any charitable trust. The provisions of this act apply regardless of any contrary provisions in the trust instrument.

Federal income tax advantages will be an important consideration for Mr. Black when he considers his alternatives for Greenacre. The tax factors involved in private land-use planning are far too exhaustive to be covered in detail in this discussion, but their importance requires at least an outline of some prominent features. It should be remembered that charitable deductions are merely one consideration in tax planning and that the donation of Greenacre, or an interest in Greenacre, may be only one of a number of charitable contributions made by Mr. Black in any one taxable period.

The Internal Revenue Code allows for deduction from income for any charitable contribution made within the taxable year.[7] Deductions are based upon a contribution base which should normally correspond to Mr. Black's adjusted income for the taxable year. Although there are two classifications of charities mentioned in the Internal Revenue Code, we will assume that Mr. Black is donating all or part of Greenacre to a governmental unit or to a conservation organization that qualifies for the highest possible deduction. Contributions to the United States, a state or political subdivision, a public school, and most private schools are deductible. Other organizations must apply and receive authorization from the Internal Revenue Service before donors will be allowed deductions for their contributions. One contemplating such a contribution should investigate the organization to which he or she seeks to donate to determine that the contribution will qualify for deduction.

The greatest advantage comes to Mr. Black where Greenacre has greatly appreciated in value. If he were to sell Greenacre, he would have to pay capital gains tax on any gain. Let us assume, for example, that Mr. Black bought Greenacre many years ago for $10,000 and it is now worth $30,000. If he were to sell it today, he would be taxed on 50 percent of the difference between the current sale price and the initial cost to him, referred to as the capital gain. This would amount to a taxable income of $10,000 ($30,000 − $10,000 = $20,000; 50 percent of $20,000 = $10,000). If Mr. Black earns $40,000 a year, this will result in a tax of about $5000.

On the other hand, if Mr. Black were to donate Greenacre to a charitable organization, he could deduct Greenacre's present fair market value from his income, limited to 30 percent of his contribution base for the taxable year. He may, however, carry over for the next 5 years the amount over the 30 percent limit. Let us assume, for example, that Mr. Black earns $40,000 a year. Thirty percent of that equals $12,000, and that figure is the maximum amount that he may deduct in any one year for the gift of Greenacre. However, Mr. Black may carry over for the next 5 years any amount over the 30 percent limit. Thus, if he gave Greenacre to Parkless Township in 1974 when it was worth $30,000, he will be able to deduct $12,000 in 1974, $12,000 in 1975, and $6,000 in 1976.

Mr. Black may take a deduction for a gift of Greenacre even though he has retained a right of entry in the event that the land is not maintained according to his wishes. Donations for interests such as the term of years

[7]Internal Revenue Code, Sec 170 (a).

or the long-term lease are not deductible, for they are less than an entire interest. It has been held, however, that certain types of conservation easements, although less than the entire interest of the land, may be deducted up to the fair market value of the property on the date of gift.

Mr. Black may realize tax advantages in other areas besides income tax. Both state law and federal law, for example, tax transfers of property upon death or in contemplation of death. In most cases, an exemption is allowed from such taxes for contributions for charitable purposes. The donor's estate would have a deduction from the gross value of the estate for the fair market value of any land devised to a charitable organization or governmental unit. Certain property tax relief may also be realized through contributions of land. Once the property is transferred, the former owner is no longer responsible for the property tax. In most cases, land held by a charitable organization would be exempt if used exclusively for charitable purposes.

Private action to maintain particular parcels of natural and open-space land allows property owners an opportunity to make a small contribution to the land planning process. Comprehensive land-use direction, however, can be practicably instituted only by public bodies, with authority to impose restrictions broadly upon all property in a given area. The legal methods that enable public bodies to perform this task are the subject of the following discussion.

Public-Initiated Land-Use Planning

ZONING

The most pervasive form of public initiated land-use planning is zoning. Zoning is defined as the "public regulation of the development and use of privately-owned land" (Crawford, 1965, p. 1). Mechanically, zoning is implemented by first isolating a particular geographic area, then dividing it into sections, assigning appropriate uses to each section. Zoning is clearly the most useful tool available to public land planners. Yet its encroachment upon the unrestricted use of private property—a concept bred deeply into Anglo-American thinking—often creates resistance to its advancement.

The first hurdle in the development of zoning law was determining whether governmental bodies possessed the authority to enact laws denying citizens the unfettered use of their land. For, is it not conceivable that zoning restrictions could so limit the use of land that the owner could make no reasonable employment of his resource? And how does this differ, from the owner's point of view, from a public taking of private land through the law of eminent domain, for which the U.S. Constitution mandates that an owner receive just and reasonable compensation.[8]

Euclid v. Ambler (Village of Euclid v. Ambler Realty Company 272 *U.S.* 303) is a cornerstone of American zoning law, for it decreed that the

[8] Constitution of the United States, Article V.

states do have constitutional authority to regulate the use of private land in a comprehensive fashion. This, the court found, is based on Article X of the U.S. Constitution, which has been construed to allow the states to legislate for the purpose of promoting and protecting the general health, safety, welfare, and morals of the people. Such authority is commonly referred to as the state's police power.

The decision, while constitutionally validating zoning, did not permit the states to regulate land use unconstrained. Police power can be exercised only in a reasonable fashion; if a court finds the restraint on the land is unreasonable, it may, despite the edict of *Euclid v. Ambler*, still strike down the ordinance. For example, an ordinance prohibiting timbering operations in a certain wooded zone might be found invalid as unreasonable if competent and reliable evidence were offered to show that management of the tract, including selective timbering, would actually produce a healthier forest. On the other hand, proof that the tract is of unique botanical significance and should not be altered, would likely persuade the court to uphold the restriction. Courts are often quick to note in such cases that they do not sit as superzoning boards. Furthermore, there is a general presumption that the ordinance is valid. Therefore, unless it appears the classification is clearly unreasonable, it will normally be preserved.

Besides meeting the test of reasonableness, a zoning ordinance cannot be oppressive to the point of confiscation. As mentioned earlier, "taking" of private property by a governmental body without just compensation is prohibited. Assume, for example, a wooded tract in the middle of a large metropolitan area is zoned to prevent its development. The reasonableness of this classification is clear—the scarcity and attractiveness of such a setting in a metropolitan area warrants its preservation. Even if the owner appreciates its value as a natural area, he or she may be financially unable to bear the loss of its development potential. Such a person may, for example, have invested in the land to sell for a retirement income. Now, for the esthetic benefit of the public in general, this income is denied. A court would likely find the zoning confiscatory, meaning that the city should purchase the land if it wishes it preserved.[9]

Confiscation determinations will turn on the facts of the case, there being no precise line of demarcation. When a zoning classification has created a marked decline in the value associated with the use of a parcel, a court will be signaled as to the possibility of a "taking" without compensation. As the spread between the former and latter value widens, the probability of the court voiding the ordinance increases. The preserving of natural areas through zoning is particularly sensitive to this attack, for typically natural lands are not economically profitable on a comparative basis with other common uses.

Thoughtful classifying of areas designed to prevent destruction of natural features, yet maintaining the economic value of the land, will help protect the validity of an ordinance. To preserve open space, for instance, an ordinance, rather than precluding any development of a large tract, may allow dwellings to be constructed from parcels not less

[9] *Industrial Land Company v. Birmingham,* 346 Mich. 667.

than 10 acres, thus allowing the owner to realize a gain, while still maintaining the essential character of the land. Another method is to isolate critical but limited areas for restriction. Floodplain and riverbank zoning is an example of this approach. Building and use restrictions are yet another important device to the same end. If the planner's goal is to preserve a scenic skyline, limiting building and structure heights often serves this purpose without rendering the land ineligible for any construction.

The master plan is an important guideline for courts contemplating the validity of an ordinance. A general or master plan is often required by statute; in fact, in many jurisdictions, courts may declare an entire ordinance void if not so grounded. Properly prepared, a general plan should clearly state the purposes of the act. By defining intent, classifications might be justified that otherwise could seem unreasonable. If the desire of the planners, for instance, is to preserve the essential agricultural character of a certain township, a 4-hectare (10-acre) minimum lot requirement is tenable. The same restriction would not be reasonable for an ordinance designed to encourage industrialization with associated high-density housing. Better plans will carefully substantiate the drafter's intent with factual documentation; if the purpose of an ordinance is the preservation of the general agricultural character of the land, statements that the region is well suited for agriculture in terms of soil type, population density, and climate would be supportive.

With this understanding of the derivation of the power to zone and the fundamental limitations upon that power, the attention of this discussion now focuses upon the enactment of zoning ordinances. *Euclid v. Ambler* declared the constitutional authority of the states to zone. Zoning legislation, however, commonly originates at the municipal or township level. Authority, therefore, must be granted from the state to subordinate governmental units by statutes known as enabling acts, which normally establish procedures for creating, amending, and administering the ordinances.

Zoning procedures vary from state to state, and, even within a state, the procedures may differ between the various levels of government. In Michigan, for example, there are separate enabling statutes for counties, townships, and cities and villages. For the purpose of this discussion, reference is to the Michigan Township Rural Zoning Act, M.S.A. 5.2963(1), pertaining solely to township ordinances.

Each enabling statute, of course, gives local units power to zone. The first step toward zoning a Michigan township is the passage of a resolution by the township board decreeing that the township intends to create a zoning ordinance. In the event that a township board is not so inclined, by statute, such a resolution may be effected by petition of registered voters. If necessary, the ordinance can be enacted through a favorable referendum vote. The Township Rural Zoning Act provides for preparation of the zoning ordinance by an independent body, the township planning commission. Furthermore, the act requires a public hearing to give interested persons an opportunity to examine the proposed ordinance and publicly comment upon its provisions. Additionally, proposed ordinances must receive approval of the county planning commission, which may make recommendations as well as pass judgment.

Once these steps are completed, the proposal is forwarded to the township legislative body, where the actual decision-making process and enactment must occur; all prior work on the ordinance is primarily preparatory and advisory. In fact, the township board could enact the ordinance in spite of recommendations to the contrary by the county planning commission. Prior to the decision of the township board, a second public hearing must be held on the ordinance.

Once a zoning law is enacted, it should not be assumed that all matters of land use are resolved. The ordinance only superimposes over the geographical unit a legal structure for land use; many new problems are bred by the institution of the ordinance. For example, what is to be done with uses of land that were formerly permitted but are inconsistent with the new ordinance? Or, how will the classification be enforced against violations? Who will decide what constitutes a given use? For instance, would a private hunting preserve be considered an "agricultural use"? If not, how does it differ from a farm where animals are raised for slaughter and upon which the farmer hunts rabbits and quail? Obviously, a zoning ordinance that does not provide a system for encountering and resolving such inquiries is not complete.

In the case of nonconforming or formerly permitted uses, the courts are often called upon to make difficult decisions in zoning and related areas of land-use law. A leading case in point, of particular interest to the environmentalist, is *Miller v. Schoene*, which is not technically a zoning case but is often cited by judges in dealing with the nonconformity problem in the context of zoning. In reading the case, note how direct legislation, as a form of zoning, may be employed to isolate and resolve particular conflicting uses. Could a zoning ordinance have treated the nuisance as effectively? Direct legislation to control land use will be discussed later in the chapter.

Miller v. Schoene, State Entomologist 276 U.S. 568
[portions of the opinion have been deleted]

Mr. Justice Stone delivered the opinion of the court:

Acting under the Cedar Rust Act of Virginia, Va. Acts 1914, chap. 36, as amended by Va. Acts 1920, chap. 260, now embodied in Va. Code (1924) as §§ 885 to 893, defendant in error, the state entomologist, ordered the plaintiffs in error to cut down a large number of ornamental red cedar trees growing on their property, as a means of preventing the communication of a rust or plant disease with which they were infected to the apple orchards in the vicinity. The plaintiffs in error appealed from the order to the circuit court of Shenandoah county, which, after a hearing and a consideration of evidence, affirmed the order and allowed to plaintiffs in error $100 to cover the expense of removal of the cedars. Neither the judgment of the court nor the statute as interpreted allows compensation for the value of the standing cedars or the decrease in the market value of the realty caused by their destruction whether considered as ornamental trees or otherwise. But they save to plaintiffs in error the privilege of using the trees when felled. On appeal the supreme court of appeals of Virginia affirmed the judgment. Miller v. State Entomologist, 146 Va. 175, 135 S. E. 813. Both in the circuit court and the supreme court of appeals plaintiffs in error challenged the constitutionality of the statute under the due process clause of the 14th Amendment, and the case is properly here on writ of error.

The Virginia statute presents a comprehensive scheme for the condemnation and destruction of red cedar trees infected by cedar rust. By § 1 it is declared to be unlawful for any person to "own, plant, or keep alive and standing" on his premises any red cedar tree which is or may be the source or "host plant" of the communicable plant disease known as cedar rust, and any such tree growing within a certain radius of any apple orchard is declared to be a public nuisance, subject to destruction. Section 2 makes it the duty of the state entomologist, "upon the request in writing of ten or more reputable freeholders of any county or magisterial district, to make a preliminary investigation of the locality . . . to ascertain if any cedar tree or trees . . . are the source of or constitute the host plant for the said disease . . . and constitute a menace to the health of any apple orchard in said locality, and that said cedar tree or trees exist within a radius of two miles of an apple orchard in said locality." If affirmative findings are so made, he is required to direct the owner in writing to destroy the trees and, in his notice, to furnish a statement of the "fact found to exist whereby it is deemed necessary or proper to destroy" the trees and to call attention to the law under which it is proposed to destroy them. Section 5 authorizes the state entomologist to destroy the trees if the owner, after being notified, fails to do so. Section 7 furnishes a mode of appealing from the order of the entomologist to the circuit court of the county, which is authorized to "hear the objections" and "pass upon all questions involved," the procedure followed in the present case.

As shown by the evidence and as recognized in other cases involving the validity of this statute, cedar rust is an infectious plant disease in the form of a fungoid organism which is destructive of the fruit and foliage of the apple, but without effect on the value of the cedar. Its life cycle has two phases which are passed alternately as a growth on red cedar and on apple trees. It is communicated by spores from one to the other over a radius of at least 2 miles. It appears not to be communicable between trees of the same species, but only from one species to the other, and other plants seem not to be appreciably affected by it. The only practicable method of controlling the disease and protecting apple trees from its ravages is the destruction of all red cedar trees, subject to the infection, located within 2 miles of apple orchards.

The red cedar, aside from its ornamental use, has occasional use and value as lumber. It is indigenous to Virginia, is not cultivated or dealt in commercially on any substantial scale, and its value throughout the state is shown to be small as compared with that of the apple orchards of the state. Apple growing is one of the principal agricultural pursuits in Virginia. The apple is used there and exported in large quantities. Many millions of dollars are invested in the orchards, which furnish employment for a large portion of the population, and have induced the development of attendant railroad and cold storage facilities.

On the evidence we may accept the conclusion of the supreme court of appeals that the state was under the necessity of making a choice between the preservation of one class of property and that of the other, wherever both existed in dangerous proximity. It would have been none the less a choice if, instead of enacting the present statute, the state, by doing nothing, had permitted serious injury to the apple orchards within its border to go on unchecked. When forced to such a choice, the state does not exceed its constitutional powers by deciding upon the destruction of one class of property in order to save another which, in the judgment of the legislature, is of greater value to the public. It will not do to say that the case is merely one of a conflict of two private interests and that the misfortune of apple growers may not be shifted to cedar owners by ordering the destruction of their property; for it is obvious that there may be, and that here there is, a preponderant public concern in the preservation of the one interest over the other.

In this instance the court showed little tolerance for the nonconforming use. Were apples of less economic importance, or red cedars greater, the court may have been of another persuasion. Generally, when no immediate injury is produced by the nonconformity, courts will allow the owner a reasonable time to eliminate the use. Michigan, by statute, has empowered the zoning unit to permit extension of nonconforming uses and, in the case of certain units, to compensate the owner for the termination of nonconformity.

Enforcement of zoning ordinances is the task of the governmental unit to which the ordinance applies. In large municipalities an agency may be created for just this function. Smaller bodies often utilize interagency cooperation for the enforcement of the zoning ordinance. A building inspector, for example, presented with plans for construction on a certain site, would not issue a building permit if the proposal violated the zoning regulation. If a use contrary to the zoning law commences, the violator is notified of the breach and ordered to abate activity inconsistent with the ordinance. Should the violator persist, the attorney for the governmental body would file a court action requesting the court to enjoin the party from continued violation. Further use of the land inconsistent with the order could result in the owner being in contempt of the court and penalized by fine and/or jail. In addition, violations of zoning ordinances are often criminal offenses for which the violator can be prosecuted through the offices of the local prosecuting attorney.

Interpretation of the zoning ordinance and relief to individuals dealt unusual hardship by the impact of the law is vested with the Zoning Board of Appeals. In Michigan, such boards are created by statute, M.S.A. 5.2963(18-23). Persons denied a permit or issued notice to terminate a use may appear before the board, and upon a presentation of the circumstances of the matter, a decision is rendered. By statute the Board of Appeals has the right to act on all questions that arise in the administration of zoning ordinances, and the authority to pass upon all controversies stemming from the application of the terms of the ordinance.

Perhaps the most important function of the Boards of Appeal is to consider application for variances. A variance is a permission to violate some provision of the zoning ordinance and may be granted upon a showing of some practical difficulty or unnecessary hardship (Crawford, 1965, p. 6). Not uncommonly, boards will issue variances but attach conditions to the grant. Zoning along a body of water, for example, may require all dwellings to be set back a certain distance from the water's edge. This might prove to be extremely harsh to an owner with a very short lot. A board confronted with this dilemma might agree to allow him to build within the setback but require that he obtain an easement from a property owner behind him and maintain his septic system beyond the restricted distance. Such a resolution offers an avenue of escape to the affected owner yet does not subject the water to possible contamination, a probable factor in establishing the setback.

One aggrieved by the decision of the zoning appeals board may have the decision reviewed by the courts. Courts will overturn decisions only under the limited situation where the board has abused its discretion and rendered an unreasonable decision or in the event that fraud or some other procedural irregularity is demonstrated. The law bestows upon the zoning appeals board considerable discretion in adjudicating zoning matters. Undoubtedly, boards acquire a measure of expertise in dealing with these problems, and courts are reluctant to substitute their discretion for that of the board where the record reveals sufficient evidence to anchor the decision. On the other hand, courts realize the political pressures at work at the board level; therefore, decisions patently offensive to the zoning scheme without corresponding hardship or those producing a calamitous situation for the landowner may be reversed.

This discussion of zoning only scratches the surface of a complex, ever-expanding topic. Nevertheless, it is important for one engaged in environmental studies to understand the fundamental concepts and framework of zoning law, for it is and shall continue to be the primary legal mechanism for land-use planning. Moreover, because zoning legislation is characteristically grass roots law, the opportunity afforded the individual to participate in the formulation and administration of a zoning ordinance is considerably greater than is possible with most other forms of law. Those with special knowledge and a sensitivity to the relationships between humans and botanical or other natural resources can render planners and legislative bodies valuable assistance. Persons in this category are encouraged to do so but are forewarned that solicitations for these perspectives are rare and that one should not hesitate to assume the initiative in presenting appropriate views and information.

One might ask how zoning laws, specifically, can preserve natural resources. The answer to this important question is limited only by our imagination and foresight. Perhaps a starting point is for the reader to acquire a copy of a zoning ordinance for a city, village, or township and examine it carefully. Consider the significance of setbacks, minimum lot sizes, conditional uses, density regulations, floodplain limitations, and the various other zoning techniques to the establishment and mainte-nance of a balanced natural system. For reasons that should be clear at this point, it is unlikely that areas the size of most national parks could be preserved in a natural state through present zoning measures. Preserva-tion in such cases can be effected best by other means, such as purchase or condemnation by government authority. Through zoning, however, thousands of lesser land masses—such as the distance between your house and your neighbor's—or the floodplain of a free-flowing river— can be preserved by restricting them to limited use.

Direct Legislation for Land-Use Planning

Perhaps the simplest form of public-initiated land planning is direct legislation. A public body with proper legislative and constitutional authority may isolate a parcel or class of property for a particular purpose. The federal government, a state, or subdivision thereof may, for example, purchase Greenacre for park purposes, thereafter develop-ing and maintaining the parcel with public funds. Even if the owner of Greenacre is reluctant to agree to the sale, the government body can, under the power of eminent domain, purchase property for the benefit of the general public.

Through direct action, a public body may, short of an actual purchase, restrict the use of a given parcel or class of property. Essentially, this is a form of zoning in a noncomprehensive fashion; consequently, the limitations of zoning law apply. Accordingly, if the constraints are too severe, the act may be construed as a "taking" of the owner's property, thus subjecting it to nullification. Direct restrictive-type legislation is often employed to preserve critical ecological areas where outright purchase is impracticable. Acts to preserve river floodplains and shore-lines are an important example of the use of this concept. A variation on this theme is to allow certain uses by permit only, providing for permits to issue upon application and approval with a department or agency which normally may issue such permits only according to prescribed regulations. To familiarize the reader with the format and content of a legislative act designed to control very particular pieces of property, portions of the Michigan Natural Beauty Roads Act are presented here. Examine the enactment carefully; attempt to pinpoint ambiguous phrases that the courts may be called upon to define. Further, assume that you are commissioned to lead a drive to establish a natural beauty road under the act. What statutory steps must be followed? What objections are likely to be encountered?

STATE OF MICHIGAN
75TH LEGISLATURE
REGULAR SESSION OF 1970
ACT NO. 150
PUBLIC ACTS OF 1970

Introduced by Senators Richardson, Rockwell and Bursley

AN ACT to designate certain roads as Michigan natural beauty roads; to provide certain powers and duties; and to provide for the development of guidelines and procedures.

THE PEOPLE OF THE STATE OF MICHIGAN ENACT:

Sec. 1.

As used in this act:

(a) "Board" means a board of county road commissioners.

(b) "Department" means the department of natural resources.

(c) "Native vegetation" means an original or indigenous plant of this state including trees, shrubs, vines, wild flowers, aquatic plants or ground covers.

(d) "Natural" means in a state provided by nature, without man-made changes, wild or uncultivated.

(e) "County local road" means a county local road as defined in section 4 of Act No. 51 of the Public Acts of 1951, being section 247.654 of the Compiled Laws of 1948.

Sec. 2.

(1) Twenty-five or more freeholders of a township may apply by petition to their board for designation of a county local road or portion thereof as a natural beauty road.

(2) Within 6 months after a petition is received, the board shall hold a public hearing to consider designating the described road as a natural beauty road. The hearing shall be held at a suitable place within the township in which the proposed natural beauty road is located. At the hearing a party or interested person shall be given an opportunity to present his support for or objections to the proposed designation. Notice of the hearing shall be given by the board by causing a notice thereof to be published at least once in each week for 2 successive weeks in a newspaper of general circulation in the county, and by posting 5 notices within the limits of the portion of the road to be designated, in public and conspicuous places therein. The posting shall be done and at least 1 publication in the newspaper shall be made not less than 10 days before the hearing.

(3) Within 30 days after the hearing, if the board deems the designation desirable, it shall file with the county clerk a true copy of its resolution designating the portion of the county local road as a natural beauty road.

Sec. 3.

(1) If the board designates a road as a natural beauty road, the property owners of record of 51% or more of the lineal footage along the natural beauty road may submit within 4 days after the road is so designated a petition to the board requesting that the natural beauty road designation be withdrawn and if the petition is valid, the designation as a natural beauty road shall be withdrawn.

(2) A designation of a natural beauty road may be withdrawn or revoked by the board after the board holds a public hearing in accordance with the procedure described in subsection (1) of section 2. Within 30 days after a hearing, if the board by majority vote determines the revocation necessary, it shall file with the county clerk a notice of its determination and publish the notice in a newspaper of general circulation, once in each week for 2 successive weeks. After publication of the notice, the road previously designated shall revert to its former status.

Sec. 4.

(1) The department shall develop uniform guidelines and procedures which may be adopted by the board to preserve native vegetation in a natural beauty road right of way from destruction or substantial damage by cutting, spraying, dusting, salting, mowing or by other means. No guidelines and procedures adopted under the authority of this act shall prohibit the application of accepted principles of sound forest management in a natural beauty road right of way.

(2) The department may advise and consult with the board on the application of the guidelines and procedures.

(3) The board shall provide for a public hearing before an act is permitted which would result in substantial damage to native vegetation in the right of way.

(4) Nothing in this act shall affect the right of a public utility to control vegetation in connection with the maintenance, repair or replacement of public utility facilities, which were constructed in a road prior to its designation as a natural beauty road, or in connection with the construction, maintenance, repair or replacement of public utility facilities crossing a natural beauty road.

Sec. 5.

The department may establish a citizen's advisory committee to assist in the formulation of proposals for guidelines and procedures.

This act is ordered to take immediate effect.

This act represents but one small page among volumes of statutes directly or indirectly contributing to the character of our land-use and planning scheme. A few among the list are mentioned to provide an idea of the scope of such laws: subdivision control laws regulating the division of property to lesser parcels and plats; property tax relief to preserve open space and farmland that is often valued for its development potential, forcing the owner to convert the land to more economically profitable uses; acts empowering local agencies to control erosion of soils; laws pertaining to such recreational use of public and private land as hunting, fishing, camping, and the operation of off-road vehicles; and restrictions on mining, drilling, and extraction of other mineral resources. A brief perusal of the compiled statutes of any state, commonly available at the public library, would provide one with an understanding of the extent of the laws affecting land use in that state.

Administration and enforcement of most land-use laws are vested with various state and local agencies; some, like the Michigan Natural Beauty Road Act, invite citizen participation. One interested in a particular land-use law will typically find the staff of the agency responsible for administering the act an excellent source of information. Also, an examination of the enactment normally identifies the legislators who introduced the bill; legislators will be able to provide further knowledge, often from a somewhat different perspective than would the agency staff. Finally, and perhaps most important for one investigating a particular law, remember that legislative acts are public records and may be readily obtained from the Legislative Services Bureau or from the enforcement agency. As the first step in any inquiry about a law, acquire a copy of the act and read it carefully. Many common misunderstandings about laws can be laid to rest by taking this simple step.

Summary

The foregoing discussion has focused upon the legal devices involved in land use and land preservation. Relative to the total body of knowledge of the law, the view is microscopic. Breadth of knowledge in the study of law, however, is not particularly critical; even the most learned students of the law can gain a clear comprehension of only a small portion of their discipline. It is far more important for one to understand the methods and sources of the law and how legal techniques are applied to a given problem. Hopefully, therefore, students of the botanical, ecological, and other environmental sciences can relate the substance of this chapter to their particular course of study, and, further, come to appreciate the processes of the law and how persons acting alone or as agents of a public body may employ the legal system toward protecting and preserving our natural resources.

Literature Cited

BRENNEMAN, R. L. 1967. *Private Approaches to Preservation of Open Land.* Conservation and Research Foundation, New London, Conn.

CRAWFORD, C. J. 1965. *Michigan Zoning and Planning.* Institute of Continuing Legal Education, Ann Arbor, Mich.

DINERSTEIN, C., J. MARKS, and R. W. SNYDER. 1973. *Legal Devices for the Preservation of Open Space and Natural Land.* Prepared under the auspices of the Environmental Council of Lenawee, Lenawee County, Mich.

Conserving Our National Heritage: The Nature Conservancy[1]
Nancy Buckingham and John W. Humke

25

The Nature Conservancy is a national conservation organization, receiving its support from the public, whose resources are solely devoted to the preservation of ecologically and environmentally significant land. The organization works in four ways to save threatened natural areas:

1. By identifying, systematically, the land that contains the best examples of all the components of the natural world. The identification is undertaken in cooperation with private and public conservation and research agencies.
2. By protecting natural areas through—in most cases—gift or purchase; by assisting or advising government or other conservation organizations.
3. By managing over 660 Conservancy-owned preserves using volunteers and staff; encouraging compatible use by researchers, educators, and the public.
4. By increasing public awareness of the need to safeguard natural areas.

Because it can act quickly, the Conservancy's successes have been many. Over 1800 areas of forests, swamps, marshes, prairies, mountains, and beaches, involving over 500,000 hectares (over a million acres) throughout the United States, have been saved from destruction through Conservancy action.

History and Development

In 1917, the Ecological Society of America established the Committee for the Preservation of Natural Conditions, thus recognizing the fact that the nation's natural areas were endangered. It is to this committee of

---

[1] The Nature Conservancy, 1800 N. Kent Street, Arlington, Virginia 22209.

scientists, as well as to its companion Committee for the Study of Plant and Animal Communities, that The Nature Conservancy traces its roots.

In 1946, the two committees became a separate entity, the Ecologists Union. After a few years of operation, during which the Union was primarily concerned with encouraging other groups to establish natural areas, the membership decided that a direct and active role in natural-areas preservation could make the group more effective. Borrowing the name of a previously established British group, the Ecologists Union became The Nature Conservancy in 1950. That same year, the Conservancy was recognized as a nonprofit association by the Internal Revenue Service. In 1951, the Conservancy was incorporated in Washington, D.C., as an organization chartered for scientific and educational purposes.

The new organization spent several years experimenting with various methods of land preservation. Then, in 1953, the first independent project was undertaken by the Conservancy at Mianus River Gorge in Westchester County, New York, (Figure 25-1), when a group of Connecticut residents asked for affiliation with the Conservancy in order to raise money to preserve and protect the Gorge. Other land projects quickly followed in the Midwest and on the West Coast.

In the late 1950s and early 1960s, as membership and project activity increased, the small organization administered by a volunteer Board of Governors and a small staff began to form chapters throughout the country. These groups of volunteers—usually composed of a mixture of scientists, attorneys, business persons, and others with professional skills—identified natural areas and sought donations or raised funds to purchase them. They also attracted many new members. The chapters normally operated in a single state or in part of a state.

In 1965, The Nature Conservancy received a 3-year $500,000 grant from the Ford Foundation to substantially expand its professional staff. Under this program the organization hired its first full-time professional president and created a number of additional staff positions, primarily in its national office.

In recent years, growth in the national office has leveled off and new resources are being devoted toward establishing a strong professional capability in the field. At the same time, a resurgence of growth and activity among Conservancy volunteer chapters and committees is being encouraged.

Present Organization and Capabilities

The Nature Conservancy is governed by a member-elected Board of Governors, which determines all matters of policy and hires the President, who is the administrative head of the organization.

The Nature Conservancy's national office is located in Arlington, Virginia. Its staff of approximately 75 persons consists of professionals with expert capability in the fields of real estate, law, finance, fund raising, public relations, preserve management, ecology, and other aspects of natural-area preservation. The functions of the national office

FIGURE 25-1.
Mianus River Gorge in Westchester County, New York. (Courtesy of The Nature Conservancy.)

include the general administration of all programs, undertaking large or unusually complex land projects, and the development of special or new programs.

The field capability of the Conservancy operates at three levels: regional offices, field offices, and chapters. It is at these levels that most land preservation projects are initiated and carried out.

There are four regional offices: the Eastern Regional Office in Boston, the Southeast Regional Office in Atlanta, the Midwest Regional Office in Minneapolis, and the Western Regional Office in San Francisco. Reporting to the regional offices are the 13 field offices. These are located in Portland, Oregon (Northwest Field Office); San Francisco, California; Denver, Colorado; Honolulu, Hawaii; Nashville, Indiana; Minneapolis, Minnesota; Lansing, Michigan; Columbus, Ohio; Washington, D.C. (Mid-Atlantic Field Office); Chapel Hill, North Carolina; Nashville, Tennessee; Miami, Florida; and Meridian, Mississippi.

The 30 chapters of The Nature Conservancy consist basically of volunteers, although 9 chapters presently employ staff. Activities of the chapters include project identification and fund raising, preserve management, and membership development. The needs of the particular chapter dictate the expertise of its staff.

FIGURE 25-2.
Blake's Reserve, South Carolina, one of the finest egret rookeries in the United States. (Courtesy of The Nature Conservancy.)

Programs

All Nature Conservancy programs either directly or indirectly serve the purpose of preserving biological diversity. Traditionally, land acquisition and financing have been the primary techniques for accomplishing this purpose. More recently, stewardship and several special programs have been developed to further this objective.

GIFTS OF LAND

People who wish to protect their land in its natural state may donate it to The Nature Conservancy. If the land is judged to be of ecological, environmental, scientific, or educational value, the Conservancy encourages such a donation for conservation purposes. Size is not necessarily a significant factor in selection. Whether $\frac{1}{2}$ hectare or thousands of hectares, each property is evaluated separately. All projects, including gifts, are reviewed by the staff and then presented to the Conservancy's Board of Governors for their approval. If approved, the transfer documents are completed, the deed recorded, and work on a preliminary inventory and management plan initiated.

Along with the satisfaction of knowing that the land will be enjoyed and studied by future generations, there are also substantial tax benefits involved in gifts of land. Through the provisions of the Internal Revenue Code, the federal government encourages donations to publicly supported nonprofit charitable organizations. Because the Conservancy is such an organization, individuals and corporations may deduct the full value of gifts they make to The Nature Conservancy, subject to certain limitations.

Two examples of large gifts will be of interest. The Santee Club in South Carolina has given what is said to be the most valuable gift ever made to conservation in the United States. The 10,118-hectare (25,000-acre) parcel donated, valued by some experts at an estimated $20 million, is located about 45 miles northeast of Charleston, South Carolina. The endangered southern bald eagle and American alligator both find refuge at Santee. Included in the gift is Blake's Reserve, considered by many field ornithologists to be one of the finest egret rookeries in the nation (Figure 25-2).

Another gift of land of great importance to conservation was made by the Union Camp Corporation, a forest products company. The Great Dismal Swamp (Figure 25-3), an area of nearly 20,235 hectares (50,000 acres) and worth $12.6 million, was then, in turn, given by the Conservancy to the Department of the Interior's Bureau of Sport Fisheries and Wildlife for use as a National Wildlife Refuge. Union Camp had decided that this was the best use for the Great Dismal, after considering the economic feasibility of logging the area or draining it for agricultural purposes.

PRIVATE PURCHASES OF LAND

The Conservancy's "private" acquisition program is a program in which the land acquired is to be retained by the Conservancy or by another private organization. These acquisitions are accomplished either with funds already identified or on hand, or through project fund raising after the land has been acquired. Obviously, where an outright gift is not possible, the Conservancy is interested in negotiating as low a price as possible (known as a "bargain sale" if a partial gift is involved) and has been very successful over the years in buying land for less than its fair market value.

An example of the Conservancy's private purchase program involves the Virginia barrier islands. Using funds provided by the Mary Flagler Cary Charitable Trust, the Conservancy has established the Virginia Coast Reserve. This program has resulted in the acquisition and preservation of barrier islands starting from Smith Island in the south to

FIGURE 25-3.
View of Great Dismal Swamp, Florida. (Courtesy of The Nature Conservancy.)

parts of Metemkin Island in the north. This prime example of a private purchase of land is creating an important link in the fragile barrier island system of Virginia. Shore development schemes have been halted by the acquisitions, thus protecting the islands' distinctive value as natural habitats, as well as the protection they afford the mainland from storms.

GOVERNMENT COOPERATION PROGRAM

Several years ago the Conservancy recognized that government was encountering problems in acquiring lands for conservation purposes that a private organization could overcome. Frequently, between the time a new park or other government project was actually made available, the resource itself was destroyed by development, or land speculation had greatly increased its cost. A private organization with flexibility and funding could step in to acquire and hold the land for repurchase by the government. The Conservancy does this through its Government Cooperative Program.

In these projects the relationship between the Conservancy and the governmental agency is based on a letter of intent requesting Conservancy assistance which does not constitute a binding contract. Borrowed funds are normally used to finance government cooperation projects. In infrequent cases, at the request of the governmental body, the Conservancy has placed use restrictions on the property via a reverter clause in the deed that would return the land to the Conservancy if it were ever used for purposes not intended in the purchase.

A Government Cooperation Program project of great significance involves the 1821-hectare (4,500-acre) Joyce Estate in Itasca County, Minnesota. In late 1973, The Nature Conservancy purchased the property from the estate of Mrs. Beatrice Joyce Kean for $2 million. The purchase was made possible through a $1.5 million loan from Northwestern National Life Insurance Company of Minneapolis. The Conservancy held the land until federal acquisition funds became available early in September 1975. The Conservancy then sold the property to the U.S. Forest Service as an addition to the Chippewa National Forest.

Funding Programs

Funds provided by individual contributors, foundations, and corporations enable The Nature Conservancy to continue to preserve natural areas by acquisition; certain of these funds also provide the monies to support the operation of the organization. The Nature Conservancy also receives gifts of natural areas from donors; however, when cash is required for purchase, four basic methods are used:

1. *Project Revolving Fund,* presently at $4 million, holds funds for the purchase of natural areas. The governing idea behind the fund is to provide money, temporarily, but quickly, where it is most needed. Upon purchase of a given area with money from the fund, it becomes the responsibility of a chapter, project committee, or sister conservation organization to raise funds to repay the Project Revolving Fund, usually within 3 years. The money allocated from the fund thus constitutes a loan. No interest is charged during the first 90 days; thereafter, however, to encourage prompt repayment so that the funds may be used again, interest is set at 1 percent below the current prime rate, or 5 percent, whichever is less.

2. *Established Lines of Credit* with several banks provide the Conservancy's acquisition program with great flexibility. The lines of credit, generally provided at the prime interest rate and necessitating a compensatory balance, are used primarily when the acquisition is for a government agency.

Unsecured credit lines of $7 million are now available to the Conservancy for national use through the State Bank of Albany, N.Y., and Manufacturers Hanover Trust Company in New York City. The Conservancy has also received assistance from several other banks and insurance companies in connection with the financing of specific project acquisitions.

3. *Guarantee and Income Fund,* a separate fund now in excess of $2 million, provides an endowment to the Conservancy and can also be used to guarantee bank loans when the Project Revolving Fund and credit lines are fully utilized.

4. *Land Preservation Fund,* a new fund that will permit the Conservancy to finance most acquisitions internally. Monies will augment the Project Revolving Fund and will be used to defray portions of the overall Conservancy program. The fundraising goal for the Land Preservation Fund is $20 million.

Stewardship

To "acquire" land is not necessarily to "preserve" it. To protect the biological and physical features of preserves, continuing stewardship must be exercised. This stewardship involves organization, evaluation, planning, and plan implementation on preserves and administration of the preserve management network.

Preserve management or stewardship places great reliance on regional, chapter, and field offices, as well as on local volunteers. The most vital component of the management system is the preserve stewardship committees. These committees are composed of interested local citizens representing a cross section of ages as well as vocational and social backgrounds. They inventory preserves, establish preserve objectives, and incorporate these data into a preserve master plan which spells out suitable uses and necessary preservation. The master plan for each preserve is prepared and sent to the chapter, the regional office, and the Director of Preserve Management at the national office for review. A preserve director is appointed by the committee to implement, on a day-to-day basis, the policies and concepts contained within the master plan.

To provide an overall program of stewardship within a chapter area and to coordinate the management of specific preserves, an intermediate level of stewardship responsibility is required. This is provided by a stewardship committee of the chapter, headed by a chairperson. Preserves are periodically visited to see that they are being adequately maintained.

The regional offices of The Nature Conservancy assist the chapters and individual preserve managers through frequent communication and in-person visits with chapter stewardship committee members. They provide the day-to-day authority for the administration of the stewardship program. In those states or geographical areas where no chapter exists, the regional offices fill the function of the chapter stewardship committee.

The national office stewardship staff is responsible for the development of stewardship policy. It is the central monitor of the preserve use and stewardship program for the organization as a whole.

Special Programs

In an effort to continue to bring innovative answers to the need to preserve and protect the natural lands, the Conservancy has initiated a number of special programs. Each of the programs is aimed at multiplying the effect of Conservancy land conservation without a correspondingly large increase in the allocation of resources.

STATE NATURAL HERITAGE PROGRAM

In an attempt to create an integrated approach to protecting natural land preservation, The Nature Conservancy has created the State Natural Heritage Program. The goal of the program is to work directly with states to assist them in providing a systematic basis for identifying ecologically significant areas, communities, species, or features, and in protecting them from adverse impacts.

In a typical program, The Nature Conservancy contracts with a state to help the state develop an overall system to coordinate the processes of ecological inventory, systematic data management and analysis, and the implementation of legal protection programs. Although the typical contract with the Conservancy lasts only 1 or 2 years, the system itself is designed to have the state operate indefinitely, adding or modifying data, revising priorities, and getting on with the business of preserving ecologically significant land.

The information compiled is also of great utility in more comprehensive land-use planning and will be likely to contribute an important body of information, ecological in nature, which would not otherwise be collected or taken into account. However, heritage programs are not comprehensive land-use planning programs, since they are primarily designed for active protection of relatively modest amounts of ecologically important landscape.

The value of cooperating closely with state governments is that it brings tremendous potential public capabilities to bear on a problem that most governments have ignored too long. Furthermore, ecological information gathering and record keeping cannot be adequately accomplished in a short period of time, and only public agencies have the resources required for the creation and long-term maintenance of a data system and for long-term application of appropriate legal protection.

Heritage contracts normally require that the Conservancy assign a staff ecologist or planner to work with the state for the term of the project. Most of this person's duties involve working in cooperation with the state to design the system and organize participants throughout the state. The Conservancy intends to rely heavily on the scientific community within the state in order to gather and evaluate information for the ecological inventory of the state. Every state contains scientists, mostly university faculty, who are interested in ecological inventories and natural-area protection. The Conservancy invites these scientists to form technical advisory committees for the heritage program, for the express purpose of using their specialized skills and knowledge to unearth leads, to classify information, and to adjust the generalized model with respect to the peculiarities of their region. In this way, the heritage program catalyzes the professional potential within the state and organizes the usually diffuse efforts of the state's scientific community. The effectiveness of this approach has been demonstrated in the Conservancy's first natural heritage program, in South Carolina, where scientists with a great variety of subject skills from many universities and federal and state government agencies have been enthusiastically helping tackle the problem of natural-area preservation.

The Heritage Program is not limited to a scientific role. States are provided with a catalog of legislative and regulatory alternatives for implementing the inventory. These, too, are developed in cooperation, where possible, with a Legal Advisory Board appointed to provide legal expertise. Currently, the program is conducting a study of state legislation and other implementation techniques pertaining to natural area protection. This study, combined with the Conservancy's practical experience in working with state natural area legislation, provides the basis for preservation/protection recommendations to the state.

Finally, once the state begins the hard work of protecting the areas identified, the field staff of the Conservancy will be ready to utilize the Conservancy's traditional financial resources and real estate skills. In most cases, the members of the field staff have a working relationship with the state agencies even before the program begins.

GOVERNMENT PROGRAM

The Government Program monitors federal legislation and administrative actions to determine which pertain to The Nature Conservancy's objectives. Legislation studied pertains to such things as natural area protection, endangered species protection, land-use planning, and the current status of 501(c)(3) organizations with respect to tax exemption. Appropriate Conservancy action is then recommended, including the identification of staff or volunteers to be responsible for the action. Other members of the conservation community also benefit from these studies by being informed on matters of mutual concern.

Three other studies were completed under contract with the Department of the Interior in conjunction with the U.S.–U.S.S.R. Agreement on Cooperation in the Field of Environmental Protection. The first of these was on U.S. federal natural area programs and activities; the second, on state programs; and the third, on private and local programs. The federal study has been published and is entitled: *Preserving Our Natural Heritage, Vol. I: Federal Activities.* It is expected that Volumes II and III will be published as well.

INTERNATIONAL PROGRAM

The Nature Conservancy has recently begun a program creating the opportunity to prevent the loss of lands containing unparalleled natural abundance and diversity in areas outside the United States. The International Program plans through a series of carefully selected acquisitions to preserve critical natural areas. Project selection focuses on precedent-setting transactions in the Caribbean, Central America, and Canada. The techniques developed by The Nature Conservancy during two decades of land conservation are being made available to other organizations and governments.

Cooperative working relationships have been established with other international and national groups to strengthen local conservation efforts, to undertake joint projects and programs, and to identify innovative programs, public and private, that may be applicable domestically and elsewhere on the globe.

A large and ecologically important tract of tropical rain forest on the Caribbean island of Dominica, about 350 miles southeast of Puerto Rico, has been given to The Nature Conservancy. The 384-hectare (950-acre) Middleham Estate, a gift of Virginia resident John D. Archbold, valued at more than $1 million, is currently being managed by the Dominion government as a national park. It is the first in what is hoped to be a series of acquisitions throughout the western hemisphere.

Article II of The Nature Conservancy's by-laws sets forth the objectives of the organization as follows:

Recognizing a dependence on natural lands for environmental stability, for essential scientific inquiry, for wholesome human life, and in fact for the survival of abundant and varied life on earth, the objectives of The Nature Conservancy are (a) to preserve natural areas for biological diversity, for the uses of science, and for the wilderness experience; (b) to preserve open land for conservation of natural features, for pleasure or recreation, and for education; (c) to restore land; (d) to improve techniques of land preservation by demonstrating to others how to do it well, by studying and trying out new ways to do it better, and by devising standards and priorities for the preservation of natural areas; (e) to advance the cause of natural area preservation in the schools, in private enterprise, in government, and in other countries; (f) to advance the foregoing objectives in cooperation with other organizations having similar and related objectives.

With these objectives in mind, the Conservancy set up a 1980 Plan. This plan is built around a Core Program of identifying, acquiring, and managing natural areas. This program will draw on the largest share of the Conservancy's resources, approximately 55 percent, and it will provide the organizing focus for other programs. To strengthen and guide the Core Program, four supporting programs are being developed: a Program of Applied Ecological Research, a Government Program (previously described in detail), a Private Enterprise Program, and a Public Awareness Program. The International Program will be further developed as an adjunct to the other programs.

The Ecological Research Program will attempt to identify major problems of programmatic significance, conduct or stimulate research on these problems, and recommend action when appropriate. Such research is in the identification and inventorying of natural areas, designing viable ecological preserves, planning and implementing preserve management, ecological land-use planning, ecosystem restoration, and conservation land development. Research will be promoted in land restoration and maintenance techniques on Conservancy preserves. The program will also attempt to advance the cause of ecosystem preservation by actively disseminating conclusions and recommendations to appropriate audiences.

Through the Government Program, federal legislative and administrative actions will continue to be monitored to determine which pertain to The Nature Conservancy's objectives. Working with the regional offices, a program of identifying and acting upon governmental matters of state, local, and regional concern to the Conservancy will be developed and promoted within the regional and field offices.

The Private Enterprise Program will serve to promote the cause of natural area preservation by demonstrating ways in which business and industry can further their own ends while protecting the ecological needs

Summary: Future Objectives and Directions

of the land. The Conservancy will use private enterprise techniques and resources to further its objectives.

The Public Awareness Program plans to develop a broad and educated public awareness of the natural land and its importance, to encourage the development of a constituency for the land, provide an additional layer of protection for natural areas, and promote an awareness of The Nature Conservancy that will create a climate conducive to its growth and effectiveness. Technical information concerning land conservation will be organized and disseminated.

Conserving Our National Heritage: The Sierra Club[1]
Eugene V. Coan

26

Questions for Consideration

1. What is the Sierra Club?
2. What has the Sierra Club accomplished in the way of preserving natural and wilderness areas?
3. What are the major battles that the Sierra Club has fought and won for the conservation movement?
4. What is the Sierra Club Legal Defense Fund?
5. What is the recreational program of the Sierra Club?
6. Why do we need organizations like the Sierra Club in the environmental movement?

Formation in the Early Years

The Sierra Club has played a leading role in the conservation movement since its organization in 1892. Founded by the naturalist John Muir, the Club's early activites centered on protection of California's 600-kilometer (375-mile) long Sierra Nevada mountain range. The Club was instrumental in defending the boundaries of Yosemite National Park, which had been established in 1890, from those who wished to use these public lands for their own purposes, and it played a critical role in seeing that the Yosemite Valley itself was added to the Park. The Sierra Club has worked consistently over the years and throughout the country to obtain protection for wild and scenic lands as national parks and monuments. Its efforts were, in part, responsible for the creation of the National Forest System. Other early milestones were the unsuccessful battle to protect the Hetch-Hetchy Valley in the Sierra Nevada and the successful creation of Kings Canyon National Park. Throughout these early years, the Club remained primarily a California organization, concerned with the problems of that western state. But even its efforts to protect California's resources required federal legislation and resulted in a considerable measure of national attention.

The turning point in the Club's role came in the 1960s with battles to preserve areas outside California. The first of these came as the result of plans announced by the Bureau of Reclamation to build dams on the upper Colorado River in the then-little-visited Dinosaur National Monument. The Club saw the battle to protect the integrity of this national monument as being crucial in ensuring the protection of all the nation's national parks and monuments. In this effort the Club came to

[1] Sierra Club, 530 Bush Street, San Francisco, California 94108.

use all the methods at its disposal to influence public policy—articles in its *Bulletin*, a film, and a book—each with a focus on the national legislation needed to accomplish the national monument's preservation. This campaign met with success when in 1956 Congress decided that no dam on the Colorado River Storage Project could be constructed within the National Park System. The second milestone was a campaign to prevent Grand Canyon National Park and Monument from similar dam construction. In this successful campaign, the Club used a wider array of methods of public persuasion than had ever been brought to bear on a single conservation issue previously: outings, books, films, exhibits, conferences, testimony, advertising, and letter-writing campaigns.

Membership and Structure

Throughout the early years, growth in the Club's membership and change in its structure was slow, but the nationwide attention generated by the Dinosaur and Grand Canyon campaigns swelled the Club's membership and broadened it geographically. Today the Club has about 150,000 members throughout the United States and Canada. It has some 46 chapters, covering the whole of the United States and about half of Canada. Most of the chapters cover one or more states. California, as a result of its still proportionately large membership, has 12 chapters. Most chapters contain still smaller units called groups, which represent local centers of activity. There are now about 230 groups in the Club. Each of the chapters and most of the local groups have their own meetings, newsletters, and outings, and they serve to focus public attention on conservation issues within their own areas, with special committees, campaigns, and sometimes their own offices. At the national level, the Club is governed by a 15-person Board of Directors elected by the membership to staggered 3-year terms. The Board, in turn, selects a five-person Executive Committee from among its members, including a president and a vice-president. There are also a multitude of other committees at the national level which recommend policy to the Board in given subject areas and which provide greater insight and coordination over particular programs than the Board could alone. The most important of these are the Sierra Club Council, which has representatives of each chapter, and focuses its attention on structural and organizational questions, and eight "regional conservation committees," which endeavor to coordinate Club conservation efforts in specific regions of the United States and Canada.

The Club's staff is relatively small, about 120, and much of it is of necessity devoted to organizational maintenance, with large membership and accounting departments. Its conservation staff, numbering about 30, is divided among the national office in San Francisco, California; a Washington, D.C., office; and regional offices in Tucson, Arizona; New York City; Anchorage, Alaska; Seattle, Washington; Los Angeles and Sacramento, California; Madison, Wisconsin; and DuBois, Wyoming. There is also an office of the International Program in New York City.

The Club has active programs of outings, book publications, advertising, and scientific research. The Sierra Club *Bulletin* is issued approximately 10 times a year, and the weekly *National News Report,* available by special subscription, is sent by first-class mail to keep Club leaders informed on the latest conservation developments and to generate timely local support on national issues.

In recent years, with environmental laws having passed Congress and been signed into law or become law in the various states, the Sierra Club has found it increasingly necessary to turn to the courts to achieve compliance with the letter and the spirit of these laws. Much of this legal work is handled by the Sierra Club Legal Defense Fund, which employs seven attorneys. The staff is aided by many public-spirited attorneys around the nation, who receive only a modest fee or only expenses for their services, as well as by cooperative arrangements with other public interest law firms, such as the Center for Law and Social Policy and the Natural Resources Defense Council. The Sierra Club Legal Defense Fund is maintained as a separate organization to ensure a healthy attorney–client relationship, and so that it may accept tax-deductible contributions.

The Sierra Club Legal Defense Fund

The Sierra Club Foundation was incorporated in 1960 as a means of guaranteeing the tax-deductibility of gifts for Club programs. In 1968, the Internal Revenue Service ruled that the Club itself could no longer receive tax-deductible contributions because of its degree of involvement in legislative affairs, and the Foundation assumed the responsibility for receipt of deductible contributions for many Club programs, particularly in its educational, legal, international, scientific, and book publishing ventures.

The Sierra Club Foundation

A study of the Sierra Club membership in 1971 demonstrated that the Club is composed chiefly of students (19 percent), teachers (18 percent), homemakers (12 percent), managers and executives (11 percent), and doctors, dentists, lawyers, and other professionals (15 percent), with only 7 percent in the blue-collar category. About 18 percent of members hold a Ph.D., law, or medical degree; and 21 percent have a Master's degree. It is a "young" organization, with 46 percent of its members under the age of 35. Many of the Club's older members joined primarily for its outings programs, while an increasing percentage of newer members join primarily to support its conservation efforts.

Who Belongs to the Sierra Club?

Over the years, the Sierra Club has considerably broadened its point of view and its goals, endeavoring in so doing to seek out and solve the basic causes of environmental problems. For example, Club concern with the preservation of scenic areas broadened into long-term protection of these areas once they had been publicly dedicated, then into overall park

Trends in Sierra Club Thinking on Environmental Issues

and forest management policies, and finally into the recognition of the need for broadly defined land- and water-use policies.

Club involvement in the key issue of energy, initially in response to aesthetic considerations of dam and power-plant siting and the obvious effects of offshore drilling, has broadened considerably to encompass the entire spectrum of energy considerations, from each of its sources to need for conservation in all its uses.

From its first statement of concern about water pollution in 1957, the Club has, in a series of policy statements and actions, become deeply involved with both air and water pollution issues, as well as with the solid waste problem and the need to recycle and reuse materials.

The Club's wildlife policies, initially concerned with the management of large game animals, now adopts a broader approach, emphasizing the necessity to maintain the diversity of ecosystems and viable populations of all species. In 1970, the Club adopted a policy on "environmentally hazardous substances" in which it emphasized the need to shift the burden of proof to new chemicals before they can be released in large-scale fashion into the environment.

An important broadening of Sierra Club thinking came in September 1971, when the Club, stating that "damage to any part of the global environment endangers the whole," launched an International Program. The Club also early recognized the need to bring our human population size into balance with our environment and resources.

Increasingly in recent years, the Club has come to recognize the need to take into account economic mechanisms in developing sound environmental goals. Policies have been adopted calling for the use of pollution emission taxes, the elimination of special subsidies, and the use of the principle of "marginal cost pricing" to ensure that the user of energy and other resources pays full social and environmental costs.

The Role of the Sierra Club

The Club has remained for all these many years on the cutting edge of the environmental movement. It very clearly recognizes its central role as an advocate. It attempts to raise crucial environmental issues, identifying the key political and legal changes required to bring about needed reform, and it takes steps to bring about the needed changes. Although the Club does recognize that political change is impossible without public commitment, it focuses less on public education than on needed political reform. There are within the environmental movement, organizations whose efforts are devoted largely to public education, and the Club works closely with them. The movement, as a whole, encompasses a full spectrum of activities needed to bring about change in public awareness and change in public policy.

Highlights of Recent Years

One of the most important actions in the last few years has been the legal case against the degradation of air quality in regions in which it is now of higher quality than would be required by existing "standards."

The Club successfully argued this case up to the U.S. Supreme Court, which refused to reverse a favorable decision of the Court of Appeals. Under this decision, the Environmental Protection Agency was ordered not to approve State Air Quality Plans that would allow air quality degradation. This has prevented industries from locating in clean air areas, such as Wyoming.

During the "energy crisis" of 1972–1974, the Club was successful in beating back industrial efforts to weaken the Clean Air Act. In another legal campaign, the Club was able to get a ruling that the U.S. Forest Service must study areas under its jurisdiction for possible inclusion in the National Wilderness Preservation System.

In California, Club efforts led to successful passage of an initiative, setting up a mechanism for planning and regulating coastal land use so that important public and environmental values would not be lost to private development. In large measure as a result of Club efforts in 1974, Congress added the Big Thicket in Texas and the Big Cypress Swamp in Florida to the National Park System, as well as added many new areas to the Wilderness System. Continuing Club concern has also prevented a large ski development from engulfing Mineral King, adjacent to Sequoia National Park.

As has often been said in the environmental movement, particularly with regard to land use: "Victories are temporary; defeats are permanent"; once an area is open to large-scale development, it cannot be wilderness. Once a disastrous project has been built, it cannot easily be unbuilt. Once bad rules for the use of public resources are adopted on behalf of special interests, they are difficult to correct. Once a nonrenewable resource is gone, it cannot be replaced.

One may draw two main conclusions from this frightening truism. First, there is a permanent need for organizations like the Sierra Club to maintain vigilance over public resources and values. Second, there is a continuing need to create comprehensive systems to maintain environmental quality, a need to get at the roots of problems and seek basic solutions—in essence, a need to engage in ever-greater long-range planning and to set up steady-state systems that will prevent environmental destruction.

A major recent innovation in the environmental movement is the recognition that economics lies at the heart of many environmental problems and is the key to the solutions of many of these problems. Whenever possible, it is desirable to create systems that make it less expensive to do the "right thing," instead of creating ever more complex regulatory systems. An example of this is the rate structure for electrical power. Under current practice, rates are often maintained by utilities and the regulatory bodies which oversee them such that people who use more electricity receive a lower rate than people who use less. Thus potential use is subsidized. If there were flat rates for domestic uses of energy, for example, then conservation would be in everyone's financial interest. In another example, the use of an emissions tax to control

Summary: Future Prospects for the Environmental Movement and the Sierra Club

pollution has a number of advantages over the creation and enforcement of specific emission standards.

There is now general recognition that the environmental problems we face are global and long-term. Unless our efforts to achieve steady-state systems become global in scope, they will ultimately be of limited value.

Environmentalists have also begun to accept the notion that our style of life will have to change; we cannot continue to use such a great proportion of the world's resources. Significant changes in values are therefore needed—changes that cannot be brought about by laws and regulations.

The environmental problems that we face are of a very large magnitude indeed, and while substantial gains have been made in recent years, there remains a very considerable amount of unaccomplished work. When this unfulfilled need is viewed, particularly at the international level, it is apparent that we are not yet in any way past the worst difficulties we face. It is my view that the Sierra Club, with its unique organizational structure, will continue to play a major role in the environmental movement for years to come.

Wild and Endangered Species of Plants
Peter B. Kaufman

27

Questions for Consideration

1. How do we distinguish between endangered and threatened species of plants?
2. Why is it important to save threatened and endangered species of plants?
3. What is the National Endangered Species Act, and how does it attempt to prevent endangered and threatened species of plants and animals from becoming extinct?
4. Why is it better to preserve threatened and endangered species of plants in their native habitats than trying to maintain them in parks, botanical gardens, and private gardens?
5. Why are the cacti one of the most endangered of all major plant groups?
6. Why are there so many endangered and threatened species of plants in California and Hawaii?
7. How can you as an individual help to protect endangered and threatened species of plants?

Plants are disappearing off the face of the earth at an alarming rate due to the disruptive activities of people. Witness the desecration of entire populations of plants over large areas of Arizona, California, parts of Appalachia, urban areas in the eastern United States, Vietnam, Brazil, and Hawaii. In Japan, many plants have become so rare that they are placed under maximum protection in botanical gardens, parks, and dedicated natural areas, with whole groves or individual trees dedicated to eminent botanists and conservationists who will protect them.

The fact of the matter is that right now 10 percent of our native higher plants are endangered, threatened, or recently have become extinct. When plants become extinct on the face of the earth, we have lost them forever—unlike rocks, minerals, and soil. These plants may be vital links in our understanding of plant evolution. They could be useful crop or ornamental plants or contain valuable drugs or other products useful to human beings. And most are essential and vital components of the total ecosystem on which other organisms depend.

In this chapter, we shall define what is meant by threatened and endangered species of plants. Classic examples of plants that have become extinct, or those which were thought to be extinct but have been found in nature, will be described. Some of the present-day threatened and endangered species of plants, such as the cacti, orchids, gentians, and insectivorous plants, will be cited and illustrated. Finally, we shall take up the main points of the National Endangered Species Act of 1973 and what you can do as an individual to help protect these plants.

Plants That Have Become Extinct in Their Natural Habitats

Have any of our native plants, other than those represented in the fossil record, become extinct in recent times? The answer is, yes. One of the prominent examples is *Franklinia alatamaha*, a shrub first discovered along the Alatamaha River in Georgia by John Bartram, King's Botanist, and his son, William, in 1765. It was last observed growing wild in this locality in 1803. Fortunately, this magnificent flowering shrub was collected by John Bartram; it is now being grown in cultivation in gardens and parks. Since *Franklinia* (named after Benjamin Franklin in 1785) is the first known species to have become extinct in the United States, it is being proposed as a symbol of endangered plants. It has already been commemorated on a postage stamp by the U.S. Post Office as an extinct plant species.

Another famous example of a plant that has become extinct in nature but is seen in cultivation is the maidenhair tree, *Ginkgo biloba. Ginkgo* is a living fossil, in that it once was a dominant part of the flora in Carboniferous times but is no longer known in wild or natural populations anywhere on earth. The only reason that we have living trees of *Ginkgo* is that they were cultivated in Chinese temple gardens, where they have survived even though they became extinct elsewhere. In the eighteenth century, westerners brought the seed of *Ginkgo* to Europe and America. *Ginkgo* became widely propagated and is now commonly grown as a street tree because of its ability to withstand air pollution and because it is relatively free of pests.

Shortia galacifolia or Oconee Bells, a shrub with pink to white bell-shaped flowers, which is native to woods of Georgia, South and North Carolina, and Virginia, was thought to be extinct. However, small populations were later discovered. Because of this, Oconee Bells is listed as endangered in the Smithsonian Institution's list of endangered plants.

Plants Thought to Be Extinct, Which Have Been Rediscovered in Nature

Our next example is *Betula uber*, or Ashe's birch, another plant thought to have become extinct but later found in wild populations. *Betula uber* was discovered by William Ashe of the U.S. Forest Service in 1914 along the banks of Dickey Creek in Smyth County in the mountains of southwest Virginia at about 850 meters (2800 feet). Ashe collected dried specimens of the plant and sent them to several well-known eastern U.S. herbaria. In 1953, collecting trips to the Dickey Creek site by Harvard horticulturist Albert Johnson failed to turn up any living plants of *Betula uber*. The plant was then considered to have become extinct. An expedition was set up in August 1975 to see if this plant could be rediscovered in the wild, based on new findings of Douglas Ogle of Virginia Highlands Community College. He found what he thought was *Betula uber*

FIGURE 27-1.
The showy lady-slipper orchid (Cyprepedium reginae) growing in Reese's bog near the University of Michigan Biological Station, Pellston, Michigan. (Courtesy of Larry Mellichamp.)

growing along Cressy Creek in the same county where it was found in 1914. The expedition team consisted of property owners Garland Ross and Ray Houslee; Warren H. Wagner, Jr., of the Department of Botany at the University of Michigan; Peter Mazzeo of the U.S. National Arboretum; and Ogle. On the appointed day, they did indeed confirm Ogle's suspicions. They rediscovered *Betula uber* in the wild along Cressy Creek in Virginia, 30 years after it had first been discovered.

Ten percent of the vascular plants native to the continental United States (including Alaska), or about 2000 species, are now listed as "endangered," "threatened," or "recently extinct." In Hawaii, a conservative estimate is that 50 percent of the native vascular plants are threatened or endangered (i.e., on the edge of extinction). This testifies to the seriousness of losing many of our valuable native vascular plants as a

Plants on the Edge of Extinction: Endangered and Threatened Species

result of the destructive activities of people and machines. The situation is likely to become aggravated in the United States and in other parts of the world where human populations and environmental destruction are increasing rapidly.

We wish to cite here a few examples of some of our native plants that are endangered or threatened, and which could become extinct in our time if the destructive activites of people continue. One of the prime examples is that of cacti native to the United States. About 72 species, or 26 percent, of the 268 species of cacti native to the United States are now either so rare or restricted in habitat localities that they are vulnerable to extinction. The entire cactus family may be one of the most endangered groups of plants in the United States (National Parks and Conservation Association, 1975). Cacti occupy very specialized niches, mostly in desert and dry areas. Some species do occur in tropical jungles of the Everglades or in cool mountain forests up to 3300 meters (11,000 feet) in elevation.

FIGURE 27-2.
Flower of the lady-slipper orchid (Cyprepedium reginae). (Courtesy of Larry Mellichamp.)

FIGURE 27-3.
The yellow lady-slipper orchid (Cyprepedium calceolus) photographed at Michigan State University, East Lansing, Michigan. (Courtesy of Larry Melli-champ.)

The problem with many species of cacti is that they are vulnerable to overzealous collectors, commercial exploitation, housing construction, grazing, fire, and agriculture. Housing developments near Tucson, Arizona, for example, have drastically reduced the native populations of the saguaro cactus. Lyman Benson (in National Parks and Conservation Association, 1975) cites case after case of rare cactus species succumbing to fire, agriculture, overgrazing, and overcollecting in desert areas of southwestern United States. He states that the "cactus habitat must be preserved if many rare native species are to survive."

Other plants are also on the edge of extinction. These include our native orchids (Figures 27-1 to 27-6), the gentians (Figures 27-7 and 27-8), the clubmosses (Figure 2-8), all our insectivorous plants (pitcher

FIGURE 27-4.

*The showy orchis (Orchis
spectabilis) growing at
Sharon Hollow near Ann
Arbor, Michigan. (Courtesy
of Larry Mellichamp.)*

plants, sundews, Venus-flytrap, butterwort, and others), and the bristle-
cone pine (Figure 27-9). Why are they on the edge of extinction? The
main reason is that these plants are avidly collected by individuals and
companies (especially biological supply houses and chain stores), just
as we have witnessed for cacti of all types. People are literally destroying the landscape to obtain these plants for their own private collections
or to sell. Our attitude toward these plants *must change.* Environmental
education must teach us that these plants are sacred parts of our ecosystem and must be left in their native habitats if they are to survive.
The National Endangered Species Act of 1973 will help us to further
this end. In the next two sections, we shall discuss what the National
Endangered Species Act aims to accomplish and what you can do as
an individual to help protect and save our rare and endangered species
of plants.

The National Endangered Species Act of 1973 as It Applies to Plants

To help you understand terms now used in the National Endangered
Species Act of 1973, and acts passed by various states, we shall define an
endangered species as "any species of fish, plant life, or wild life which is in
danger of extinction throughout all or a significant part of its range other
than a species of insecta determined by the commission or the Secretary

of the Interior to constitute a pest whose protection under this act would present an overwhelming and overriding risk to man." A *threatened species* means "any species which is likely to become an endangered species within the foreseeable future throughout all or a significant part of its range" (Michigan Public Act Number 203, passed July 11, 1974). In the United States, we have already emphasized that 10 percent of our native higher plants have become endangered, threatened, or recently extinct. Table 27-1 indicates the number of threatened and endangered species of native higher plants in each of the states as of this writing. It is clear from this list that most of these plants are in Hawaii and California, states where "people pressure" and human "developments" have had par-

FIGURE 27-5.
The putty-root orchid (Aplectrum hyemale) growing in Mack County, North Carolina. (Courtesy of Larry Mellichamp.)

ticularly drastic and detrimental impact on natural and wilderness areas. It is probable that many more plants will be added to this inventory because of continuing pressures by people on their natural environment. It is very fortunate that now these endangered species of plants are officially recognized and can be protected under the National Endangered Species Act of 1973. The big question that remains is how effectively this act will be enforced.

The National Endangered Species Act of 1973 has three primary purposes: (1) to provide means of preserving ecosystems on which endangered and threatened species depend, (2) to preserve endangered species of plants and animals, and (3) to carry out the purposes of international treaties regarding endangered and threatened species. The Secretary of the Interior is the principal administrator of the law. The Secretary of Agriculture is responsible for enforcement of protecting endangered and threatened plants. The Smithsonian Institution in Washington, D.C., has provided the original list of plants with about 2000 total for the United States; of this total, 761 are endangered, 1238 are threatened, and 100 are extinct.

Among the provisions of the National Endangered Species Act, it authorizes the acquisition of land to carry out the needs of the law. To accomplish this, it provides monies from the Water and Conservation

FIGURE 27-6.
Flowers of the putty-root orchid (Aplectrum hyemale). (Courtesy of Larry Mellichamp.)

FIGURE 27-7.
The fringed gentian (Gentiana procera), growing in Radrick fen, University of Michigan Matthaei Botanical Gardens. (Courtesy of Larry Mellichamp.)

Funds of the U.S. Department of the Interior. It also enlists the cooperation of the states by authorizing the use of federal funds to be used with matching funds from the states. And it provides foreign financial aid, especially for protecting species that are taken in other countries and imported into the United States.

The *National Parks and Conservation Magazine* (1975) lists several practical ways in which we, as individuals, can help to protect and save our rare and endangered species of plants. These include the following:

How Can We Help Save Endangered Plants?

1. "By urging your state conservation agencies to provide adequate protection of endangered and threatened plants under their jurisdiction."

FIGURE 27-8.
The bottled gentian (Gentiana andrewsii), growing
along Huron River near
Ann Arbor, Michigan.
(Courtesy of Larry Melli-
champ.)

2. "By informing yourself regarding endangered and threatened species in your area." These are listed, state by state, in the *Federal Register*, Part IV, Endangered and Threatened Species of Plants (June 16, 1976).

3. "By arranging for the long-term protection of any endangered or threatened species on your own land and encouraging others to do the same."

4. "By assuming the moral obligation not to harm any endangered or threatened species by your own actions and to teach by example your family, friends, and neighbors a conservation ethic."

5. "Do NOT TRANSPLANT endangered species of plants. If they are in imminent danger of unavoidable destruction—such as bulldozing—then obtain advice from knowledgeable botanists as to proper action to take."

6. "By being alert for infractions of protective regulations and reporting them to responsible authorities."

FIGURE 27-9.
A rare and endangered plant, the bristlecone pine (Pinus aristata) growing in the Sierra Nevada Mountains in California. Some of these trees are over 4000 years old. (Courtesy of Larry Mellichamp.)

Summary

In this chapter we have spelled out the nature of the crisis concerning the rapid disappearance of many of our rare and endangered species of plants. The same crisis applies to many species of animals. With the advent of the National Endangered Species Act of 1973, we now have at last begun to take some positive steps to prevent this tragic loss from continuing unabated. The best that will come from this act is that people will begin to recognize the problem, will educate themselves as to which

TABLE 27-1 / Endangered and Threatened Species of Native Higher Plants by State in the United States[a]

State	Endangered	Threatened	State	Endangered	Threatened
Alabama	27	46	Montana	2	8
Alaska	9	21	Nebraska	—	1
Arizona	64	58	Nevada	43	84
Arkansas	5	17	New Hampshire	6	4
California	242	393	New Jersey	4	9
Colorado	23	17	New Mexico	15	26
Connecticut	3	6	New York	3	14
Delaware	2	5	North Carolina	16	48
District of Columbia	—	2	North Dakota	—	3
Florida	84	128	Ohio	3	12
Georgia	23	65	Oklahoma	5	6
Hawaii	637	202	Oregon	43	135
Idaho	21	52	Pennsylvania	4	13
Illinois	5	16	Rhode Island	1	4
Indiana	1	9	South Carolina	9	35
Iowa	1	2	South Dakota	—	1
Kansas	—	2	Tennessee	25	31
Kentucky	7	22	Texas	95	135
Louisiana	1	8	Utah	56	101
Maine	4	6	Vermont	4	4
Maryland	1	8	Virginia	11	32
Massachusetts	1	6	Washington	16	72
Michigan	5	7	West Virginia	—	11
Minnesota	3	7	Wisconsin	3	10
Mississippi	—	12	Wyoming	3	18
Missouri	7	17			

[a] Adapted from listing in *National Parks and Conservation Magazine* (1975).

species of plants and animals are endangered, and will help take definitive action in concert with the states and federal government to help protect and save our endangered and threatened species of plants and animals. We must learn to cherish and protect *all* our native plants, just as peoples in China, Japan, and our American Indians have done for centuries.

Selected References

Department of Interior Fish and Wildlife Service. 1976. Endangered and Threatened Species. Plants. Part IV, *Federal Register* **41**(117), 24,524–24,572.

KINKEAD, E. 1976. The search for *Betula uber*. *The Explorer* **18**(2), 12–22.

National Parks and Conservation Association. 1975. Help Save Our Endangered Plants. *National Parks and Conservation Magazine*, January.

PRANCE, G. T., and T. S. ELIAS. 1977. *Extinction Is Forever*. The New York Botanical Garden. Bronx, N.Y.

Can One Love a Plastic Tree?[1]
Hugh H. Iltis

28

Every planner, landscape architect, or human ecologist should read Martin Krieger's "What's wrong with plastic trees?" (*Science*, **179**, 446–455, Feb. 2, 1973) if he wishes to catch a glimpse of the nightmare future that technology is preparing for man and nature. His article discusses the titanic events of the environmental crisis, of man vs. nature, totally outside the framework of biological reference; hence, one of his conclusions—that plastic trees and all sorts of nature substitutes have a valid place in planning—reads like a bad fairy tale. If he had only contemplated Hans Christian Andersen's "The Emperor's Nightingale," in which a mechanical nightingale is given the emperor to substitute for the real one whose song the emperor had loved. Eventually, of course, the clockwork breaks. Death comes and sits on the emperor's bed. But the real nightingale reappears and sings so sweetly that the emperor recovers. It is an old moral—you cannot make a real nightingale out of wheels and diamonds, an idea quite lost on our author.

If there is nothing wrong with plastic trees, if plastic trees can "give most people the feeling that they are experiencing nature," why not invent plastic dogs instead of live ones? Why not plastic corsages with synthetic perfumes, instead of orchids or gardenias? Why not plastic

---

[1] This chapter has been reprinted with minor corrections from *Landscape Architecture Quarterly* 63(4), 361–363, 1973, where it appeared under the title "Down the Technological Fix." The author is Director of the Herbarium, and Professor of Botany at the University of Wisconsin, Madison. Used with permission of Hugh H. Iltis.

dolls, which need no diapers, instead of babies? Why not 3000 giant Disneylands, one in each county, and then develop the rest of the country to grow more food and build more cities?

Why worry about the extinction of the African giant sable antelope or the Indian tiger? Or the preservation of the weedy Mexican grasses ancestral to corn or Peruvian wild potatoes? Why protect the Amazonian rain forest, or preserve the arctic tundras? According to Krieger, such proposals are "imperialistic at worst, unrealistic at best." But if biologists and ecologists or, for that matter, planners, won't concern themselves about the fate of Nature, who is there that will? And since most ecologists and planners are in the "developed" countries, should they remain uninvolved to satisfy misguided notions of what it is to be "imperialistic"? Of course, we are all against "imperialism" and for "social justice"! But we are also against stupidity and misinformation.

What, then, is a socially concerned teacher and biologist to do when he reads such misconceptions? What are we to think of *Science*, that editorially confused journal which proclaims its adherence to social justice and the scientific comprehension of the environmental crisis, yet publishes, regularly now for years, the unenlightened "optimism" of the technological bamboozlers (to use Theodore Rozshak's apt expression): Spilhaus, Doxiadis, Weinberg, Handler, Buckminster Fuller, and Seaborg? And now, as a final insult, these gratuitous environmental opinions of a biologically innocent planner sanctified, as it were, by publication in *Science*.

One wonders why *Science* publishes this author who values flowers by cost/benefit ratios, and argues preservation of nature only in the framework of rarity and the free market, in apparent ignorance of the vast and complex ecological arguments as to why nature and its diversity must be protected.

Why, indeed, must nature be preserved? This question has been answered in detail so many times by others—biological diversity as a basis of long-range ecological stability; genetic diversity as the necessary concomitant of continuing evolution (including gene preservation for future crop breeding options); and that vast uncharted New World of esthetic diversity; of human genetic needs for natural pattern, for natural beauty, for natural harmony, all the results of natural selection over the illimitable vistas of evolutionary time—of the complementary coadaptations of man to nature, of man and woman, of mother and child.

Do plastic trees have mycorrhizae? Produce oxygen? Transpire and cool the air? Have fragrant flowers visited by bees and produce fruits that feed the birds? Do they have leaves that decompose into a rich humus? But further, in contemplating plastic trees as economically inexpensive nature substitutes, one may well ask the question: Can one love a plastic tree? Or the sound of wind in a plastic pine? Is, indeed, "the demand for a rare [read natural] environment . . . a learned one"? Is the love of a *living* tree or flower truly taught only by *culture,* or is it due to the interaction of culture and evolution? With such wonderful plastic surrogates, will this love eventually become obsolete? Will mail-order plastic women filled with warm water and greased with Vaseline sufficiently satisfy our human needs? Will the false harmony of false trees or of surrogate sex be able to produce feelings of affection? Will all these makeshift substitutes send us screaming into the night for the satisfying totality of the emotions that evolution has led us to expect? Has our innocent apologist never heard of Charles Darwin?

And what of the special needs of children? Suppose that they have *biological imperatives* for wilderness, for natural beauty, for natural harmony. If these are not satisfied, what will happen to their orderly and adapted ontogeny? Suppose that, for the sake of social justice, *all* children, not only those of the rich, should have a chance to experience the feeling of untouched wilderness (in order to grow up to be happy, healthy, and wise). What if, long *after* all of nature has finally been ground up in the garbage disposal of the technologic sink (with bamboozlers like the author at the switch), it becomes suddenly clear that there are indispensable genetic needs for many of these components of nature? But by then it would be way too late.

All planners should be human ecologists. They enunciate and illuminate what an alive, evolved, and evolving man *must* have to remain human, with human biological needs foremost on their minds; with the needs of the technological colossus in proper perspective. And what does it mean to put human needs first? "Not until man places man second, or, to be more precise, not until man accepts his dependency on nature and puts himself in place as part of it, not until then does man put man first! This is the greatest paradox of human ecology" (H. H. Iltis, *BioScience* **20,** 820, 1970).

But what, in fact, does our present school of planners think its duty is? Is it to offer frivolity of choice to a human population uniformly programmed to genetically determined and culturally influenced needs? Thus, Krieger offers genuine, unspoiled nature only to those rich enough to rent a plane to visit it, and small city parks for the poor masses who can afford only to ride a streetcar. What brand of social justice gives the poor a tiny city park, the rich a giant wilderness? "A summum bonum of preserving trees has no place in an ethic of social justice"—indeed! It should of course be obvious that there can never be any meaningful social justice without "preserving trees."

The counterculture is bad enough in its simplistic insistence on the *Greening of America.* On having its car and driving it, too. On living simply, in affluence. Are we now to be blessed with a countercounterculture, which will hasten the destruction of most of what is biologically sacred, a destruction, while begun by a mindless technocratic profit-oriented capitalism, is now to be completed in the guise of social justice and relevancy by a pack of technologically optimistic liberal planners?

No matter what Harry Harlow's experiments might suggest, to the affection-starved baby monkey, a terrycloth, wire female with only a light bulb heart does not much of a loving mother make! And, likewise, plastic trees or tiny city parks do not a healthy landscape make. We cannot condition humans to be happy and *human* with the surrogates of technology—we can only make them happy and human with what they, biologically, have been selected to experience.

The planner who maneuvers himself into becoming an apologist for our cultural derelictions, including the virtues of plastic trees; the planner who encourages the faked and denatured environment, no matter how good his intentions may be, becomes himself an addict of the "technological fix,"[2] a technological junkie, hooked on growth, hooked on profit and hooked on the propagation of that one, grand, and damnable lie (the lie that makes the absurdly destructive extremes of the technological revolution possible): that *man can adapt to anything*, even plastic trees; that man doesn't really need the matrix of nature to exist in; that "the way in which [man] experiences nature is conditioned by . . . society"; and that, therefore, society can de-condition man from wanting to experience the "real thing," the real nature, that lives and blooms and flies and sings.

Whatever the finer points of man's existence may be, the French sociologist and lay theologian Jacques Ellul has put it well in *The Technological Society* (1964, p.325):

The milieu in which man lives is no longer his. He must adapt himself, as though the world were new, into a universe for which he was not created . . . He was made to have contact with living things, and he lives in a world of stone.

It cannot be, then, that our affection, our apparently overwhelming need for flowers, trees, and wild land is fortuitous, a mere accidental cultural fixation. We may expect, as a matter of fact, that science will furnish the objective proofs of suppositions about man's needs for a living environment which we, at present, can only guess at through timid intuition; that one of these days we shall find the intricate neurological bases of why a leaf or a lovely flower affects us so very differently from a broken beer bottle.[3]

Meanwhile, modern technological civilization continues in its accelerative growth and with unprecedented speed, magnitude, and complexity, which are so great that most people, in fact, do not have the faintest notion what is good for them, for their families, for their society, or for humanity as a whole. The problems are simply too complex, involved, and removed for anyone but an occasional highly sophisticated specialist to understand. And that in itself is fraught with danger. Who will judge? And how? And even if we understood a problem and wished to effect a change, the momentum of technological civilization is so great that, like the sorcerer's apprentice, it is often quite beyond any rational control.

[2] A.M. Weinberg argues persuasively [*BioScience* **23**(1), 41–45, 1973] that for every "technological fault" there is a "technological fix," ironically forgetting that the latter phrase was coined by environmentalists to designate a cure of an ill analogous to the "fix" of a drug user hooked on heroin; that is, it is precisely because so many "technological fixes" are bad that they do not represent a valid solution (e.g., the green revolution and unlimited food, atomic energy and unlimited power, etc.).

[3] At the University of Wisconsin, Sharon Decker, with Hugh Iltis, is compiling an annotated bibliography on "Man's needs for Nature," sponsored by the Horticultural Research Institute, which may represent a small beginning toward such an understanding. [cf. H. H. Iltis, 1966, The meaning of human evolution to conservation, *Wisconsin Academy Review* **13**(2), 16–23.]

Krieger may, in fact, correctly describe what he observes in a horde of Disneyland visitors. But many of these may have been conditioned by their megalopolis environment and upbringing to blindly deny their own biological well-being, an increasingly prevalent phenomenon, especially of big cities. The general biological ignorance bodes ill for democratic decisions on environmental issues. Except for a deliberate expansion of public understanding of biology and evolution, I don't know how else the public will ever understand or realize its own condition. Meanwhile, we shall pay a terrible price in environmental damage for keeping evolution out of the schools and out of planning, and continuing the public's ignorance. Since we are, and always will be, biological creatures, the planner as well as the public should be biologically sophisticated if it is not to make erroneous assumptions. The lack of evolutionary input into the environmental crisis, and the mere existence of the "teaching of evolution vs. creation" controversy in California (not in 1873, but in 1973) all point to the great need for reevaluating priorities in teaching and planning.

What, then, is a planner to be; what, then, is he to do? As a socially responsible individual, where must he lead? He must, above all, be a biologist and human ecologist, sensitive to man's evolution and its holistic implications, *whose principal job it is to preserve the evolutionary harmony and diversity of this earth.* Nothing really matters more than this—no cleverness, no "fix," no good intentions. There can never be a healthy humanity, both physically and socially, without its ancient evolutionary and ecological base.

Thus, neither the planner, the physician, the teacher, nor the landscape architect can compromise the evolutionary nature of man: he must accept it, because it is. He must accept the basic principle that the optimum environment for all organisms (including man) is that in which they evolved, because *it,* in fact, selected *them,* and in a dynamic sense still continues to do so. No experimentation is necessary to show this; it is true.

Summary

Let us therefore demand that the future of the human environment, the only environment to which man is genetically adapted, be left to those enlightened planners who, in prudence, humility, and biological understanding wish to protect and preserve it.

It is time for men and women to commit themselves to a contemplative study of nature, however hard that may be for us to begin. . . . We are far from knowing all the facts. We need more information. It is too easy to say that people prefer their landscapes humanized and that we adore wilderness only after it no longer howls. The presumed fact that people like to tame wilderness does not prove that people are well off without wilderness. We are still ignorant of what people, in the deepest levels of their brains, need from the world.

DANIEL MCKINLEY

Glossary

Ablation rate   The speed of decrease of volume, primarily as a result of melting and evaporation.

Acaricide   A chemical used to kill acarids, such as mites, ticks, and water mites.

Activated carbon   An absorbent form of carbon used to filter toxic gases from the air and to remove dissolved organic matter from wastewater.

Activated sludge   Secondary sewage treatment product that has been treated with air so that aerobes can live in it, thus purifying it so that it can be used to make methane gas or fertilizer.

Aerobic   Living or active only in the presence of oxygen.

Afterripening   A complex enzymatic process occurring in seeds, bulbs, tubers, and fruits after harvesting and often necessary for subsequent germination.

Agouti   *Dasyprocta agouti,* a large diurnal seed-eating rodent that lives on the forest floor of much of the lowland New World tropics.

Air foil   An object, the shape and orientation of which can control stability of another object to which it is attached.

Air layering   Vegetative propagation of a plant by enclosing a branch or shoot in a moist medium (such as sphagnum or soil) until roots have formed.

Aleurone   Protein matter in the form of minute granules concentrated in a special peripheral layer of endosperm cells in grain seeds.

Alkaloid   Any of a very large group of organic bases containing nitrogen and usually oxygen that occur in seed plants in the forming of acids.

Allelopathy   The reputed detrimental influence of one living plant upon another due to secretion of toxic substances.

Alluvial   Clay, silt, sand, gravel, or similar material deposited by running water, especially during recent geologic time, the deposits usually occurring on floodplains of streams.

Alternator   An electric generator or dynamo producing alternating current.

Ampere   A unit of electric current, or flow; equal to the current produced by 1 volt of energy at a resistance of 1 ohm; amperage is the strength of an electric current in amperes, in coulombs per second.

Amylase   Any of the enzymes that accelerate the hydrolysis of starch and glycogen.

Anaerobic   Living or active in the absence of free oxygen.

Analog   A species in one group corresponding in some particular characteristics with a member of another group.

Anionic   Having the character of a negatively charged ion.

Anticlone   A high-pressure area.

Aphotic zone   An area without light in a body of water.

Apiary   A collection of hives or colonies of bees kept for honey.

Asbestos   A mineral that readily separates into long flexible fibers suitable for uses where incombustible, nonconducting, or chemically resistant material is required.

Assimilation   To take in and absorb as one's own.

Association   A major unit, often taken to be the fundamental unit in an ecological community.

Autecology   The study of the details of how an individual or a species interacts with its environment.

Autotrophic   Capable of self-nourishment; capable of using carbon dioxide and carbonates as the sole source of carbon, and a simple inorganic nitrogen compound for metabolic synthesis.

Auxin   Any organic substance characterized by its ability in low concentrations to promote growth of plant shoots along the longitudinal axis and to produce various other effects, such as root formation and bud inhibition.

Axil   The distal, usually upper, angle or point of divergence between a branch or leaf and the axis from which it arises.

Biennial   Plants that require two growing seasons in which to go from seed to seed.

Biodegradable   Capable of being readily decomposed by biological means, such as by bacterial action.

Biological magnification   The accumulation and concentration of a persistent toxic compound in the environment.

Biomass   The weight of living material contained in a sample.

Birth rate   The number of births per year per 1000 population in a given group.

BOD   Biochemical oxygen demand; the amount of oxygen used in meeting the metabolic demands of aerobic microorganisms in water rich in organic matter.

Bog   An inadequately drained area rich in plant residues, usually acidic, frequently surrounding a body of open water and having a characteristic flora.

Bole   The trunk of a tree, especially the lower portion of such a trunk.

Boreal   Of or relating to the northern biotic area, characterized by dominance of coniferous forests and tundra.

Bract   A somewhat modified leaf associated with the reproductive structures of a plant.

Cacodyl oxide   A heavy oily liquid $(AsC_2H_6)_2O$ that has a repulsive odor and is obtained by distilling arsenic trioxide with potassium acetate.

Cacodylic acid   A crystalline deliquescent compound $(CH_3)_2AsOOH$ obtained by oxidizing cacodyl oxide; a herbicide.

Calcareous   Growing on limestone or in soil impregnated with lime.

Caliche   A succession of crusts of calcium carbonate that form within, or on top of, the stony soil of arid or semiarid regions.

Callus   A thickened area or protuberance on the surface of a plant.

Canopy   The uppermost spreading branch layer of a forest.

Carbon adsorption   A filtering process utilizing carbon.

Carnivore   A meat-eating animal.

Catabolic   A process that involves destructive metabolism involving release of energy and resulting in true excretion products.

Catalyst   A substance, usually present in small amounts relative to the reactants, that modifies (usually increases) the rate of a chemical reaction without being consumed in the process.

Chaparral   A dense impenetrable thicket of stiff or thorny shrubs or dwarf trees.

Chutney   A condiment that has the consistency of jam and is made of acid fruits with added raisins, dates, and onions, and seasoned to taste.

Clear-cutting   The removal of all timber from an area.

Cleft graft   A plant graft made by cutting the stock squarely across, splitting the cut end, and inserting one or two scions in the split so that cambia of stock and scion are in contact.

Climate   The prevailing weather conditions of an area.

Cloning   The aggregate of the asexually produced progeny of an individual whether natural or by plant cuttings.

Cloud seeding   The process by which dry ice or other compounds are dropped onto clouds for modifying the weather.

COD   Chemical oxygen demand.

Community   A body of individuals organized into a unit or normally manifesting a unifying trait. An ecological term applied to plant or animal communities.

Composite   A member of the plant family Compositae (or Asteraceae), which contains artichokes, daisies, dandelions, and sunflowers.

Compost   A mixture of decomposing organic material used as a fertilizer.

Coniferous   Bearing cones; refers to vegetation made up mainly of conifers, which are gymnosperms.

Conspecific   Of the same species.

Convection   The circulatory motion that occurs in a fluid or gas at a nonuniform temperature owing to the variation of its density and the action of gravity.

Convective currents   Mass movements of hot air, usually rising; and cool air, usually sinking.

Coriolis force   The force created by the earth's rotation, causing the winds to curve to the right in the northern hemisphere and to the left in the southern hemisphere.

Corm   A rounded, thick, modified underground stem base bearing membranous or scaly leaves or buds and acting as a vegetative reproductive structure.

Cotyledon   The seed leaf of an embryonic sporophyte.

Coulomb   A unit of electrical current quantity, equal to the amount of electricity provided by a current of 1 ampere flowing for 1 second.

Crown   The highest part; the head of foliage in a tree or shrub.

Crozier   The young unfolding fern frond.

Crustose   Having a thin thallus adhering closely to the substratum of rock, bark, or soil.

Cucurbitaceous A member of the plant family Cucurbitaceae, which contains cucumbers, gourds, melons, and squash.

Cull To separate one group from another; usually refers to removal of undesirable individuals or objects from a group.

Cultigen A plant found only in cultivation, not known to be indigenous to any area.

Cuticle A thin, continuous noncellular layer of fatty substances secreted by epidermal cells on the external surface of plant organs, functioning in the prevention of desiccation.

Damping-off Falling over and death of seedlings, usually caused by fungal attack.

Death rate The number of deaths per year per 1000 population in a given group.

Decibel A unit used to express relative differences in power, usually between acoustic or electric signals; a measurement of sound intensity.

Deciduous Falling off or shedding of leaves at the end of the growing period.

Defoliation To strip off leaves; to cause the leaves of a plant to fall prematurely.

Detrivore Feeding on animal wastes.

Diatom Any of the unicellular or colonial algae constituting a class having a silicified cell wall (frustule) that persists as a skeleton after death.

Diesel An internal combustion engine that uses the heat of highly compressed air to ignite a spray of fuel introduced after the start of the compression stroke.

Dioecious Having the male reproductive organs in one individual and the female in another.

Dipterocarp A tree, chiefly of tropical Asia, yielding valuable wood, aromatic oils, and resins and distinguished by having two-winged fruit.

Diversified habitat A habitat distinguished by a large variety of plant species.

Doldrums The equatorial ocean region in which the air is heated and rises, characterized by very light, fluctuating winds.

Dormancy Existing in a latent form or in a minimum degree of metabolism, but capable of bursting into full activity.

Ebb tide The period of a tide between high water and succeeding low water.

Ecdysis The act of shedding off a dead skin or integument, as by insects in metamorphosis.

Ecology A branch of science concerned with the interrelationship of organisms with their environment.

Ecotone A transition area between two adjacent ecological communities, usually showing competition between organisms common to both areas.

Ectotherm A cold-blooded animal.

Edaphic   Referring to the drainage, slope, soil chemistry, and so on, with respect to where a plant is found.

Effluent sewage   Raw, untreated sewage.

Electrolysis   The decomposition of a compound into its ions by the action of an electric current.

Electrolyte   A substance that dissociates into ions in solution or when fused, thereby becoming electrically conducting.

Electrostatic precipitator   A device that removes particles suspended in gas by electrostatic charging and subsequent precipitation onto a collector in a strong electric field. Used to control air pollution in smokestacks.

Embryonic plant   The young sporophyte of a seed plant, resulting from union of the egg and one of two sperm nuclei, usually comprising a rudimentary plant/with plumule, radicle, and cotyledons and typically embedded in endosperm that provides nutriment for the developing plant upon germination.

Emergent   Being above water or above the level of forest.

Emulsion   A suspension of small globules of one liquid in a second liquid with which the first will not mix.

Endocarp   The inner layer of the pericarp when it consists of two or more layers of different texture or consistency (as in the apple or orange).

Endosperm tissue   A nutritive tissue in seed plants formed around the embryo within the embryo sac by division of the endosperm nucleus.

Endotherm   A warm-blooded animal.

Entropy   The measure of the amount of energy in a system not available for doing work.

Epidermal cells   A layer of primary tissue in higher plants that is commonly one cell thick and provides protection; also, the outer epithelial layer of the external integument of the animal body.

Epilimnion   The layer of water that overlies the thermocline of a lake and is subject to the action of wind.

Epiphyte   A plant that grows upon another plant nonparasitically, deriving its moisture and nutrients from the air.

Equitant   Said of leaves whose bases overlap one another, as in the Iris family.

Ericaceous   Adjective referring to a member of the heath family, or Ericaceae, which also contains azaleas, blueberries, cranberries, and rhododendron.

Estuary   A water passage where the tide meets the current of the stream.

Ethnobotany   The study of the interrelationship between human and plant populations.

Ethylene   A colorless, flammable gaseous olefin hydrocarbon, $CH_2{=}CH_2$, used in organic synthesis, and promoting growth of plants and ripening of fruits; a plant hormone.

Eukaryotic   Having a visibly evident nucleus.

Eulittoral   A landward subdivision of the littoral zone of a body of water.

Eutrophication   Rich in dissolved nutrients but shallow and oxygen-deficient.

Evapotranspiration rate   The rate of water vapor loss through a leaf's stomata.

Fauna Animals collectively—especially the animals of a particular region or time.

Fen Low, peaty land covered wholly or partially with water, usually alkaline in pH.

Fermentation An anaerobic process in which an organic compound (usually a carbohydrate) is broken down anaerobically into other compounds, often producing ethyl alcohol or methane and CO_2.

Ferrous Containing iron, and therefore capable of rusting and eventual decomposition.

Fission A nuclear reaction in which an atomic nucleus is split into fragments and energy is released.

Fixation The metabolic assimilation of atmospheric chemicals (e.g., nitrogen) by heterotrophic bacteria into the plant's growth cycle.

Flat plate collector A device in which a fluid (usually water) flows over a metal sheet that is exposed to sunlight, the heated fluid being used for various purposes, including home and hot water heating.

Flood tides Periods of rising water levels, caused by the moon's orbit.

Flora The plants or plant life characteristic of, peculiar to, or adapted for living in a particular region.

Fodder Livestock feed made of hay or straw.

Foliose Leafy or resembling a leaf, as with certain lichens.

Formation The largest unit in ecological community organization, comprising two or more associations together with the successional communities that lead to their establishment.

Fossil fuel A fuel obtained from mineral deposits in the earth; oil, coal, and natural gas.

Frond A fern leaf.

Front The leading edge of a large mass of air.

Frost line The level to which ground frost will penetrate, usually about 1.5 to 1.8 meters (5 to 6 feet).

Frugivorous An organism that eats fruit.

Fungicide A chemical used to kill fungi or to check the growth of spores.

Fusion The union of atomic nuclei to form heavier nuclei, resulting in the release of enormous quantities of energy.

Gas turbine An air-breathing internal combustion engine, consisting essentially of an air compressor, a combustion chamber, and a turbine wheel.

Generic Characteristic of a genus, or referring to a large inclusive group, as opposed to specific individuals.

Genus A taxonomic category ranking below a family and above a species, used in taxonomic nomenclature, either alone, or followed by a species epithet, to form the name of a species.

Geothermal Pertaining to the internal heat of the core of the earth.

Germination The resumption of growth by the embryo in a seed after planting; involves the development of a young plant from an embryo after a period of dormancy or quiescence.

Germ plasm The hereditary material of the germ cells; genes.

Gesneriad Belonging to the family Gesneriaceae, a large family of tropical herbs or (rarely) woody plants having chiefly opposite leaves and strongly zygomorphic flowers; includes the African violets.

Gibberellin Any of several plant-growth regulators that are produced by a fungus (*Gibberella fujikuroi*), and higher plants, which promotes growth of shoots when applied in low concentrations.

Greenhouse effect The heating effect caused by CO_2 absorption and reradiation back to earth of infrared rays.

Guild An ecological group of plants distinguished from ordinary herbs, shrubs, and trees by a special mode of life, usually involving some degree of dependence upon other plants.

Habitat The place where a plant or animal species naturally lives and grows.

Head The upper or anterior part of an engine.

Head In a hydroelectric power dam, the usable difference in elevation between the incoming water and the level at which the water leaves the dam.

Heat engine A device to produce electrical energy by means of the temperature difference between two bodies.

Heath A plant of the family Ericaceae, typically growing on open barren, rather acidic, and (frequently) poorly drained soil.

Hectare A metric unit of area equal to 2.471 acres.

Herbicide A chemical used to kill plants, especially weeds.

Herbivore A plant-eating animal.

Hermaphrodite Having both sexes in the same organism; more specifically in plants, having functional male and female parts in the same flower.

Heteromorphic alternation of generations Offspring deviation from the usual form of the preceding generation.

Heterosis A greater vigor or capacity for growth, frequently displayed by crossbred animals or plants as compared with those resulting from inbreeding.

Heterotrophic Obtaining nourishment from outside sources; most animals, and those plants that do not carry on photosynthesis.

Hinterland The land directly adjacent to and inland from a coast.

Horse latitudes Two belts of calm light winds and high barometric pressure located around 30° north and 30° south latitude.

Hummock A slight rise of ground above a level surface.

Humus The organic component of the soil resulting from the partial decay of leaves and other vegetable matter.

Hydrologic cycle The water cycle, including rain, soil percolation, runoff, water flow, and evaporation.

Hydrolysis A chemical reaction, in which the water ions (H^+ and OH^-) react with a compound to produce a weak acid, a weak base, or both.

Hydrolyzer A device that decomposes a chemical compound by reaction with water.

Hydrophyte A plant, especially a perennial aquatic plant, having its overwintering buds underwater and growing wholly or partly in water.

Hypha   One of the individual threads that make up the mycelium of a fungus.

Hypolimnion   The part of a lake below the thermocline made up of water that is stagnant and of essentially uniform temperature except during the period of overturn.

Inarch   To form an approach graft (e.g., use of sucker sprouts to bridge rabbit damage on apple trees).

Indole ring   A crystalline compound (C_8H_7N) found especially in jasmine oil and coal tar, which may be formed by reductive distillation of indigo and zinc. Also, in auxin-type plant hormones and many hallucinogenic compounds called "indole alkaloids."

Inflorescence   The mode of development and arrangement of flowers on an axis.

Infrared   Thermal radiation of wavelengths longer than those of visible light.

Infrutescence   The structure bearing the fruits; the infrutescence develops from the inflorescence.

Intercrop   To grow two or more crops simultaneously in the same ground.

Interpretive naturalist   A person who interprets specific natural elements and relates them to the whole environment.

Inverter   A device for converting direct current into alternating current.

Joule   A unit of electrical energy, or work; equivalent to the work done in raising 1 coulomb of electricity 1 volt.

Juvenile hormone   A hormone in the metamorphosis process, bringing on maturation and development of the insect larvae.

Kinetic energy   The energy of a system produced by its motion, equal to $\frac{1}{2}mv^2$, where m is the mass of the moving system and v is its velocity.

Krummholz   Stunted forest characteristic of most alpine regions.

Lateritic   Lateritic soils are red-to-yellow clays rich in iron and aluminum oxides and very low in silicon. They are formed by perpetual leaching of ordinary soils at high temperatures, until the silicates and many other minerals have been removed. The red cotton-field soils of Georgia are midlatitude lateritic soils. They are also known as latosols.

Lauraceous   A member of the plant family Lauraceae, which contains avocados, bay, and cinnamon.

Leach   To remove nutritive or harmful elements from the soil by percolation.

Leaf trace   The vascular supply of a leaf or branch consisting of one or more vascular bundles that are extensions of the central vascular cylinder.

Legume   Dry, dehiscent fruit developed from a simple superior ovary and usually dehiscing into two valves, with the seeds attached to the ventral structure.

Littoral zone   The area of, or relating to, or on, or near a shore, especially of the sea.

Lupine A plant of the genus *Lupinus,* a genus of herbs with palmately compound leaves and white, yellow, blue, or purple flowers in long racemes. Member of the legume family, Leguminosae (Fabaceae).

Macrophyte A member of the plant life of a body of water that is large enough to be observed by the naked eye.

Mariculture Producing food from the ocean, compared to agriculture.

Mesophytic Growing in, or adapted to, a moderately moist environment.

Mesotrophic Requiring either a single amino acid, or ammonia and an organic acid as a source of metabolic nitrogen.

Methane CH_4, a colorless, odorless, inflammable gas formed by the decomposition of vegetable matter, as in a methane generator.

Microwave Electromagnetic wavelengths less than 1 meter in length.

Migration The move from one region to another with the change in seasons.

Mixing depth The distance between the earth and the altitude, where rising warm air becomes the same temperature as the surrounding air.

Mordant A chemical (as a salt or hydroxide of chromium or aluminum or tin) that serves to fix a dye in or on a substance (as a textile fiber) by combining with the dye to form an insoluble compound.

Morphology The features comprised in the form and structure of an organism or any of its parts; also, the study of the forms, relations, metamorphoses, and phylogenetic development of organs apart from their functions.

Mulch A protective plant cover made of straw, leaves, or other materials.

Mycelium Network of filamentous cells or hyphae, forming the typical vegetative structure of fungi.

Mycorrhizal The symbiotic association of the mycelium of a fungus (as various Basidiomycetes and Ascomycetes) with the roots of a seed plant (as various conifers, beeches, heaths, and orchids), in which the fungal hyphae form an interwoven mass of threads on or within the roots.

Nematocide A chemical used to kill nematodes.

Nematodes Very small cylindrical unsegmented worms commonly found in soils.

Neotropical Of, relating to, or constituting a biogeographic region that includes South America, the West Indies, and tropical North America.

Niche The role of an organism in an ecological community, involving especially its way of life and its effect on the environment.

Node A thickened or swollen enlargement as in a stem, where a leaf is inserted.

Nonferrous Not containing iron, and therefore incapable of rusting and decomposing.

Nuclear fusion The combining of hydrogen nuclei to produce energy.

Oligocene The geological epoch extending from about 40 to 30 million years ago.

Ongotrophic   A lake deficient in plant nutrients and usually having abundant dissolved oxygen with no marked stratification.

Ontogeny   The biological development or course of development of an individual organism.

Organophosphate   An organic compound (containing carbon) with phosphate (PO_4) groups attached. Used in some insecticides.

Orographic   Dealing with the mountains.

Outcrossing   A mating of individuals of different strains but usually of the same breed.

Ozone layer   The layer of ozone gas (O_3) in the stratosphere, believed to be important in filtering out ultraviolet radiation from the sun.

Paca   *Cuniculus paca,* a large nocturnal seed- and seedling-eating rodent that lives on the forest floor of much of the lowland New World tropics.

Palynologist   A person practicing the branch of science concerned with the study of pollen and spores, whether living or fossil.

Parasitic   An organism that lives at the expense of another organism in order to obtain its food, with no recompense to its host.

Parenchyma   A tissue of higher plants, consisting of thin-walled living cells that remain capable of cell division when mature, that may function in photosynthesis and storage, and that make up much of the ground tissue of leaves, stems, and roots.

Parthenocarpic   The production of fruits without fertilization (seedless).

Patch-budding   Plant budding in which a small rectangle of bark bearing a scion bud is fitted into a corresponding opening in the stock.

Peduncle   A slender plant stalk, especially one that supports a fruiting structure.

Pelagic zone   Of the open sea or ocean.

Percolation   To filter slowly through porous soil or stone.

Perennial   Continuing or lasting for several years; a plant that dies back seasonally and produces new growth from a perenniating part (e.g., *Delphinium, Asparagus*).

Perianth   The external envelope of a flower or floral leaves; the calyx and corolla collectively.

Pesticide   A chemical used to kill insects and other undesired pests.

Petiole   A slender stem that supports the blade of a foliage leaf.

Phenology   The pattern in time of the production of plant parts, such as leaves or flowers.

Phenols   Aromatic hydrocarbons built on a basic chemical structure of C_6H_5OH or combinations thereof; phenols are usually poisonous to animals and plants.

Pheromone   A substance produced by one organism for communication with another; e.g., sex hormones produced by female insects.

Phloem   A complex tissue in the vascular system of higher plants, consisting mainly of sieve tube elements and companion cells, functioning chiefly in the translocation or transfer of organic solutes from one part of a plant body to another.

Photic zone   The uppermost layer of the sea or other body of water receiving sufficient light from the sun to affect living organisms, especially by permitting the occurrence of photosynthesis.

Photochemical smog   Air pollution caused by the interaction of radiant energy and chemical compounds in the atmosphere.

Photosynthesis   The synthesis of carbohydrates in plants from water and carbon dioxide by the action of sunlight on chlorophyll.

Physiography   The description of the relief features of the earth's surface.

Phytotoxicity   Being poisonous to plants.

Plankton   The passively floating or weakly swimming animal and plant life of a body of water consisting chiefly of minute plants (such as diatoms and blue-green algae) and of minute animals (such as protozoans, entomastracans, and various larvae) but including also larger forms (such as jellyfishes) that have only weak powers of locomotion.

Ploidy   Degree of replication of chromosomes or genomes.

Plumule   The primary bud of a plant embryo, usually situated at the apex of the hypocotyl and consisting of leaves and an epicotyl that elongates to extend the axis as a primary stem.

Pocosino   A swamp or marsh, especially in an upland on interfluvial area of the coastal plain of the southeastern United States.

Podzolic soils   Soil that has an organic mat and a thin, usually acidic, organic mineral layer.

Polar front   The boundary between the cold polar easterly winds and the warm tropical westerlies.

Polder   A tract of low land reclaimed from the sea or other body of water (as by dikes).

Polymorphs   Organisms having or assuming various forms, characters, styles, or functions.

Polyploid   Having a chromosome number that is a multiple greater than two of the haploid number.

Progeny   Offspring of animals or plants.

Propagation   To increase the number of individuals, that is, by sexual or vegetative reproduction.

Propagules   A propagatable shoot (e.g., bulbs, tubers, and runners).

Prune   To cut down or reduce by cutting off the superfluous parts, branches, or shoots of a plant for better shape or more fruitful growth.

Pulse seed   The edible seeds of plants having pods (the legumes), including peas, beans, and lentils.

Pyrolysis   Chemical decomposition by heat.

Quantum   The very smallest increments or parcels into which many forms of energy are subdivided, and which are always associated with a frequency v such that the quantum is equal to frequency multiplied by Planck's constant.

Quincunx system   A planting in the form of a series of squares or rectangles with a plant at the center of each.

Rain shadows   A localized area of rainfall.

Rarify   To make thin or less dense.

Ratoon   A stalk or shoot arising from the root or crown of a perennial plant.

Resin   Any of various substances of different textures obtained from tropical trees or pine trees or made synthetically, used chiefly in varnishes, ink, and medicine.

Respiration   The processes by which a living organism or cell takes in oxygen, distributes and utilizes it in oxidation, and gives off products, especially carbon dioxide and energy as ATP and heat.

Roguing   To weed out inferior, diseased, or abnormal individuals from a crop.

Rootstock   A stock for grafting, consisting of a root or a piece of root.

Rosaceous   Adjective referring to a member of the plant family Rosaceae, which contains apples, cherries, plums, and roses.

Rotary engine   An engine, such as a turbine, in which power is supplied directly to vanes or other rotary parts.

Saprophytic   Obtaining nourishment from nonliving biological materials, as for example, fungi that live on dead trees or on lumber.

Savanna   A treeless plain; an open, level region.

Scarification   To treat hard-coated seeds by mechanical abrasion or with acid to facilitate water absorption and hasten germination.

Scion   A detached living portion (as a year-old shoot) of a plant prepared for union with a stock in grafting and usually supplying aerial parts to a graft.

Secondary succession   The plants that grow up on a piece of ground that has been partially cleared of vegetation by fire, people, a tree fall, or other means.

Sedge   Plants belonging to the family Cyperaceae, a large family of monocotyledonous plants distinguished chiefly by having achenes, solid stems, and three-ranked stem leaves.

Selective cutting   The cutting out of trees that are mature or defective to encourage the growth of remaining trees in a forest.

Senesce   To wither or to grow old.

Shield budding   Plant budding in which an oval or shield-shaped piece of bark bearing a scion bud is fitted into an approximately T-shaped opening in the bark of the stock.

Sibs   Other members of the offspring from one parent plant.

Side graft   A plant graft in which the scion is inserted into the side of the stock and permitted to grow until union is established between stock and scion.

Silica gel   Extremely absorbent colloidal silica that is made by coagulation of hydrated silica and is used as a drying agent, chromatography absorbent, and as a catalyst or catalyst carrier; $SiO_2 \cdot nH_2O$, the natural form of silica in plants.

Solar cell   A device used to convert sunlight into electrical energy.

Solar panel   A great number of solar cells connected together.

Solstice   The time at which the sun is at its greatest distance from the equator, when the sun appears to stop or cease to recede from the equator. In the northern hemisphere, the summer solstice is on June 21, the winter on December 22.

Somatic cell hybridization The joining together of two differentiated cells of different species to form one large cell with two nuclei with different genomes.

Spore A minute unicellular reproductive or resistant resting body; morphologically a mass of protoplasm, usually with a single definite nucleus, capable of producing a new individual.

Standing crop A synonym of "biomass"; this term is most frequently used in studies of the amount of food available in a habitat for a predator or herbivore.

Steppe Arid land characterized by xerophytic vegetation and found usually in large tracts and in regions of extreme temperature range.

Stock A plant stem into which a graft is inserted; the basal half of a graft.

Stratification The placing of seeds in damp sand, peat moss, or sawdust at low temperature (0 to 4°C) to facilitate germination; a necessary procedure for seeds requiring moisture or low temperature or both during their resting period or after-ripening period.

Stratified charge engine A type of automobile engine with two combustion chambers, noted for more complete combustion of hydrocarbon fuels, and as a result, causes less air pollution.

Stratopause The outer edge of the stratosphere.

Stratosphere The portion of the atmosphere containing the ozone layer, at an altitude between 15 and 50 km above the earth's surface.

Style The part of the carpel lying between the ovary and the stigma.

Sublittoral Situated on the watery side of a shoreline; also, the region in a lake between the deepest growing rooted vegetation and the hypolimnion.

Submergent Partly or completely under water.

Subsistence farming Farming in which there is produced only enough food for the farmer to eat; there are no surpluses raised for sale.

Substrate The material or substance upon which an enzyme acts; or the surface on which a plant or animal grows.

Substratum The layer of atmosphere closest to the earth's surface.

Succession The process of change in the biological population of an area; the sequence of identifiable ecological states or communities in the process from barrenness to climax.

Sucker A shoot originating from the roots or lower part of the stem of a plant and usually developing rapidly, often at the expense of the parent plant; also, an accessory propagation shoot.

Supercooled Cooled below its freezing point without solidification.

Swidden An impermanent agricultural plot produced by cutting trees and burning off vegetative cover.

Symbiotic The intimate living together of two dissimilar organisms in any of various mutually beneficial relationships.

Sympatric Two species capable of occupying the same range without loss of identity due to interbreeding.

Synecology A branch of ecology that deals with the structure, development, and distribution of ecological communities in relation to environment.

Tannin A substance from the bark and fruit of many plants (especially oak and sumac) that has many commercial and medicinal uses, commonly used in leather tanning and other preservative applications because of its antibacterial and antifungal properties.

Taproot   A root having a prominent central portion, growing vertically downward and giving off small lateral roots in succession.

Taxa   The name applied to a naturally related group in a formal system of nomenclature.

Taxonomy   The science of classification, as of plants or animals.

T-budding   *See* Shield budding.

Temperature inversion   An atmospheric condition in which pollutants are trapped near the earth's surface by a layer of warm air lying over a layer of cold air.

Tenacious bracts   Bracts that cling to one another.

Tendril   A portion or the whole of a leaf, stipule, or stem that is modified into a slender, spirally coiling organ serving to attach a plant to its support.

Terpenes   Any of the various unsaturated hydrocarbons $C_{10}H_{16}$ found in essential oils and oleoresins of plants and used in organic synthesis. Main component of natural "smog" given off by vegetation.

Thalloid   A combination of cells presenting no differentiation of leaf and stem vegetative or assimilative parts.

Tidal generator   A device to produce electricity from the movement of the ocean tides.

Tolerance   The natural, or acquired, ability to be resistant to the effects of a toxic compound.

Topography   The configuration of surface, including its relief and the position of its natural and man-made features.

Torque   The moment of a force, a measure of its tendency to produce torsion and rotation about an axis.

Trade winds   Winds that blow from the northeast to the southwest toward the equator in the northern hemisphere and from the southeast to the northwest toward the equator in the southern hemisphere.

Transpiration   The loss of water vapor from the surfaces of leaves or other parts of plants via stomata, the cuticle, and lenticels.

Tropopause   The outer edge of the troposphere.

Troposphere   The portion of the atmosphere that is below the stratosphere; extends 7 to 10 miles from the earth's surface; and is the portion in which temperature rapidly decreases with altitude, clouds form, and convection is active.

Tuber   A short, thickened fleshy stem or terminal portion of a stem or rhizome, usually formed underground, that bears buds capable of developing into a new plant.

Tundra   A level of undulating treeless plain that is characteristic of arctic and subarctic regions; marks the limit of arborescent vegetation.

Turbidity   Measure of thickness or denseness.

Turbogenerator   A machine that converts mechanical energy into electrical energy using a turbine-driven compressor.

Turnover   To reverse the layers of water in a lake or the ocean.

Understory   A foliage layer lying beneath and shaded by the main canopy of a forest.

Unflagellated   Without the long tapering structure, a flagellum, that projects singly or in groups from a cell or microorganism and is often used as a means of locomotion.

Vegetation   Plant life or total plant cover within a given area.
Vortex   Fluid flow involving rotation about an axis.

Watt   A unit of electric power; wattage is the amount of current (in amperes) multiplied by the voltage; expressed in joules per second.
Weather   The condition of the atmosphere with regard to temperature, moisture, winds, and so on.
Wind generator   An electrical generator powered by the wind.

Xerophytic   Adapted to living in a dry habitat.
Xylem   A complex tissue in the vascular system of higher plants, which functions chiefly in conduction of water and mineral nutrients. The primary conducting elements in the xylem are tracheids and vessel members.

Index

E

Easements, 465
Ecklonia, 158
Edel, M., 173
Edible wild foods
  recipes, 367–69
Edible wild plants
  asparagus, 327, 330
  blueberries, 364
  cattails, 352–53
  common sorrel, 343–44
  cranberries, 364–65
  dandelion, 338
  environmental damage, 366
  groundnuts, 353–55
  habitats, 323–25, 345–46, 352, 361
    dry, 325
  Jerusalem artichoke, 327, 331
  juniper, 339–40
  lamb's quarters, 339–41
  maple, 346–47
  mayapple, 348
  milkweed, 331–33
  mints, 355
  nettles, 355–58
  New Jersey tea, 341–42
  nutritional value, 323
  ostrich fern, 357–58
  pawpaw, 359–60
  poisons, 365
  pokeweek, 334–35
  purslane, 342
  reasons for collecting, 323
  roadsides and railroad banks, 326
  sassafras, 348–49
  strawberries, 344–45
  sumac, 336
  violets, 349–50
  watercress, 361–62
  wet habitats, 326, 328–30
  wild chives and onions, 345
  wild fruits, 336–37
  wild ginger, 350–51
  wild leeks, 351–52
  wild nuts, 337–38
  wild rice, 362–63
  wintergreen, 352
Eichhornia crassipes, 45, 161
Eisenia, 158
Emergent aquatic plants, 141
Emigration rate, 258
Emphysema, 182

Endangered species of plants
  extinct plants, 502
  National Endangered Species
    Act, 506
  plants near extinction, 503
  plants rediscovered, 502
  saving endangered species, 509
Endosperm, 40–42
Endothia parasitica, 51
Enteromorpha, 159
Energy
  alternative sources, 419–40
  geothermal, 437–39
  solar, 429–35
  water, 419–29
  wind power, 435–37
Energy transformation, 229–30
Environmental education
  adult education, 459
  community nature centers, 455
  elementary education, 457
  secondary education, 459
  specialized education, 458
Environmental impact statement,
    241–42
Environmental issues, 497
Ephedra, 27
Epilimnion, 152
Epiphytes, 134–36
Equisetum, 22–23
Equisetum giganteum, 23
Ergot, 14–15
Esposito, J. C., 172–73
Ethnobotany, 281
  definition, 273
  middle America, 287–88
Euglenoids, 18
Euglenophyta, 18
Eutrophication, 156
Eutrophic lakes, 156
Evolution
  by natural selection, 264–65
Extinct plants, 502

F

Federal Noise Act, 186
Federal Water Pollution Control
    Act, 232
Federal wilderness area, 243
Fen, 150–51
Fermentation, 14